Periodensystem der Elemente mit Gmelin-Systemnummern

1 H 2																1 H 2	2 He 1
3 Li 20	4 Be 26											5 B 13	6 C 14	7 N 4	8 O 3	9 F 5	10 Ne 1
11 Na 21	12 Mg 27											13 Al 35	14 Si 15	15 P 16	16 S 9	17 Cl 6	18 Ar 1
19 * K 22	20 Ca 28	21 Sc 39	22 Ti 41	23 V 48	24 Cr 52	25 Mn 56	26 Fe 59	27 Co 58	28 Ni 57	29 Cu 60	30 Zn 32	31 Ga 36	32 Ge 45	33 As 17	34 Se 10	35 Br 7	36 Kr 1
37 Rb 24	38 Sr 29	39 Y 39	40 Zr 42	41 Nb 49	42 Mo 53	43 Tc 69	44 Ru 63	45 Rh 64	46 Pd 65	47 Ag 61	48 Cd 33	49 In 37	50 Sn 46	51 Sb 18	52 Te 11	53 J 8	54 Xe 1
55 Cs 25	56 Ba 30	57** La 39	72 Hf 43	73 Ta 50	74 W 54	75 Re 70	76 Os 66	77 Ir 67	78 Pt 68	79 Au 62	80 Hg 34	81 Tl 38	82 Pb 47	83 Bi 19	84 Po 12	85 At	86 Rn 1
87 Fr	88 Ra 31	89*** Ac 40	104 71	105 71													

* NH_4 23

**Lanthanide 39	58 Ce	59 Pr	60 Nd	61 Pm	62 Sm	63 Eu	64 Gd	65 Tb	66 Dy	67 Ho	68 Er	69 Tm	70 Yb	71 Lu
***Actinide	90 Th 44	91 Pa 51	92 U 55	93 Np 71	94 Pu 71	95 Am 71	96 Cm 71	97 Bk 71	98 Cf 71	99 Es 71	100 Fm 71	101 Md 71	102 No(?) 71	103 Lr 71

Reihenfolge der Gmelin-Systemnummern siehe Innenseite des hinteren Deckels

Gmelin Handbook of Inorganic Chemistry

8th Edition

Gmelin Handbook of Inorganic Chemistry

8th Edition

Gmelin Handbuch der Anorganischen Chemie

Achte, völlig neu bearbeitete Auflage

Prepared and issued by

Gmelin-Institut für Anorganische Chemie
der Max-Planck-Gesellschaft
zur Förderung der Wissenschaften
Director: Ekkehard Fluck

Founded by — Leopold Gmelin

8th Edition — 8th Edition begun under the auspices of the Deutsche Chemische Gesellschaft by R. J. Meyer

Continued by — E. H. E. Pietsch and A. Kotowski, and by Margot Becke-Goehring

Springer-Verlag Berlin Heidelberg GmbH 1988

STAFF GMELIN HANDBOOK

Dr. G. Bär, D. Barthel, Dr. N. Baumann, Dr. W. Behrendt, D. Benzaid, Dr. L. Berg, Dipl.-Chem. E. Best, Dr. R. Bohrer, K. D. Bonn, P. Born-Heck, Dipl.-Ing. V. A. Chavizon, E. Cloos, Dipl.-Phys. G. Czack, I. Deim, Dipl.-Chem. H. Demmer, Dipl.-Geol. R. Ditz, R. Dowideit, Dipl.-Chem. M. Drößmar, Dr. D. Düster, U. Duggen, Dr. P. Eigen, M. Engels, Dr. H.-J. Fachmann, Dr. J. Faust, B. Fischer, G. Funk, Dipl.-Ing. N. Gagel, Dr. U. W. Gerwarth, Dipl.-Phys. D. Gras, Dipl.-Bibl. W. Grieser, Dr. I. Haas, Dr. R. Haubold, B. Heibel, Dipl.-Min. H. Hein, G. Heinrich-Sterzel, A. Heise-Schuster, H.-P. Hente, H. W. Herold, U. Hettwer, Dr. R. Heumüller, Dr. G. Hönes, Dr. W. Hoffmann, Dr. W. Huisl, R. Jonuschat, Dr. R. Jotter, Dr. J. von Jouanne, Dr. B. Kalbskopf, Dipl.-Chem. W. Karl, H.-G. Karrenberg, Dipl.-Phys. H. Keller-Rudek, B. Kirchner, A. Klusch, Dipl.-Chem. C. Koeppel, Dipl.-Chem. H. Köttelwesch, R. Kolb, E. Kranz, Dipl.-Chem. I. Kreuzbichler, Dr. A. Kubny, Dr. M. Kunz, Dr. W. Kurtz, M. Langer, Dr. U. Lanzendörfer, Dr. B. Ledüc, Dr. A. Leonard, Dipl.-Chem. H. List, H. Mathis, E. Meinhard, C. Metz, K. Meyer, Dr. M. Mirbach, Dipl.-Chem. B. Mohsin, Dr. U. Neu-Becker, K. Nöring, Dipl.-Chem. R. Nohl, Dipl.-Min. U. Nohl, Dr. U. Ohms-Bredemann, Dr. W. Petz, I. Rangnow, Dipl.-Phys. H.-J. Richter-Ditten, Dipl.-Chem. H. Rieger, E. Rieth, A. Rosenberger, E. Rudolph, G. Rudolph, Dipl.-Chem. S. Ruprecht, Dr. B. Sarbas, Dr. H. Schäfer, Dr. D. Schiöberg, V. Schlicht, B. Schmitt, Dipl.-Chem. D. Schneider, Dipl.-Min. P. Schubert, Dr. K. Schücke, A. Schwärzel, Dr. B. Schwager, Dipl.-Ing. H. M. Somer, Dr. C. Strametz, Dr. G. Swoboda, Dr. U. Taubald, M. Teichmann, Dr. D. Tille, Dipl.-Ing. H. Vanecek, Dipl.-Chem. P. Velić, Dipl.-Ing. U. Vetter, H.-M. Wagner, Dipl.-Phys. J. Wagner, R. Wagner, Dr. E. Warkentin, Dr. C. Weber, Dr. A. Wietelmann, Dr. M. Winter, Dr. B. Wöbke, K. Wolff

STAFF GMELIN ONLINE DATA SYSTEM

Dr. P. Kuhn, Dr. G. Olbrich,

Gmelin Handbook of Inorganic Chemistry

8th Edition

Th
Thorium

Supplement Volume C 7

Compounds with Carbon: Carbonates, Thiocyanates, Alkoxides, Carboxylates

With 12 illustrations

AUTHOR Kenneth W. Bagnall, Chemistry Department, University of Manchester, Manchester, U.K.

CHIEF EDITORS Karl-Christian Buschbeck, Gmelin-Institut, Frankfurt am Main

Cornelius Keller, Supervising Scientific Coordinator for the Thorium Supplement Volumes, Schule für Kerntechnik, Kernforschungszentrum Karlsruhe

System Number 44

Springer-Verlag Berlin Heidelberg GmbH 1988

LITERATURE CLOSING DATE: MID OF 1986
IN MANY CASES MORE RECENT DATA HAVE BEEN CONSIDERED

Library of Congress Catalog Card Number: Agr 25-1383

ISBN 978-3-662-06317-0 ISBN 978-3-662-06315-6 (eBook)
DOI 10.1007/978-3-662-06315-6

Originally published by Springer-Verlag Berlin Heidelberg New York London Paris Tokyo in 1988
Softcover reprint of the hardcover 8th edition 1988

Preface

The present volume of the Gmelin thorium series describes the solid thorium-carbon compounds with the exception of the carbides and coordination compounds of the type $ThA_n \cdot x\,B$, where B is a neutral ligand. The complex equilibria of the Th^{4+} ion with C-containing complexing agents are treated in the ThD1 volume.

A first look into this volume demonstrates that a very large number of Th^{IV} complexes has been prepared. This may be explained by the fact that the Th^{4+} ion is the largest tetravalent ion of the periodic table. Therefore, the preparation of complexes with, e.g., multidentate ligands can give a well-established picture of the coordination number as a function of charge and ionic radius. However, there are very few modern and updated comprehensive treatments of such data. Many compounds described in this volume are characterized by no other means than analytical composition and IR spectra (whereby IR spectra of organic Th^{IV} salts mostly give information only on the ligand). Besides thorium carbonate and carbonato complexes, which are relevant for the environmental behaviour of this radioactive element and some organic complexes like oxalates, which are used in the field of analytical and separation chemistry nearly all other compounds described here are practically only of scientific interest. On the other hand in order to have scientifically reliable data, a very large part of these compounds needs further investigation and characterization. A close look into this volume, therefore, can provide many interesting ideas for modern research topics in the field of thorium chemistry in general and in special and complex chemistry as well.

I thank Prof. Bagnall, Manchester, for his excellent contribution and the very good cooperation. Thanks also to the Gmelin Institute, especially Dr. K.-Chr. Buschbeck, the supervising editor and Prof. Dr. E. Fluck, the director for their support and excellent cooperation.

Karlsruhe,
February 1988

Cornelius Keller

Volumes published on "Radium and Actinides"

Ac Actinium

Main Volume		– 1942
Suppl. Vol. 1: Element and Compounds		– 1981

Np, Pu, ... Transuranium Elements

Main Volume

Part A: The Elements		
A 1, I	History, Occurrence, Properties of Atomic Nuclei	– 1973
A 1, II	Nuclides: Manufacture, Recovery, Enrichment	– 1974
A 2	General Properties, Uses, Storage, Biology	– 1973
Part B: The Metals		
B 1	Metals	– 1976
B 2	Binary Alloy Systems 1	– 1976
B 3	Binary Alloy Systems 2	– 1977
Part C: The Compounds		
C	Compounds	– 1972
Part D: Chemistry in Solution		
D 1	Aqueous Solutions. Coordination Chemistry	– 1975
D 2	Extraction, Ion Exchange. Molten Salts	– 1975
Index		– 1979

Pa Protactinium

Main Volume		– 1942
Suppl. Vol. 1: Element		– 1977
Suppl. Vol. 2: Metal. Alloys. Compounds. Chemistry in Solution		– 1977

Ra Radium

Main Volume		– 1928
Suppl. Vol. 1: History, Cosmochemistry. Geochemistry		– 1977
Suppl. Vol. 2: Element. Compounds		– 1977

Th Thorium

Main Volume		– 1955
Suppl. Vol. Part A: The Element		
A 1	Thorium Deposits. Geochemistry (in preparation)	
A 2	History. Isotopes. Recovery of Thorium from Ores	– 1986
Suppl. Vol. Part C: The Compounds		
C 1	Compounds with Noble Gases, Hydrogen, Oxygen	– 1978
C 2	Ternary and Polynary Oxides	– 1976

Table of Contents

15 Thorium Carbonates, Cyanides, Thiocyanates, Alkyl(Aryl) Oxides, and Carboxylates

Kenneth W. Bagnall
Chemistry Department, University of Manchester
Manchester, U.K.

This volume covers the available information on the carbonates, carbon-containing pseudohalides such as cyanides and thiocyanates, and the much more extensively documented alkyl oxides, aryl oxides, and carboxylates. A considerable number of carbon-containing thorium compounds is also described in "Thorium" Suppl. Vol. E, 1985, both as complexes with neutral ligands such as alcohols, amines or pyridine, and in combination with ligands such as diketones and Schiff bases. Examples are the oxodiacetate compound, $Th(OOCCH_2OCH_2COO)_2 \cdot 3C_5H_5NO \cdot 3H_2O$, described on pp. 69 and 74, and the alkoxide compound $Th[OC(CCl_3)(CH_3)_2]_4 \cdot 2py$, described on pp. 12 and 14. The present volume deals with the compounds which contain only the "acidic" constituent, such as the "pure" oxodiacetates and alkoxides.

Unless otherwise stated, all of the compounds described in this volume are white or colourless solids.

15.1 Thorium Carbonates and Carbonato Complexes

15.1.1 Introduction

Since the last edition of the "Thorium" volume was published in 1955, several books and reviews of the chemistry of the actinides have appeared, some of which include information on thorium carbonates and carbonato complexes [1 to 3], but no publications dealing exclusively with this area of thorium chemistry are available.

Mixed carbonato oxalato and carbonato ethylenediaminetetraacetato complexes are described later in this volume on pp. 98/102 and 124/5, respectively.

Abbreviations. In the tables the following abbreviations are used: b.p. = boiling point, and m.p. = melting point; d_{calc} = calculated density and d_{obs} = observed density. In referring to the thorium coordination compounds volume its citation "Thorium" Suppl. Vol. E, 1985 is abbreviated as Th E, preferably in the tables.

References for 15.1.1:

[1] Pascal, P. (Nouveau Traité de Chimie Minérale, Vol. 9, Masson, Paris 1963, pp. 1141/2).
[2] Bagnall, K. W. (The Actinide Elements, Elsevier, New York 1972, pp. 1/272).
[3] Brown, D. (in: Bailar, J. C.; Emeléus, H. J.; Nyholm, R. S.; Trotman-Dickenson, A. F., Comprehensive Inorganic Chemistry, Vol. 5, Pergamon, Oxford 1973, pp. 277/86).

15.1.2 Thorium Carbonates

The available information on thorium carbonate and the basic carbonates is summarised in Table 1, p. 2.

Table 1
Thorium Carbonates.

$Th(CO_3)_2$	decomposes at ca. 390°C [16]
$Th(CO_3)_2 \cdot 0.5H_2O$	IR spectrum: ν_3(CO), 1550, 1380, ν_1(CO), 1085, ν_2(CO), 830, ν_4(CO), 752, 680 cm^{-1} [4]
$Th(CO_3)_2 \cdot 3.00$ to $3.57H_2O$	thermal decomposition diagram given in [1]
$ThO(CO_3)$	[5]
$ThO(CO_3) \cdot H_2O$	[6]
$ThO(CO_3) \cdot 2H_2O$	see "Thorium" 1955, p. 301
$ThO(CO_3) \cdot 3H_2O$	diamagnetic susceptibility, $\chi_m = -74$ to -77×10^{-6} [14]
$ThO(CO_3) \cdot 8H_2O$	solubility product, $S = [ThO^{2+}]\ [CO_3]^{2-} = 9 \times 10^{-9}$ in H_2O [10] see also "Thorium" 1955, p. 301
$ThO_2 \cdot ThO(CO_3) \cdot xH_2O$	x=1.5 or 4, see "Thorium" 1955, p. 301
$3ThO_2 \cdot ThO(CO_3) \cdot H_2O$	see "Thorium" 1955, p. 301
$6ThO_2 \cdot ThO(CO_3)$	see "Thorium" 1955, p. 301
$Th(OH)_2(CO_3)$	[7]
$Th(OH)_2(CO_3) \cdot 1.5H_2O$	diamagnetic susceptibility, $\chi_m = -58 \times 10^{-6}$ [14]
$Th(OH)_2(CO_3) \cdot 2H_2O$	IR spectrum (4000 to 400 cm^{-1}) illustrated in [5]; ν_1(CO), 1535, ν_5(CO), 1380, ν_2(CO), 1070, ν_6(CO), 840, δ(CO)+δ(ThO_2C), 730 cm^{-1} [5] diamagnetic susceptibility, $\chi_m = -67$ to -71×10^{-6} [14]
$Th(OH)_2(CO_3) \cdot 4H_2O$	diamagnetic susceptibility, $\chi_m = -82$ to -86×10^{-6} [14]
$Th(OH)_3(HCO_3)$(?)	[8]

The anhydrous carbonate, $Th(CO_3)_2$, has only been reported as an intermediate in the pyrolysis of $Th(C_2O_4)_2$ below 320°C [12, 15, 16], but two hydrates are known. $Th(CO_3)_2 \cdot 0.5H_2O$ has been prepared by heating thorium hydroxide at 100 to 150°C under CO_2 at 1800 to 3000 atm pressure for 36 to 720 h; the product was finally dried over phosphorus pentoxide at 100°C. When the preparation is carried out at temperatures above 150°C, substoichiometric carbonates are obtained [4]. Some details of its infrared spectrum [4] are included in Table 1. The other hydrate, $Th(CO_3)_2 \cdot 3.00$ to $3.57H_2O$, has been obtained by treating a mixture of $Th(CH_3COO)_4$ (p. 47) and water with CO_2 under pressure; the yields are 91% at 850 lb/in² and 94% at 950 lb/in² [1 to 3].

The basic carbonate, $ThO(CO_3)$, is apparently formed when $Th(OH)_2(CO_3) \cdot 2H_2O$ (see below) is heated at 500 to 600°C [5]. However, $ThO(CO_3) \cdot H_2O$, which is reported to be the product when higher hydrates are heated to ca. 200°C, is said to decompose to ThO_2 on further heating and the thermal decomposition of the monohydrate has been studied by thermogravimetry, differential thermal analysis, and Hahn's emanation technique [6]. The dihydroxide carbonate, $Th(OH)_2(CO_3)$, obtained as a precipitate in a solution chemistry study of the reaction of aqueous $Th(NO_3)_4$ with Na or NH_4 carbonates by high frequency and pH titration techniques, followed by drying at 250°C [7], is probably the same compound as $ThO(CO_3) \cdot H_2O$. A hydrate, $Th(OH)_2CO_3 \cdot xH_2O$, is precipitated when an aqueous solution of $(NH_4)_6[Th(CO_3)_5] \cdot 3H_2O$ (p. 10) at ca. pH 7 is heated [13]. Other authors have reported that a product of composition close to $ThO(CO_3) \cdot 2H_2O$ is precipitated when 0.2M aqueous $Th(NO_3)_4$ is treated with two equivalents of 0.2 M sodium carbonate solution; the precipitate was washed with water and dried at ca. 60°C [8]. Continued washing, or digestion with hot water, decreases the CO_2 content of this product and it has been suggested [8] that the compound should be formulated as $Th(OH)_3(HCO_3)$ or even $Th(OH)_4 \cdot CO_2$.

The solubility of $ThO(CO_3) \cdot 8H_2O$ in aqueous 3M $NaClO_4$ is 46.8 mg/L as against 41.6 mg/L in water, and the solubility product, $[ThO^{2+}][CO_3]^{2-}$, has been reported to be 9×10^{-9} in water [10].

The IR spectrum of an unspecified thorium carbonate revealed peaks corresponding to ν_1(CO), 1081; ν_2(CO), 859, 840; ν_3(CO), 1527, 1460, and ν_4(CO), 752 cm^{-1} [11].

Satellite phenomena in the X-ray photoelectron spectrum of $Th(OH)_2(CO_3) \cdot xH_2O$ have been reported. The structure of this hydroxide-carbonate is unknown; the given formula is the most probable one [9].

The thermal decomposition at 600, 800, and 1000°C of an unspecified Th carbonate (precipitated from aqueous $Th(NO_3)_4$) has been reported in an investigation of the powder characteristics of the ThO_2 produced by a variety of methods [17].

References for 15.1.2:

[1] Head, E. L. (CONF-670502 [1967] 366/73; 6th Rare Earth Res. Conf., Gatlinburg, Tenn., 1967; N.S.A. **22** [1968] No. 23000).

[2] Head, E. L. (Australian 426165 [1972]; N.S.A **28** [1973] No. 20856).

[3] Head, E. L. (U.S. 3374069 [1968]; C.A. **68** [1968] No. 97197).

[4] Ehrhardt, H.; Schweer, H.; Seidel, H. (Z. Anorg. Allgem. Chem. **462** [1980] 185/98).

[5] Kharitonov, Yu. Ya.; Molodkin, A. K.; Balakaeva, T. A. (Zh. Neorgan. Khim. **14** [1969] 2761/7; Russ. J. Inorg. Chem. **14** [1969] 1453/6).

[6] Saito, Y. (Funtai Oyobi Funmatsu Yakin **17** [1971] 295/9 [Japan.]; C.A. **76** [1972] No. 132295).

[7] Nguyen-Dinh-Ngo; Martynenko, L. I. (Zh. Neorgan. Khim. **14** [1969] 1541/5; Russ. J. Inorg. Chem. **14** [1969] 807/9).

[8] Mirza, E. B.; Karkhanavala, M. D. (J. Indian Chem. Soc. **40** [1963] 903/4).

[9] Allen, G. C.; Tucker, P. M. (Chem. Phys. Letters **43** [1976] 254/7).

[10] Zakharov-Nartsissov, O. I.; Mikhailov, G. G. (Izv. Vysshikh Uchebn. Zavadenii Khim. Khim. Tekhnol. **3** [1960] 45/8; C.A. **1960** 16116).

[11] Ross, S. D.; Goldsmith, J. (Spectrochim. Acta **20** [1964] 781/4).

[12] Osinovik, E. S.; Yanchuk, A. F. (Vestsi Akad. Navuk Belarussk. SSR Ser. Khim. Navuk **1966** No. 3, pp. 131/3; C.A. **66** [1967] No. 14389).

[13] Chernyaev, I. I.; Golovnya, V. A.; Molodkin, A. K. (Zh. Neorgan. Khim. **6** [1961] 394/9; Russ. J. Inorg. Chem. **6** [1961] 200/2).

[14] Belova, V. I.; Syrkin, Ya. K.; Molodkin, A. K.; Ivanova, O. M.; Shiporina, L. M. (Zh. Neorgan. Khim. **13** [1968] 1458/60; Russ. J. Inorg. Chem. **13** [1968] 766/7).

[15] D'Eye, R. W. M.; Sellman, P. G. (J. Inorg. Nucl. Chem. **1** [1955] 143/8).

[16] Dell, R. M.; Wheeler, V. J. (Reactivity Solids 5th. Intern. Symp., Munich, FRG, 1965, pp. 395/408).

[17] Moorthy, V. K.; Kulkarni, A. K. (Trans. Indian Ceram. Soc. **22** [1963] 116/29).

15.1.3 Thorium Carbonato Complexes

The known compounds, together with their properties, are listed in Table 2.

Table 2
Thorium Carbonato Complexes.
Syntheses, structures, and chemical behaviour are discussed in Sections 15.1.3.1 to 15.1.3.3, pp. 7/12.

tricarbonato complexes	
$(NH_4)_2Th(CO_3)_3\cdot 6H_2O$(?)	reported as $(NH_4)_2CO_3\cdot Th(CO_3)_2\cdot 6H_2O$ in "Thorium" 1955, p. 339 and as $(NH_4)_2[Th(CO_3)_3(H_2O)_3]\cdot 3H_2O$ in [4]
$(CN_3H_6)_2[Th(CO_3)_3]$	CN_3H_6 = guanidinium; decomposes at >200°C [18]
$(CN_3H_6)_2[Th(CO_3)_3]\cdot 4H_2O$	loses $4H_2O$ below 100°C; thermal decomposition curve illustrated in [18]; IR spectrum: ν(Th–O(CO_2)), 310, 250 cm^{-1} [18]
tetracarbonato complexes	
$Na_4[Th(CO_3)_4]$	decomposes at >300°C [18]
$Na_4[Th(CO_3)_4]\cdot 7H_2O$	loses $7H_2O$ at 100°C; thermal decomposition curve illustrated in [18] IR spectrum: ν(Th–O(CO_2)), 310, 250 cm^{-1} [18], 295, 265 cm^{-1} [26]; full details of the spectrum (1560 to 265 cm^{-1}) are given in [26]
$[CN_3H_6]_4[Th(CO_3)_4]$	decomposes at >200°C [18]
$[CN_3H_6]_4[Th(CO_3)_4]\cdot 6H_2O$	loses $6H_2O$ below 100°C; thermal decomposition curve illustrated in [18] IR spectrum: ν(Th–O(CO_2)), 310, 250 cm^{-1} [18], 244, 230 cm^{-1} [26]; full details of the spectrum (1506 to 230 cm^{-1}) are given in [26]
pentacarbonato complexes	
$Na_6[Th(CO_3)_5]$ [a)]	decomposes at >290°C [10], >300°C [12, 18], >330°C [9], 350°C (−2CO_2) [2, 4], 550°C [23] IR spectrum: ν(coordinated CO_3), 1055, 725 cm^{-1}; ν(ionic[a)] CO_3), 1080, 865, 855, 705 to 700 cm^{-1}; ν(coordinated + ionic[a)] CO_3), 1460, 1345 cm^{-1} [11]; ν_1(CO), 1570 cm^{-1}; ν_5(CO), 1372, 1350 cm^{-1}; ν_2(CO), 1063, 1048 cm^{-1}; ν_6(CO), 858 cm^{-1}; δ(CO) + δ(ThO_2C), 718, 713 cm^{-1} [14][b)] IR spectrum (4000 to 400 cm^{-1}) illustrated in [14]
$Na_6[Th(CO_3)_5]\cdot H_2O$	loses H_2O at 100°C [24] or 150°C [9]
$Na_6[Th(CO_3)_5]\cdot 2H_2O$	loses $1H_2O$ at 70°C [2] or 100°C [9]
$Na_6[Th(CO_3)_5]\cdot 3H_2O$	IR spectrum: ν_1(CO), 1570 cm^{-1}; ν_5(CO), 1375, 1350 cm^{-1}; ν_2(CO), 1046 cm^{-1}; ν_6(CO), 856 cm^{-1}; δ(CO) + δ(ThO_2C), 718 cm^{-1} [14] IR spectrum (4000 to 400 cm^{-1}) illustrated in [14] diamagnetic susceptibility, χ_m = −208 to −219 × 10^{-6} [33]
$Na_6[Th(CO_3)_5]\cdot 5H_2O$	[4]

Table 2 (continued)

$Na_6[Th(CO_3)_5]\cdot 10H_2O$	[4]
$Na_6[Th(CO_3)_5]\cdot 11H_2O$	[4]
$Na_6[Th(CO_3)_5]\cdot 11.3H_2O$	diamagnetic susceptibility, $\chi_m = -307$ to -344×10^{-6} [33]
$Na_6[Th(CO_3)_5]\cdot 12H_2O$	reported as $3Na_2CO_3\cdot Th(CO_3)_2\cdot 12H_2O$ in "Thorium" 1955, p. 326; loses $12H_2O$ at 50 to 170°C [10], ca. 100°C [18, 23], 100 to 150°C [4], loses $10H_2O$ at 50°C [2] or 75 to 80°C [9], loses $11H_2O$ at 70°C [2, 4]; thermal decomposition curve illustrated in [9, 18] crystals are biaxial negative [2]; crystal optical properties, refractive indices $n_\gamma=1.503$, $n_\alpha=1.490$, $n_\beta=1.479$ [9] or $n_\gamma=1.504$, $n_\alpha=1.490$, $n_\beta=1.472$ [2] X-ray crystallographic data: 2 crystal modifications [10], triclinic, lattice parameters, a = 9.6, b = 9.9, c = 21.4 Å, $\alpha=60°$, $\beta=51°$, $\gamma=86°$ and a = 9.6, b = 13.2, c = 17.0 Å, $\alpha=50.7°$, $\beta=71.3°$, $\gamma=49°$; density, $d_{obs}=2.32(1)$ g/cm³, Z = 2 [10]; triclinic, space group $P\bar{1}$-C_i^1 (No. 2); lattice parameters, a = 13.71(2), b = 9.89(2), c = 9.60(2) Å, $\alpha=95.90(10)°$, $\beta=104.80(10)°$, $\gamma=91.40(10)°$; density, $d_{calc}=2.35$ g/cm³, $d_{obs}=2.31$ g/cm³, Z = 2 [18]; a = 9.60(2), b = 9.92(2), c = 13.64(3) Å, $\alpha=90.47(18)°$, $\beta=104.38(21)°$, $\gamma=95.52(19)°$; density, $d_{calc}=2.35$ g/cm³, $d_{obs}=2.31$g/cm³ [21] IR spectrum illustrated in [14, 25], ν(coordinated CO_3), 1055, 815, 735, 675 to 650 cm⁻¹; ν(ionic[a] CO_3), 1070, 870 cm⁻¹; ν(coordinated + ionic[a] CO_3), 1460, 1330 cm⁻¹ [11]; ν_1(CO), 1535 cm⁻¹; ν_5(CO), 1378, 1340 cm⁻¹; ν_2(CO), 1067, 1054, 1050 cm⁻¹; δ(CO) + δ(ThO_2C), 735, 725 cm⁻¹ [14]; ν(Th–O(CO_2)), 300, 265 cm⁻¹ [26], full details of the spectrum (1564 to 265 cm⁻¹) are given in [26] solubility product in neutral aqueous carbonate solution, $S=\{[Th(CO_3)_5]^{6-}\}[Na^+]^6=1.37$ [19] diamagnetic susceptibility, $\chi_m=-283$ to -316×10^{-6} [33]
$Na_6[Th(CO_3)_5]\cdot 14H_2O$	diamagnetic susceptibility, $\chi_m=-312$ to -341×10^{-6} [33]
$Na_6[Th(CO_3)_5]\cdot 20H_2O$	loses $8H_2O$ in air [4, 10] or at 25°C [2], loses 8 to $9H_2O$ at 20 to 30°C [2, 4] and loses $19H_2O$ at 70°C [4] crystal optical properties: refractive indices, $n_\gamma=1.476$, $n_\beta=1.470$, $n_\alpha=1.462$ [2, 4] crystallographic data: monoclinic, a:b:c = 1.461:1:1.495, $\beta=106.2°$ [2, 4]
$K_6[Th(CO_3)_5]\cdot 8H_2O$	probably triclinic [10]
$K_6[Th(CO_3)_5]\cdot 10H_2O$	reported as $3K_2CO_3\cdot Th(CO_3)_2\cdot 10H_2O$ in "Thorium" 1955, p. 334
$K_6[Th(CO_3)_5]\cdot 13H_2O$	[4]
$Tl_6[Th(CO_3)_5]$	[1, 13], reported as $3Tl_2CO_3\cdot Th(CO_3)_2$ in "Thorium" 1955, p. 352

References for 15.1.3 on pp. 13/4

Table 2 (continued)

$Tl_6[Th(CO_3)_5]\cdot H_2O$	[4, 6], decomposes at 180 to 380°C [13], thermal decomposition curve illustrated in [6]
$Tl_6[Th(CO_3)_5]\cdot 2H_2O$	loses $2H_2O$ at 50 to 180°C [13]
$(NH_4)_6[Th(CO_3)_5]\cdot 3H_2O$	[4], thermal decompositon curve illustrated in [6]
$(NH_4)_3(CN_3H_6)_3[Th(CO_3)_5]\cdot 3H_2O$ (CN_3H_6 = guanidinium)	[6]
$(CN_3H_6)_6[Th(CO_3)_5]$	decomposes above 150°C [7], 180 to 200°C [4], above 200°C [18], 230 to 310°C [23] IR spectrum: ν_1(CO), 1590(?) cm^{-1}; ν_5(CO), 1350 cm^{-1}; ν_2(CO), 1059, 1047 cm^{-1}; ν_6(CO), 859 cm^{-1}; δ(CO) + $\delta(ThO_2C)$, 723 cm^{-1}; ν(unidentate CO_3)(?), 1485 cm^{-1} [14] IR spectrum illustrated in [14] diamagnetic susceptibility, $\chi_m = -318$ to -324×10^{-6} [33]
$(CN_3H_6)_6[Th(CO_3)_5]\cdot H_2O$	loses H_2O at 70°C [4]
$(CN_3H_6)_6[Th(CO_3)_5]\cdot 3H_2O$	[6], loses $3H_2O$ at 50 to 80°C [7] crystal optical properties: refractive indices, $n_\gamma = 1.853$, $n_\alpha = 1.530$ [7]
$(CN_3H_6)_6[Th(CO_3)_5]\cdot 4H_2O$	reported as $3(CN_3H_6)_2CO_3\cdot Th(CO_3)_2\cdot 4H_2O$ in "Thorium" 1955, p. 343; loses $3H_2O$ below 70°C [4], loses $4H_2O$ below 100°C [18, 23]; thermal decomposition curve illustrated in [7, 18], see also [6]
$(CN_3H_6)_6[Th(CO_3)_5]\cdot 4H_2O$	crystallographic data: monoclinic, space group $P2_1/a$-C_{2h}^5 (No. 14), lattice parameters, a = 16.81(2), b = 12.91(1), c = 8.13(1) Å, β = 108.8(2)°; Z = 2, density, $d_{calc} = 1.91$, $d_{obs} = 1.94$ g/cm³ [17]; monoclinic, space group, $B2/b$-C_{2h}^6 (No. 15) or Bb-C_s^4 (No. 9), lattice parameters, a = 16.28(2), b = 16.81(2), c = 12.91(2) Å, β = 108.8(1)°; Z = 4, density, $d_{calc} = 1.92$, $d_{obs} = 1.94$ g/cm³ [18]; monoclinic, space group Bb-C_s^4 (No. 9), lattice parameters, a = 16.15(3), b = 16.70(3), c = 13.23(3) Å, β = 108.41(22)°; Z = 4, density, $d_{calc} = 1.92$, $d_{obs} = 1.94$ g/cm³ [20] crystal optical properties: refractive indices, $n_\gamma = 1.583$, $n_\alpha = 1.539$ [7] IR spectrum: ν_1(CO), 1600(?) cm^{-1}; ν_5(CO), 1370 cm^{-1}; ν_2(CO), 1056, 1048 cm^{-1}; ν_6(CO), 860 cm^{-1}; δ(CO) + $\delta(ThO_2C)$, 723 cm^{-1} [14]; $\nu(ThO(CO_2))$, 310, 250 cm^{-1} [18]; ν(unidentate CO_3(?)), 1480 cm^{-1} [14] IR spectrum illustrated in [14, 25] solubility product in neutral aqueous carbonate solution, $S = [\{Th(CO_3)_5\}^{6-}][(CN_3H_6)^+]^6 = 0.109$ [19]; binding energies, $Th5d_{5/2} = 87.4$ eV, N1s = 400.4 eV [34] diamagnetic susceptibility, $\chi_m = -272$ to -407×10^{-6} [33]
$Ca_3[Th(CO_3)_5]\cdot 7H_2O$	[2, 4]
$Ba_3[Th(CO_3)_5]\cdot 7H_2O$	[2, 4]

Table 2 (continued)

$[Co(NH_3)_6]Th(CO_3)_{3.5}\cdot 7H_2O$(?)	whitish pink or pink [8]
$[Co(NH_3)_6]_2[Th(CO_3)_5]\cdot H_2O$	decomposes above 110°C [8]
$[Co(NH_3)_6]_2[Th(CO_3)_5]\cdot 4H_2O$	reported as $[Co(NH_3)_6]_2[Th(CO_3)_5(H_2O)]\cdot 3H_2O$ in [15, 16], pale yelow [14], orange [15, 16], pink [8] IR spectrum: ν_1(CO), 1665, 1641 cm^{-1}; ν_5(CO), 1345 cm^{-1}; ν_2(CO), 1053 cm^{-1}; ν_6(CO), 870 cm^{-1}; δ(CO)+$\delta(ThO_2C)$, 778(?), 782 cm^{-1}; ν(unidentate CO_3(?)), 1441 cm^{-1} [14] IR spectrum (4000 to 400 cm^{-1}) illustrated in [14] crystal optical properties: refractive indices, $n_\gamma \simeq n_\alpha = 1.597$ [8] X-ray powder diffraction data are given in [15]
$[Co(NH_3)_6]_2[Th(CO_3)_5]\cdot 6H_2O$	orange [8], loses $5H_2O$ at 110°C, thermal decomposition curve illustrated in [8] crystal optical properties: refractive indices, $n_\gamma = 1.601$, $n_\alpha = 1.587$ [8]
$[Co(NH_3)_6]_2[Th(CO_3)_5]\cdot 7H_2O$(?)	thermal decomposition curve illustrated in [8]
$[Co(NH_3)_6]_2[Th(CO_3)_5]\cdot 8H_2O$	orange [8]
$[Co(NH_3)_6]_2[Th(CO_3)_5]\cdot 9H_2O$	orange or ruby red [8], dark yellow [14], loses $3H_2O$ on standing in air or at 90°C [8], thermal decomposition curve illustrated in [6, 8] IR spectrum: ν_1(CO), 1500 cm^{-1}; ν_5(CO), 1370 cm^{-1}; ν_2(CO), 1051 cm^{-1}; ν_6(CO), 860 cm^{-1}; δ(CO)+$\delta(ThO_2C)$, 726 cm^{-1} [14] IR spectrum (4000 to 400 cm^{-1}) illustrated in [14] diamagnetic susceptibility, $\chi_m = -210\times10^{-6}$ [33]
$[Co(NH_3)_6]_2[Th(CO_3)_5]\cdot 9$ to $10H_2O$	ruby red [8]
$[Co(NH_3)_6]_2[Th(CO_3)_5]\cdot 10H_2O$	orange [4]
$[Co(NH_3)_6]_2[Th(CO_3)_5]\cdot nH_2O$	orange [6]

[a] In [11] incorrectly formulated as $Na_4[Th(CO_3)_4]\cdot Na_2CO_3$ [14].
[b] Some adsorbed water present [14].

15.1.3.1 Tricarbonatothorates

The hydrated ammonium salt, reported as $(NH_4)_2CO_3\cdot Th(CO_3)_2\cdot 6H_2O$ in "Thorium" 1955 (p. 339), is described as $(NH_4)_2[Th(CO_3)_3(H_2O)_3]\cdot 3H_2O$ in [4]. However, there is no evidence to support the presence of water within the coordination sphere, and this product requires further investigation, for the attempted preparation of this compound by the published method [27] yielded $(NH_4)_6[Th(CO_3)_5]\cdot 3H_2O$ [2].

The anhydrous guanidinium salt, $(CN_3H_6)_2[Th(CO_3)_3]$, is obtained when the tetrahydrated salt is heated at up to 100°C; it decomposes above 200°C [18]. The tetrahydrate has been prepared by shaking a small quantity (5×10^{-4} mol) of $(CN_3H_6)_6[Th(CO_3)_5]\cdot 4H_2O$ (p. 10) with aqueous (<0.1 M, 50 mL) $(CN_3H_6)_2CO_3$ or $(CN_3H_6)HCO_3$ for 72 hours, and by shaking $(CN_3H_6)_4[Th(CO_3)_4]\cdot 6H_2O$ (p. 8, 2.5×10^{-3} mol) with water (50 mL) [18]. The sodium salt could not be prepared in this way.

References for 15.1.3 on pp. 13/4

15.1.3.2 Tetracarbonatothorates

The anhydrous sodium and guanidinium salts remain on heating the hydrates at up to 100°C; the former decomposes above 300°C and the latter above 200°C [18]. The hydrated salts, $Na_4[Th(CO_3)_4]\cdot 7H_2O$ and $(CN_3H_6)_4[Th(CO_3)_4]\cdot 6H_2O$, have been obtained by shaking small quantities (5×10^{-4} mol) of the corresponding pentacarbonatothorates(IV) (see below) with an aqueous solution (0.10 to 0.20 M) of the appropriate carbonate or acid carbonate, the product being dried in air [18]. The IR spectra of the hydrates indicate that the carbonate groups are bidentate and it has been suggested that the complex anion is probably dodecahedral (D_{2d}) [26].

Studies of the solubilities of the tetracarbonatothorates(IV) in aqueous carbonate or hydrogen carbonate show that the major species in solution are pentacarbonatothorates(IV) (see Section 15.1.3.3, below) in equilibrium with the solid tetracarbonato complex salt [19].

15.1.3.3 Pentacarbonatothorates

The anhydrous sodium [2, 4, 9, 10, 12, 18, 23], thallium [13] and guanidinium [4, 7, 14, 18, 23] salts, $M^I_6[Th(CO_3)_5]$, have been obtained by heating the appropriate hydrated salts at a variety of temperatures (see Table 2, pp. 4/7). There is also some disagreement concerning the reported decomposition temperatures of the anhydrous salts. The morphological properties of the thallium salt have been reported [1].

The IR spectra of $Na_6[Th(CO_3)_5]\cdot xH_2O$ (x = 0 and 12) have been interpreted as indicating that the salts have the composition $Na_4[Th(CO_3)_4]\cdot Na_2CO_3\cdot xH_2O$ [11], but this is incorrect [14]; the spectra indicate that the CO_3 groups are bridging or bidentate [14]. Other authors [5] have proposed that the dodecahydrate could be regarded as either $Na_4[Th(CO_3)_4]\cdot Na_2CO_3\cdot 12H_2O$ or as $Na_2[Th(CO_3)_3]\cdot 2Na_2CO_3\cdot 12H_2O$, with the latter favoured on the grounds that the thorium atom would be 6-coordinate, but both formulations are incorrect [14].

Studies of the thermal dehydration of $Na_6[Th(CO_3)_5]\cdot 12H_2O$ indicate that the last molecule of water is lost at about 70°C and that this final step is strongly endothermic. These observations led to the suggestion that one molecule of water is bonded to the thorium atom in the complex anion and that the hydrates should be written as $Na_6[Th(CO_3)_5(H_2O)]\cdot(x-1)H_2O$, with an octahedral anion involving monodentate CO_3 groups [2]; the tetrahydrated guanidinium salt has also been reported as $(CN_3H_6)_6[Th(CO_3)_5(H_2O)]\cdot 3H_2O$ [7]. However, X-ray crystallographic structure determinations (pp. 9 and 11) show that this view is incorrect and that water is not bonded to the thorium atom [21].

$Na_6[Th(CO_3)_5]\cdot 20H_2O$ has been prepared [2] by the method originally reported [27] for the dodecahydrate and also [2] by adding an aqueous solution of $Th(NO_3)_4$ to a 4- to 8-fold excess of aqueous 1M Na_2CO_3 or 1M $NaHCO_3$. Crystals of the hydrate separate after standing for several days, but these slowly lose water to form $Na_6[Th(CO_3)_5]\cdot 12H_2O$ [2]. Both hydrates crystallise simultaneously if an insufficient excess of the sodium carbonate is present. Crystals of $Na_6[Th(CO_3)_5]\cdot 20H_2O$ are also obtained when a solution of $Th(C_2O_4)_2\cdot 5H_2O$ (p. 82) in aqueous 1M Na_2CO_3 is allowed to stand [32].

$Na_6[Th(CO_3)_5]\cdot 12H_2O$ is obtained as a powder when the stoichiometric quantity of solid Na_2CO_3 is added to aqueous $Th(NO_3)_4$ [2]. In an alternative preparation of the dodecahydrate, aqueous 1M $Th(NO_3)_4$ is added to an aqueous solution of Na_2CO_3 or $NaHCO_3$ ($[CO_3^{2-}]$ or $[HCO_3^-] > 0.3$M) with vigorous stirring until the onset of persistent crystallisation [18]. The complex salt is reported to be rather easily hydrolysed and if the solution containing the reactants is heated instead of allowing it to evaporate at room temperature, hydrates with 5, 10 or 11 H_2O are obtained [4].

Studies of the solubility of thorium in the systems $ThOCO_3 - Na_2CO_3-NaClO_4-H_2O$ [5] and $Th(CO_3)_2-Na_2CO_3-H_2O$ [9] (for $ThO(CO_3)$ and $Th(CO_3)_2$, see p. 2) suggest that the solid phase is $Na_6[Th(CO_3)_5]\cdot12H_2O$ and solubility data for thorium in aqueous solutions containing Na_2SO_4 and Na_2CO_3 indicate that the solid phase is $Na_6[Th(CO_3)_5]\cdot xH_2O$ [3]. The crystallisation field of $Na_6[Th(CO_3)_5]\cdot12H_2O$ has been partially defined in a study of the solubility isotherm in the aqueous system $Th(C_2O_4)-Na_2CO_3-Th(CO_3)_2-Na_2C_2O_4$ at 25°C [29]. It has been reported that the dodecahydrate dissolves incongruently in water [9] and that the water in this hydrate is zeolitic in character [12]. The salt loses water to form the dihydrate at 50°C [2] or 75 to 80°C [9], whereas the monohydrate is said to be formed at 70°C [2]; the IR spectrum of the trihydrate has been reported [14], but it is not clear how this hydrate was obtained. The IR spectra of products obtained by heating $Na_6[Th(CO_3)_5]\cdot12H_2O$ at ~130 to 150°C, ~500 to 600°C, and ~900°C have also been reported [14].

Crystals of $Na_6[Th(CO_3)_5]\cdot12H_2O$ [4] and $Na_6[Th(CO_3)_5]\cdot20H_2O$ [2, 4] were originally reported to possess monoclinic symmetry; subsequently two forms of the dodecahydrate were reported, both possessing triclinic symmetry, with the two cells related [10] and triclinic symmetry has been confirmed [18]. The structure of the complex anion [21] consists of five bidentate CO_3 groups bonded to the thorium atom, which is surrounded by ten oxygen atoms at the vertices of an irregular decahexahedron [21, 23] (**Fig. 1**). It is isomorphous with $Na_6[Ce(CO_3)_5]\cdot12H_2O$ [22]. Interatomic distances in $[Th(CO_3)_5]^-$ are shown in Table 3.

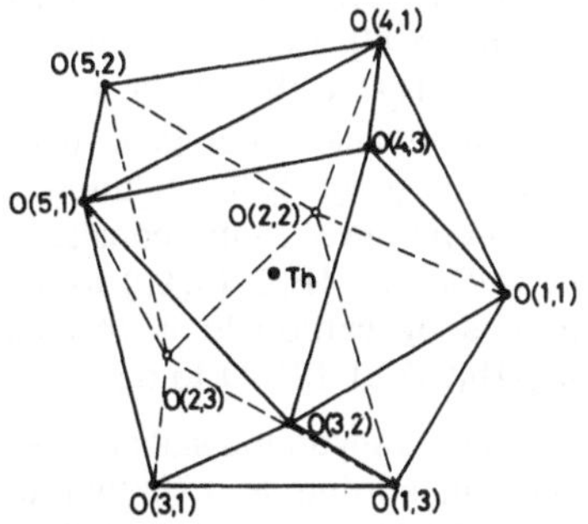

Fig. 1. The $[Th(CO_3)_5]^{6-}$ irregular decahexahedron formed by the O atoms of the carbonate groups around the Th atom in $Na_6[Th(CO_3)_5]\cdot12H_2O$ [21].

Table 3
Interatomic Distances (in Å) of $[Th(CO_3)_5]^{6-}$ (see Fig. 1) in $Na_6[Th(CO_3)_5]\cdot12H_2O$ [21].

Th–O(1,1)	2.467(9)	Th–O(3,2)	2.453(9)
Th–O(1,3)	2.515(9)	Th–O(4,1)	2.483(9)
Th–O(2,2)	2.463(9)	Th–O(4,3)	2.542(10)
Th–O(2,3)	2.547(8)	Th–O(5,1)	2.469(11)
Th–O(3,1)	2.561(9)	Th–O(5,2)	2.553(9)

The potassium and thallium salts have been less extensively studied. $K_6[Th(CO_3)_5]\cdot13H_2O$ is mentioned in [4], and the octahydrate is reported to be formed when aqueous $Th(NO_3)_4$ and K_2CO_3 are mixed; the initial precipitate redissolves on stirring and, on addition of ethanol to the solution, an oily liquid separates which finally yields the product as a powder [10].

The thallium compound, $Tl_6[Th(CO_3)_5]\cdot H_2O$, is precipitated when saturated aqueous Tl_2SO_4 or $TlNO_3$ is added to an aqueous solution of thorium hydroxide in $(NH_4)_2CO_3$ [4, 6]. Thermal

References for 15.1.3 on pp. 13/4

dehydration studies indicate that the thallium compound has the greatest thermal stability of all the known pentacarbonato salts [6]. A thermogravimetric study of $Tl_6[Th(CO_3)_5]\cdot 2H_2O$, apparently obtained in the same way as the monohydrate, led the authors to the surprising conclusion that only two CO_3 groups were bonded to the thorium atom and that the formula should be written as $3Tl_2CO_3\cdot[Th(CO_3)_2(H_2O)_2]$ [13], which seems highly unlikely.

The only recorded ammonium salt is the trihydrate, $(NH_4)_6[Th(CO_3)_5]\cdot 3H_2O$, prepared by adding ethanol [4] or acetone [4, 6] to a solution of freshly precipitated thorium hydroxide in aqueous 2 M $(NH_4)_2CO_3$. The complex salt separates as an oil which is gradually converted to a white powder on careful agitation with acetone [6]. This salt is very unstable, losing NH_3, CO_2 and H_2O in air. When a solution of the salt at pH ca. 7 is heated, $Th(OH)_2CO_3\cdot xH_2O$ (p. 2) is formed and hydrated ThO_2 results on heating a solution at pH 5 to 6 [6].

A mixed ammonium guanidinium salt, $(NH_4)_3(CN_3H_6)_3[Th(CO_3)_5]\cdot 3H_2O$, is precipitated when aqueous $(CN_3H_6)NO_3$ or $(CN_3H_6)Cl$ is added to a solution of thorium hydroxide in aqueous $(NH_4)_2CO_3$, but products with other ratios of the cations NH_4^+: $(CN_3H_6^+)$ are also obtained [6] and it is possible that these are mixtures of ammonium and guanidinium salts.

The tetrahydrated guanidinium salt, $(CN_3H_6)_6[Th(CO_3)_5]\cdot 4H_2O$, has been prepared by the addition of aqueous $(CN_3H_6)Cl$ or $(CN_3H_6)NO_3$ to a solution of thorium hydroxide in aqueous ammonium carbonate [6] and this salt is precipitated when thorium hydroxide, nitrate, sulfate, oxalate (p. 78) or basic carbonate (p. 2) is treated with at least a twofold excess of aqueous $(CN_3H_6)_2CO_3$ [7]. A simple preparative route consists in dissolving solid $Th(C_2O_4)_2\cdot 6H_2O$ (p. 82) in saturated $(CN_3H_6)_2CO_3$ solution, then leaving the solution to stand overnight in air to crystallise [7]. When a mixture of $Th(C_2O_4)_2\cdot 6H_2O$ and solid $(NH_4)_2C_2O_4$ is heated in 30 to 100 cm^3 H_2O with 2 to 5 moles of $(CN_3H_6)_2CO_3$ until dissolution is almost complete the carbonatooxalato complex, $(CN_3H_6)_6[Th_2(CO_3)_5](C_2O_4)_2$ is obtained, but with an excess of saturated $(CN_3H_6)_2CO_3$ solution, $Th(C_2O_4)_2\cdot 6H_2O$ yields $(CN_3H_6)_6[Th(CO_3)_5]\cdot 4H_2O$ [32]. A similar preparative route consists in addition of solid $(CN_3H_6)_4[Th(C_2O_4)_4]\cdot 2H_2O$ (p. 94) or $K_4[Th(C_2O_4)_4]\cdot 4H_2O$ (p. 92) to aqueous 0.5 M $(CN_3H_6)_2CO_3$ [30]. The replacement of oxalate groups by carbonate is reversible in that addition of solid $(CN_3H_6)_6[Th(CO_3)_5]\cdot 4H_2O$ or $Na_6[Th(CO_3)_5]\cdot 12H_2O$ to aqueous $K_2C_2O_4$ yields $K_4[Th(C_2O_4)_4]\cdot 4H_2O$ [30].

The tetrahydrated salt has also been prepared by adding aqueous 1 M $Th(NO_3)_4$ slowly, with vigorous stirring, to a solution of $(CN_3H_6)_2CO_3$ or $(CN_3H_6)HCO_3$ (concentration >0.3 M) up to the onset of persistent crystallisation [18] and it is precipitated when $CN_3H_6NO_3$ is added to a solution of $Th(NO_3)_4$ in saturated aqueous $(NH_4)_2CO_3$ [30] and by adding a saturated aqueous solution of $(CN_3H_6)_2CO_3$ to a solution of $Th(NO_3)_4$ in aqueous $(NH_4)_2CO_3$ until the onset of crystallisation [30]. The effect of the presence of NH_4Cl on the precipitation of the salt has also been studied [31].

The monohydrate, $(CN_3H_6)_6[Th(CO_3)_5]\cdot H_2O$, results when the tetrahydrate is heated at up to 70°C [4]. The hydrates slowly effloresce in air and are soluble in saturated aqueous $(CN_3H_6)_2CO_3$ solution, particularly on warming, and in aqueous alkali metal or ammonium carbonate, oxalate, and halides [7]. The binding energies for the Th $5d_{5/2}$ and guanidine N1s electrons for the tetrahydrate (Table 2, p. 6) were obtained from the X-ray electron spectra [34] under high vacuum, so that some, or all, of the water of crystallisation may have been lost.

When aqueous $Th(NO_3)_4$ is added to a large excess of aqueous $(CN_3H_6)_2CO_3$ the initial precipitate redissolves, and then $(CN_3H_6)_6[Th(CO_3)_5]\cdot 4H_2O$ precipitates; when the proportion of $Th(NO_3)_4$ is doubled, the initial precipitate again redissolves and, on stirring, the tetrahydrated salt separates as before. However, when the tetrahydrate is left in contact with the mother liquor for two or more hours, the trihydrated salt is obtained. This last is also obtained by stirring an aqueous solution of $Th(NO_3)_4$ (1 mol) with saturated aqueous $(CN_3H_6)_2CO_3$ (ca.

7.65 mol), when an oily mass separates; dilution of the mother liquor with one third of its volume of water with stirring then yields the solid trihydrate. This is slowly hydrolysed in water [7].

In the structure of the anion of the tetrahydrate all five carbonate groups are bidentate and the thorium atom is surrounded by ten oxygen atoms at the vertices of an irregular decahexahedron [20, 23], as in the corresponding guanidinium cerium(IV) salt [28] and the sodium salt (p. 9) [21, 23]. The structure of the $[Th(CO_3)_3]^{6-}$ entity in the sodium salt is shown in Fig. 1, p. 9. Th–O distances and angles of the (CN_3H_6) salt are shown in Table 4.

Table 4
Distances and Angles in the $[Th(CO_3)_5]^{6-}$ Polyhedron of $[CN_3H_6]_6[Th(CO_3)_5]\cdot 4H_2O$ [23].

Th–O distances in Å

Th–O(1,2) = 2.47(1)	Th–O(3,1) = 2.53(1)	Th–O(4,2) = 2.50(1)
Th–O(1,3) = 2.48(1)	Th–O(3,2) = 2.50(1)	Th–O(5,1) = 2.47(1)
Th–O(2,1) = 2.49(1)	Th–O(4,1) = 2.54(1)	mean 2.49(1)
Th–O(2,3) = 2.54(1)		

angles relative to the pyramid edges

O(1,2)–Th–O(1,3) = 52.4(3)°	O(3,2)–Th–O(3,1) = 51.8(3)°
O(1,2)–Th–O(4,2) = 71.0(3)°	O(3,2)–Th–O(4,1) = 75.6(3)°
O(1,2)–Th–O(2,1) = 68.8(3)°	O(3,2)–Th–O(2,3) = 66.3(2)°
O(1,2)–Th–O(5,2) = 70.9(3)°	O(3,2)–Th–O(5,1) = 68.7(3)°
mean 65.8(3)°	mean 65.7(3)°

angles relative to the prism edges

O(1,3)–Th–O(5,1) = 70.0(3)°	O(4,1)–Th–O(4,2) = 53.3(3)°
O(5,1)–Th–O(5,2) = 51.1(3)°	O(4,2)–Th–O(3,1) = 79.8(3)°
O(5,2)–Th–O(2,3) = 78.0(3)°	O(3,1)–Th–O(1,3) = 77.6(3)°
O(2,3)–Th–O(2,1) = 52.0(3)°	mean 66.9(3)°
O(2,1)–Th–O(4,1) = 73.9(3)°	total mean 72.1(3)°
	O(1,2)–Th–O(3,2) = 171.8(2)°

angles relative to the base edges

O(4,2)–Th–O(1,3) = 74.7(3)°	O(4,1)–Th–O(2,3) = 74.7(3)°
O(1,3)–Th–O(5,2) = 80.0(3)°	O(2,3)–Th–O(5,1) = 98.0(3)°
O(5,2)–Th–O(2,1) = 81.1(3)°	O(5,1)–Th–O(3,1) = 71.2(3)°
O(2,1)–Th–O(4,2) = 88.0(3)°	O(3,1)–Th–O(4,1) = 80.4(3)°
mean 80.9(3)°	mean 81.1(3)°

The replacement of Na^+ by $CN_3H_6^+$ in the last leads only to a slight deformation of the polyhedron [23].

The calcium and barium salts, $M^{II}_3[Th(CO_3)_5]\cdot 7H_2O$, have been prepared by metathesis of aqueous $Na_6[Th(CO_3)_5]\cdot 12H_2O$ (p. 8) with aqueous $CaCl_2$ [4], $Ca(NO_3)_2$ [2], or $BaCl_2$ [2, 4]. Little is known about them.

References for 15.1.3 on pp. 13/4

A solid of composition $[Co(NH_3)_6]Th(CO_3)_{3.5} \cdot 7H_2O$ has been reported to be formed when a solution of thorium(IV) hydroxide or nitrate in aqueous $(NH_4)_2CO_3$ is treated with an aqueous solution of $[Co(NH_3)_6]Cl_3$ [8]; it is also obtained as a precipitate when equal volumes of solutions of thorium(IV) salts and $[Co(NH_3)_6]Cl_3$ in saturated aqueous $(NH_4)_2CO_3$ are mixed and the clear solution is diluted with three times its volume of water [8]. The nature of this product is uncertain.

Several hydrates of the hexammine cobalt(III) salt, $[Co(NH_3)_6]_2[Th(CO_3)_5]$ have been recorded. The formation of the monohydrate has been observed in the thermal dehydration of the higher hydrates and the tetrahydrate is precipitated as an orange solid by adding aqueous $[Co(NH_3)_6]Cl_3$ gradually to a solution containing thorium(IV) in aqueous 2M $(NH_4)_2CO_3$ at pH 9.0 [15] or 9.0 to 9.25 [16]. The formula is written as $[Co(NH_3)_6]_2[Th(CO_3)_5(H_2O)] \cdot 3H_2O$ in [15, 16]. The tetrahydrate is also obtained when the octahydrate (see below) is washed with acetone [8]. Studies of the precipitation of the tetrahydrate have led to the proposal of $[Co(NH_3)_6]Cl_3$ as a reagent for the gravimetric and radiometric determination of thorium [16]. The tetrahydrate can also be used as a carrier for trace level amounts of Eu^{III}, Am^{III}, Cm^{III}, and Np^{V} [24].

The orange hexahydrate is reported to be formed when the stoichiometric quantity of $[Co(NH_3)_6]Cl_3$ as a saturated solution in aqueous $(NH_4)_2CO_3$ is added to a solution of $Th(NO_3)_4$ in aqueous $(NH_4)_2CO_3$; rhombic crystals of the product separate after dilution of the mixture with twice its volume of water. These crystals gradually change to pyramids of the same composition on standing in the mother liquor [8]. The hexahydrate is also formed when $[Co(NH_3)_6]_2$-$[Th(CO_3)_5] \cdot 9H_2O$ (see below) effloresces in air [8].

The heating curve for a heptahydrate is reported in [8], but no preparative details are given.

The octahydrate has also been recorded; when a solution of $[Co(NH_3)_6]Cl_3$ in 1 to 5% aqueous $(NH_4)_2CO_3$ is added to a solution of $Th(NO_3)_4$ in saturated aqueous $(NH_4)_2CO_3$, acicular orange crystals separate which change rapidly to a fine powder of composition $[Co(NH_3)_6]_2[Th(CO_3)_5] \cdot 9$ to $10H_2O$. Attempts to isolate the acicular crystals yielded an orange powder which, after drying in air, analysed as the octahydrate [8].

$[Co(NH_3)_6]_2[Th(CO_3)_5] \cdot 9H_2O$ has been obtained by adding a saturated solution of $(NH_4)_2CO_3$ to aqueous $Th(NO_3)_4$, followed by addition of a saturated solution of $[Co(NH_3)_6]Cl_3$, also in saturated aqueous $(NH_4)_2CO_3$; ruby red crystals of the enneahydrate separate on standing for 1 day. These are readily soluble in aqueous $(NH_4)_2CO_3$, so that the yield of the product depends on the final concentration of $(NH_4)_2CO_3$ in the solution. Thus, the yield increases as the concentration of $(NH_4)_2CO_3$ used to dissolve the cobalt(III) salt decreases and is 100% at concentrations close to 2.5%, when the product separates as orange crystals [8]. The large crystals of the enneahydrate are hydrolysed on prolonged standing in water or, more rapidly, on heating [8]. See also [6].

$[Co(NH_3)_6]_2[Th(CO_3)_5] \cdot 9$ to $10H_2O$ (see above) has also been prepared by slow crystallisation of the solution obtained by adding 1 volume of an aqueous $(NH_4)_2CO_3$ solution of thorium hydroxide to 1 to 3 volumes of a saturated solution of $[Co(NH_3)_6]Cl_3$. The product separates as ruby red crystals after standing for several days. Precipitation is said to be accelerated by stirring or by diluting with water, when the composition approaches that of the enneahydrate [8], although the same authors also report the formation of a product of composition $[Co(NH_3)_6]Th(CO)_{3.5} \cdot 7H_2O$ (see above) on dilution with water [8].

The orange decahydrate has been reported to be obtained by treating an aqueous $(NH_4)_2CO_3$ solution of thorium hydroxide or nitrate with a solution of $[Co(NH_3)_6]Cl_3$ [4] and this is presumably the same as the salt with 9 to $10H_2O$.

References for 15.1.3:

[1] Burakova, T. N. (Uch. Zap. Leningrad. Gos. Univ. No. 178 Ser. Geol. Nauk No. 4 [1954] 157/95; C.A. **1955** 8730).
[2] Chernyaev, I. I.; Golovnya, V. A.; Molodkin, A. K. (Zh. Neorgan. Khim. **3** [1958] 2671/86; Russ. J. Inorg. Chem. **3** No. 12 [1958] 100/19).
[3] Ingles, J. C.; Kelley, F. J. (Can. Dept. Mines Tech. Surv. Mines Branch Res. Rept. R-32 [1958]; C.A. **1959** 10932).
[4] Chernyaev, I. I.; Golovnia, V. A.; Molodkin, A. K. (Proc. 2nd Intern. Conf. Peaceful Uses At. Energy, Geneva 1958, Vol. 28, pp. 203/9).
[5] Zakharov-Nartsissov, O. I.; Mikhailov, G. G. (Izv. Vysshikh Uchebn. Zavadenii Khim. Khim. Tekhnol. **3** [1960] 45/8; C.A. **1960** 16116).
[6] Chernyaev, I. I.; Golovnya, V. A.; Molodkin, A. K. (Zh. Neorgan. Khim. **6** [1961] 394/9; Russ. J. Inorg. Chem. **6** [1961] 200/2).
[7] Chernyaev, I. I.; Molodkin, A. K. (Zh. Neorgan. Khim. **6** [1961] 587/92; Russ. J. Inorg. Chem. **6** [1961] 298/301).
[8] Chernyaev, I. I.; Molodkin, A. K. (Zh. Neorgan. Khim. **6** [1961] 809/15; Russ. J. Inorg. Chem. **6** [1961] 413/6).
[9] Luzhnaya, N. P.; Kovaleva, I. S. (Zh. Neorgan. Khim. **6** [1961] 1440/2; Russ. J. Inorg. Chem. **6** [1961] 738/40).
[10] Ryabchikov, D. I.; Volynets, M. P.; Zarinskii, V. A.; Ivanov, V. I. (Zh. Analit. Khim. **18** [1963] 348/56; J. Anal. Chem. [USSR] **18** [1963] 307/13).

[11] Karyakin, A. V.; Volynets, M. P. (Zh. Strukt. Khim. **3** [1962] 714/6; J. Struct. Chem. [USSR] **3** [1962] 689/90).
[12] Ivanov, V. I. (Sovrem. Metody Anal. **1965** 50/8; C.A. **64** [1966] 9019).
[13] Karkhanavala, M. D.; Daroowalla, S. H. (J. Indian Chem. Soc. **46** [1969] 729/35).
[14] Kharitonov, Yu. Ya.; Molodkin, A. K.; Balakaeva, T. A. (Zh. Neorgan. Khim. **14** [1969] 2761/7; Russ. J. Inorg. Chem. **14** [1969] 1453/6).
[15] Ueno, K.; Hoshi, M. (J. Inorg. Nucl. Chem. **32** [1970] 3817/22).
[16] Ueno, K.; Hoshi, M. (Radiochem. Radioanal. Letters **4** [1970] 221/9).
[17] Voliotis, S.; Faucherre, J.; Dervin, J. (Compt. Rend. C **274** [1972] 1163/5).
[18] Dervin, J.; Faucherre, J.; Herpin, P. (Bull. Soc. Chim. France **1973** 2634/7).
[19] Dervin, J.; Faucherre, J. (Bull. Soc. Chim. France **1973** 2930/3).
[20] Voliotis, S.; Rimsky, A. (Acta Cryst. B **31** [1975] 2612/5).

[21] Voliotis, S.; Rimsky, A. (Acta Cryst. B **31** [1975] 2615/20).
[22] Voliotis, S.; Rimsky, A. (Acta Cryst. B **31** [1975] 2620/2).
[23] Voliotis, S.; Fromage, F.; Faucherre, J.; Dervin, J. (Rev. Chim. Minérale **14** [1977] 441/6).
[24] Saito, A.; Morimoto, T.; Ueno, K. (Radiochem. Radioanal. Letters **43** [1980] 203/14).
[25] Jolivet, J.-P.; Thomas, Y.; Taravel, B.; Lorenzelli, V.; Busca, G. (J. Mol. Struct. **79** [1982] 403/8).
[26] Jolivet, J.-P.; Thomas, Y.; Taravel, B.; Lorenzelli, V.; Busca, G. (J. Mol. Struct. **102** [1983] 137/43).
[27] Rosenheim, A.; Samter, V.; Davidsohn, J. (Z. Anorg. Allgem. Chem. **35** [1903] 424/53).
[28] Voliotis, S.; Rimsky, A.; Faucherre, J. (Acta Cryst. B **31** [1975] 2607/11).
[29] Kovaleva, I. S.; Luzhnaya, N. P. (Zh. Neorgan. Khim. **7** [1962] 1693/8; Russ. J. Inorg. Chem. **7** [1962] 873/5).
[30] Molodkin, A. K.; Ivanova, O. M.; Skotnikova, G. A. (Zh. Neorgan. Khim. **9** [1964] 295/307; Russ. J. Inorg. Chem. **9** [1964] 162/8).

[31] Hoshi, M.; Ueno, K. (J. Nucl. Sci. Technol. **15** [1978] 585/8).
[32] Molodkin, A. K.; Skotnikova, G. A. (Zh. Neorgan. Khim. **9** [1964] 555/61; Russ. J. Inorg. Chem. **9** [1964] 308/11).
[33] Belova, V. I.; Syrkin, Ya. K.; Molodkin, A. K.; Ivanova, O. M.; Shiporina, L. M. (Zh. Neorgan. Khim. **13** [1968] 1458/60; Russ. J. Inorg. Chem. **13** [1968] 766/7).
[34] Nefedov, V. I.; Molodkin, A. K.; Salyn, Ya. V.; Ivanova, O. M.; Porai-Koshits, M. A.; Balakaeva, T. A.; Belyakova, Z. V. (Zh. Neorgan. Khim. **19** [1974] 2628/31; Russ. J. Inorg. Chem. **19** [1974] 1435/7).

15.1.4 Hydroxocarbonatothorates

The recorded hydroxocarbonatothorates(IV) are listed in Table 5. These compounds are mentioned in [1] without preparative details and their formulae were written to include water molecules bonded to the thorium atom, which is probably incorrect.

Table 5
Hydroxocarbonatothorates.

$Na[Th(CO_3)_2(OH)(H_2O)_3]\cdot 3H_2O$	[1]
$Na_2[Th(CO_3)_2(OH)_2(H_2O)_2]\cdot 8H_2O$	[1]
$Na_5[Th(CO_3)_4(OH)(H_2O)]\cdot 8H_2O$	[1]
$K_2[Th(CO_3)_2(OH)_2(H_2O)_2]\cdot 3H_2O$	[1]
$K_3[Th(CO_3)_3(OH)(H_2O)_2]\cdot 3H_2O$	[1]
$(CN_3H_6)_5[Th(CO_3)_3(OH)_3]\cdot 5H_2O$	[1] CN_3H_6 = guanidinium

Reference for 15.1.4:

[1] Chernyaev, I. I.; Golovnya, V. A.; Molodkin, A. K. (Proc. 2nd Intern. Conf. Peaceful Uses At. Energy, Geneva 1958, Vol. 28, pp. 203/9).

15.1.5 Fluorocarbonatothorates

Information on the known fluorocarbonatothorates(IV) is summarised in Table 6.

Table 6
Fluorocarbonatothorates.

$(NH_4)(CN_3H_6)_4[ThF_3(CO_3)_3]$	[1] (CN_3H_6 = guanidinium)
$(CN_3H_6)_5[ThF_3(CO_3)_3]$	decomposes to ThO_2 at 650°C [2] crystallographic data: orthorhombic, space group $P2_12_12_1-D_2^4=V^4$ (No. 19), lattice parameters, a = 9.53(2), b = 29.79(3), c = 9.11(2) Å; Z = 4, density, d_{calc} = 1.97, d_{obs} = 1.95 g/cm³ [3]; for solubility data, see [2]

The ammonium guanidinium salt, $(NH_4)(CN_3H_6)_4[ThF_3(CO_3)_3]$, is reported to crystallise from the solution obtained by dissolving hydrated $(CN_3H_6)_6[Th(CO_3)_5]$ in saturated aqueous NH_4F. When this product was left in the mother liquor, the composition changed to a product in which the ratio Th : CO_3 was approximately 2 : 3. The ammonium guanidinium salt is stable in air but is hydrolysed by water [1].

$(CN_3H_6)_5[ThF_3(CO_3)_3]$ has been prepared by heating a solution containing a mixture of $(CN_3H_6)F$ and $(CN_3H_6)_2CO_3$ ($1<[F^-]/[CO_3^{2-}]<6$) to 50 to 60°C, followed by very slow addition of a concentrated solution of $Th(NO_3)_4$ until the precipitate failed to redissolve on shaking. The filtrate was then left to crystallise for 20 min to 2 h [2]. The solubility of the salt in solutions containing F^- and CO_3^{2-} has been studied, and for $[F^-]<0.25$ M the solid was found to be the pentacarbonatothorate(IV) (p. 10) whereas with $[CO_3^{2-}]<0.05$ M it was ThF_4 [2]. The solubility (s) curves (log s = f log $[F^-]$) are illustrated in [2]. For fluoride and carbonate concentrations, respectively, $0.25 \leqq [F^-] \leqq 0.30$ M and $0.05 \leqq [CO_3^{2-}] \leqq 0.30$ M, the solubility is constant ($s = 3.2 \times 10^{-3}$ M) and in both cases the equilibria involve the $[ThF_3(CO_3)_3]^{5-}$ ion [2]. The thermal decomposition of the salt was also studied but no data are given in [2] other than that ThO_2 was formed at 650°C.

The structure of the complex anion in the guanidinium salt (**Fig. 2**) is a monocapped square antiprism with two of the three bidentate carbonate groups occupying two opposite edges of the square face and the third an edge of the pyramidal cap on the other square face [3]. Distances and angles are given in Table 7.

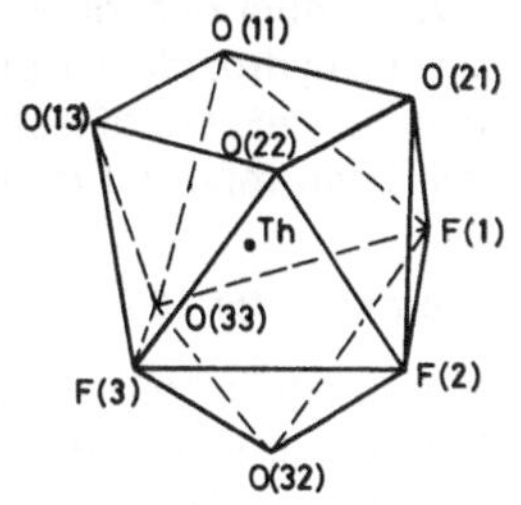

Fig. 2. The coordination polyhedron around the Th atom in the anion of the salt $(CN_3H_6)_5[ThF_3(CO_3)_3]$ [3].

Table 7
Interatomic Distances (in Å) and Angles (in °) of the $[ThF_3(CO_3)_3]^{5-}$ Polyhedron (see Fig. 2) of $(CN_3H_6)_5[ThF_3(CO_3)_3]$ [3].

distances					
Th–O(11)	2.51(2)	Th–O(22)	2.48(1)	Th–F(1)	2.27(1)
Th–O(13)	2.52(2)	Th–O(32)	2.54(2)	Th–F(2)	2.30(1)
Th–O(21)	2.56(2)	Th–O(33)	2.43(1)	Th–F(3)	2.29(1)
		mean Th–O	2.51(2)	mean Th–F	2.29(1)
angles					
pyramid edges		edges of the bases		edges of the prism	
O(32)–Th–O(33)	51.6(5)	F(1)–Th–F(2)	84.7(5)	O(33)–Th–O(11)	76.9(5)
O(32)–Th–F(1)	79.1(5)	F(2)–Th–F(3)	84.8(5)	O(11)–Th–F(1)	73.6(5)
O(32)–Th–F(2)	72.2(5)	F(3)–Th–O(33)	82.3(5)	F(1)–Th–O(21)	72.2(5)
O(32)–Th–F(3)	74.7(5)	O(33)–Th–F(1)	83.8(5)	O(21)–Th–F(2)	76.6(5)
mean	69.4(5)	O(11)–Th–O(13)	52.0(5)	F(2)–Th–O(22)	76.0(5)
		O(13)–Th–O(22)	73.9(5)	O(22)–Th–F(3)	77.4(5)
		O(22)–Th–O(21)	51.4(5)	F(3)–Th–O(13)	73.6(5)
		O(21)–Th–O(11)	75.3(6)	O(13)–Th–O(33)	79.5(5)
		mean	73.5(5)	mean	75.7(5)

References for 15.1.5 on p. 16

References for 15.1.5:

[1] Molodkin, A. K.; Ivanova, O. M.; Skotnikova, G. A. (Zh. Neorgan. Khim. **9** [1964] 295/307; Russ. J. Inorg. Chem. **9** [1964] 162/8).
[2] Dervin, J.; Fromage, F. (Bull. Soc. Chim. France **1975** 133/7).
[3] Voliotis, S. (Acta Cryst. B **35** [1979] 2899/904).

15.2 Thorium Cyanides and Cyano Complexes

15.2.1 Thorium Cyanides

Uncharacterised cyanide-containing products are obtained when $ThCl_4$ or $ThBr_4$ is treated with an alkali metal cyanide in anhydrous liquid ammonia, a reaction which yields $UCl_3CN \cdot 4NH_3$ (see "Uranium" Suppl. Vol. E 1, 1979, p. 14) in the case of UCl_4 [1]. Similarly, cyanide-containing products have resulted from the reaction of anhydrous HCN with dialkylamides, $Th(NR_2)_4$, in hexane or benzene; these are presumably of the form $(R_2N)_xTh(CN)_{4-x}$. Their IR spectra exhibit features at 2120 to 2150 cm^{-1} due to ν(C≡N) [2].

A product described as $Th[Th(CN)_8]$ (presumably $Th(CN)_4$) is said to be obtained when a thorium hydroxide gel, prepared by adding aqueous NH_3 to aqueous $Th(NO_3)_4$, is treated with a 10 to 20% excess of a stabilised solution of HCN in an autoclave [3]. No analyses or IR spectra have been reported for this product, which requires further investigation.

References for 15.2.1:

[1] Bagnall, K. W.; Baptista, J. O. (J. Inorg. Nucl. Chem. **32** [1970] 2283/5).
[2] Bagnall, K. W.; Baptista, J. O. (from Bagnall, K. W., Proc. 10th Rare Earth Res. Conf., Carefree, Ariz., 1973, Vol. 2, pp. 856/60).
[3] Paris, R. A.; Amblard, P. A.; Rousset, A. C. (Fr. 2157203 [1973]; C.A. **80** [1974] No. 49880).

15.2.2 Thorium Cyanometallates

The available information on compounds of this type is summarised in Table 8; all of the recent work has been concerned with the cyanoferrates(II).

Conductimetric and potentiometric titration of aqueous $Th(NO_3)_4$ against $K_4[Fe(CN)_6]$ indicates that $Th[Fe(CN)_6]$ is formed [1], but the compound was not isolated in this work. Thorium(IV) interferes with the precipitation of $K_4(UO_2)_4[Fe(CN)_6]_3$ [2] (see "Uranium" Suppl. Vol. C 13, 1983, p. 31).

The hexahydrate, $Th[Fe(CN)_6] \cdot 6H_2O$, is precipitated when aqueous 1M $Th(NO_3)_4 \cdot 6H_2O$ is added to aqueous 1M $K_4[Fe(CN)_6] \cdot 3H_2O$, the precipitate being subsequently washed and dried at room temperature [8]. Thermogravimetry and differential thermal analysis indicate that 2 molecules of water are lost at 150°C and at 200°C some $ThFe_2O_4$ is formed [8]. The Mössbauer spectrum also suggests that decomposition begins at 150°C [8]. A study of the thermal decomposition of the hydrated cyanoferrate(II) indicates that the process involves hydrolysis by the hydration water [4, 5]. The hydrolysis apparently begins during dehydration at room temperature, leading to a product of composition $Th_6(OH)_6Fe[Fe(CN)_6]_5 \cdot 24H_2O$ and further hydrolysis occurs at 160°C, with the formation of $Th_6(OH)_{12}Fe_2[Fe(CN)_6]_4 \cdot 3H_2O$ and, at 230°C, $Th_6(OH)_{18}Fe_3[Fe(CN)_6]_3$ appears to be formed, with HCN being lost at each stage [5]. It is not clear whether these hydrolysis products are genuine compounds or mixtures. There is also solution chemistry evidence for the formation of a basic thorium cyanoferrate(II) involving the ThO^{2+} ion [6].

Table 8

Thorium Cyanometallates.

$Th[Fe(CN)_6]\cdot 4H_2O$	decomposes at 200 to 400°C [8], see "Eisen" B, 1932, p. 1129; "Thorium" 1955, p. 301
$Th[Fe(CN)_6]\cdot 6H_2O$	loses $2H_2O$ at 150°C [8] IR spectrum, ν(CN), 2075 cm^{-1} [8]; Mössbauer spectrum reported in [7, 8]
$Th[Fe(CN)_6]\cdot xH_2O$	thermal decomposition curve illustrated in [5]; solubility product, $pK_s = 8.17$ at 25°C (1.0 M KNO_3, $\mu = 1.04$); for other solubility data, see [10]
$Na_{12}Th[Fe(CN)_6]_4\cdot 3H_2O$	water lost up to 450°C, decomposition at >450°C [9] IR spectrum, ν(CN), 2040 cm^{-1} [9]; X-ray powder diffraction data [9]
$Th_6(OH)_6Fe[Fe(CN)_6]_5\cdot 24H_2O$	[5]
$Th_6(OH)_{12}Fe_2[Fe(CN)_6]_4\cdot 3H_2O$	[5]
$Th_6(OH)_{18}Fe_3[Fe(CN)_6]_3$	[5]
$Th[Pt(CN)_4]_2\cdot 16H_2O$	see "Platin" C, 1940, p. 310 and "Thorium" 1955, p. 301

The isomer shifts for Fe^{II} in the Mössbauer spectrum of the hexahydrate do not appear to be affected very appreciably by the high charge of the thorium(IV) cation [7]. The ζ potentials have been determined for solutions of the thorium compound which is precipitated when 0.05 M aqueous solutions of $Th(NO_3)_4$ and $K_4[Fe(CN)_6]$ are mixed [3].

A product of composition $Na_{12}Th[Fe(CN)_6]_4\cdot 3H_2O$ is said to be precipitated when aqueous 0.03 M $Th(NO_3)_4$ is added to aqueous 0.015 M $Na_4[Fe(CN)_6]$ in the presence of 0.01 M HNO_3 at 80°C [9]. The white product became blue after drying at 80°C [9]. This compound has been studied for its use as an ion-exchange medium and the distribution coefficients for a variety of metal cations, including Th^{4+} and La^{3+}, have been reported [9].

References for 15.2.2:

[1] Gaur, J. N.; Bhattacharya, A. K. (J. Indian Chem. Soc. **31** [1954] 467/70).
[2] Sochevanov, V. G.; Shmakova, N. V.; Volkova, G. A. (Zh. Analit. Khim. **15** [1960] 77/83; J. Anal. Chem. [USSR] **15** [1960] 79/85).
[3] Sanyal, N. N.; Kundu, N. (J. Indian Chem. Soc. **43** [1966] 485/8).
[4] Seifer, G. B.; Makarova, Z. A. (Dokl. Akad. Nauk SSSR **169** [1966] 358/60; Dokl. Chem. Proc. Acad. Sci. USSR **166/171** [1966] 702/4).
[5] Seifer, G. B.; Makarova, Z. A. (Zh. Neorgan. Khim. **11** [1966] 1056/62; Russ. J. Inorg. Chem. **11** [1966] 570/3).
[6] Kourim, V.; Laznicek, M.; Dolezal, J. (J. Radioanal. Chem. **21** [1974] 355/62).
[7] Brar, A. S.; Date, S. K.; Sandhu, H. S.; Sandhu, S. S. (Indian J. Chem. Sect. A **19** [1980] 165/6).

[8] Brar, A. S.; Sandhu, H. S.; Sandhu, S. S. (Thermochim. Acta **41** [1980] 253/6).
[9] Srivastava, S. K.; Pal, N.; Agarwala, S.; Singh, R. J. (Ann. Chim. [Rome] **74** [1984] 461/9).
[10] Bellomo, A.; De Marco, D.; Casale, A. (Talanta **22** [1975] 197/9).

15.3 Thorium Thiocyanates and Thiocyanato Complexes

15.3.1 Introduction

A review of actinide thiocyanate compounds is available [1]. A considerable number of papers on nitrogen and oxygen donor complexes of thorium(IV) thiocyanate have been published and these are treated in "Thorium" Suppl. Vol. E, 1985.

The available X-ray crystallographic and vibrational spectroscopic evidence indicates that the thiocyanate group is N-bonded to the thorium atom. The formulae of all of the thiocyanates and thiocyanato complexes reported in this section are therefore written in this way, including those of the compounds for which such evidence is lacking.

Reference for 15.3.1:

[1] Bagnall, K. W. (MTP [Med. Tech. Publ. Co.] Intern. Rev. Sci. Inorg. Chem. Ser. One **7** [1972] 139/56).

15.3.2 Thorium Thiocyanates

The available information on thorium thiocyanates is summarised in Table 9.

Table 9
Thorium Thiocyanates.

$Th(NCS)_4$(?)	[8], decomposes to $ThO(NCS)_2$(?) in air at 280 to 450°C [11], 520 to 650°C [10], 560°C [9] or ca. 650°C [7]
$Th(NCS)_4\cdot4H_2O$	[1], loses H_2O at 100°C (partial decomposition) [4] or above 135°C [2] IR spectrum: ν(CN), ca. 2070, ν(CS), 800, 784, δ(NCS) 480 cm^{-1} [5], density, $d_4^{25}=2.413$ g/cm³ [3] crystal optical properties: refractive indices, $n_g=1.760$, $n_m=1.740$, $n_p=1.716$, mean refractive index = 1.739; molar refraction, $R_D=$ 89.2 cm³ [3] diamagnetic susceptibility, $\chi_m=-276$ to -335×10^{-6} [6]
$Th(NCS)_4\cdot4C_5H_5N$	see Th E, pp. 11, 13
$Th(NCS)_4\cdot4(2\text{-}CH_3C_5H_4N)$	see Th E, pp. 11, 13
$Th(NCS)_4\cdot4(2\text{-}H_2NC_5H_4N)$	see Th E, pp. 11, 13
$Th(NCS)_4\cdot4$(2,4- or 2,6-$(CH_3)_2C_5H_3N$)	see Th E, pp. 11, 13
$Th(NCS)_4\cdot4C_9H_7N$	C_9H_7N = quinoline or isoquinoline, see Th E, p. 15
$Th(NCS)_4\cdot2L$	L = bipyridine or 1,10-phenanthroline, see Th E, pp. 16, 17 and 16, 21, respectively

Table 9 (continued)

$Th(NCS)_4 \cdot 6C_4H_{12}N_4 \cdot 4H_2O$	$C_4H_{12}N_4$ = 1,4-diaminopiperazine, see Th E, pp. 19, 22
$Th(NCS)_4 \cdot 2C_{12}H_9N_3$	$C_{12}H_9N_3$ = 2-(2-pyridyl)benzimidazole, see Th E, pp. 20, 23
$Th(NCS)_4 \cdot 4(CO(NH_2)_2) \cdot 4H_2O$	see Th E, pp. 38, 40
$[Th(NCS)_4\{(CH_3)_2NCON(CH_3)_2\}_4]$	see Th E, pp. 42/4
$Th(NCS)_4 \cdot 4CH_3CON(CH_3)_2$	see Th E, pp. 49, 52
$Th(NCS)_4 \cdot 4C_2H_5CON(CH_3)_2$	see Th E, p. 54
$Th(NCS)_4 \cdot 4C_2H_5CON(C_2H_5)_2$	see Th E, p. 54
$Th(NCS)_4 \cdot 4(CH_3)_2CHCON(CH_3)_2$	see Th E, p. 55
$Th(NCS)_4 \cdot x(CH_3)_3CCON(CH_3)_2$	x = 3 or 4, see Th E, p. 55
$Th(NCS)_4 \cdot xC_{11}H_{12}N_2O$	x = 2 or 4, $C_{11}H_{12}N_2O$ = antipyrine, see Th E, pp. 61, 62
$Th(NCS)_4 \cdot xC_{11}H_{13}N_3O$	x = 1,2 or 4, $C_{11}H_{13}N_3O$ = 4-aminoantipyrine, see Th E, pp. 62, 63
$Th(NCS)_4 \cdot 3C_{23}H_{24}N_4O_2$	$C_{23}H_{24}N_4O_2$ = diantipyrylmethane, see Th E, pp. 62, 63
$Th(NCS)_4 \cdot 4C_5H_5NO$	C_5H_5NO = pyridine-N-oxide; see Th E, pp. 68, 73
$Th(NCS)_4 \cdot 4CH_3C_5H_4NO$	$CH_3C_5H_4NO$ = 2,3 or 4-methyl pyridine-N-oxide, see Th E, pp. 69, 73
$Th(NCS)_4 \cdot 2C_7H_9NO$	C_7H_9NO = 2,6-dimethyl pyridine-N-oxide, see Th E, pp. 70, 73
$Th(NCS)_4 \cdot 4C_9H_7NO$	C_9H_7NO = quinoline or isoquinoline-N-oxide, see Th E, pp. 71, 74
$Th(NCS)_4 \cdot 2C_{10}H_8N_2O$	$C_{10}H_8N_2O$ = 2,2′-bipyridyl-N-oxide, see Th E, pp. 71, 75
$Th(NCS)_4 \cdot C_{10}H_8N_2O_2$	$C_{10}H_8N_2O_2$ = 2,2′-bipyridyl-N,N′-dioxide, see Th E, pp. 72, 75
$Th(NCS)_4 \cdot 2C_{12}H_8N_2O$	$C_{12}H_8N_2O$ = 1,10-phenanthroline-N-oxide, see Th E, pp. 72, 76
$Th(NCS)_4 \cdot 2C_{12}H_8N_2O_2$	$C_{12}H_8N_2O_2$ = 1,10-phenanthroline-N,N′-dioxide, see Th E, pp. 72, 76
$Th(NCS)_4 \cdot 6(CH_3)_3PO$	see Th E, pp. 77, 80
$Th(NCS)_4 \cdot 4\{(CH_3)_2N\}_3PO$	see Th E, pp. 85, 86
$Th(NCS)_4 \cdot 2\{[(CH_3)_2N]_2PO\}_2O$	see Th E, p. 88
$Th(NCS)_4 \cdot 4(C_6H_5)_3PO$	see Th E, pp. 90, 91
$Th(NCS)_4 \cdot 4(C_6H_5)_2SO$	see Th E, pp. 105, 108
$Th(NCS)_4 \cdot 2C_4H_8OS$	C_4H_8OS = tetrahydrothiophene-1-oxide, see Th E, pp. 110, 111
$Th(NCS)_4 \cdot 3L$	L = $C_{14}H_{13}NO$, $C_{14}H_{13}NO_2$ or $C_{15}N_{15}NO_2$ (Schiff bases), see Th E, pp. 178, 182
$Th(NCS)_2(C_{14}H_{10}NO_2) \cdot 2H_2O$	$C_{14}H_{12}NO_2$ (Schiff base), see Th E, pp. 186, 193

References for 15.3.2 on pp. 20/1

Table 9 (continued)

$Th(NCS)_4 \cdot 2L$	$L = C_{16}H_{16}N_2O_2$ or $C_{12}H_{20}N_2O_2$ (Schiff bases), see Th E, pp. 187, 192 and 191, 192, respectively
$Th(NCS)_4 \cdot L$	$L = C_{18}H_{21}N_3O_2$ or $C_{20}H_{26}N_4O_2$ (Schiff bases), see Th E, pp. 188, 193
$ThO(NCS)_2$(?)	decomposition to ThO_2 in air at 490 to 690°C [11] or ca. 700°C [7]

The anhydrous thiocyanate, $Th(NCS)_4$, has been reported to be an intermediate in the thermal decomposition of the complexes with N-oxides, $Th(NCS)_4 \cdot 4L$ (L = pyridine-N-oxide, quinoline- and isoquinoline-N-oxide, 2-, 3- and 4-picoline-N-oxide [10]), $Th(NCS)_4 \cdot 2L$ (L = lutidine-N-oxide [10, 11], 2,2'-bipyridyl-N-oxide [8, 10], 1,10-phenanthroline-N-oxide [8, 10], 1,10-phenanthroline-N,N'-dioxide [10], $Th(NCS)_4 \cdot L$ (L = 2,2'-bipyridyl-N,N'-dioxide [9, 10]), with a sulfoxide, $Th(NCS)_4 \cdot 2tmso$ (tmso = $(CH_2)_4SO$ [11]) and with Schiff bases, $Th(NCS)_4 \cdot 3L$ ($L = C_{14}H_{13}NO$ from salicylaldehyde and toluidine; $C_{14}H_{13}NO_2$ from salicylaldehyde and anisidine, and $C_{15}H_{15}NO_2$ from salicylaldehyde and *p*-phenetidine) [7]. In these reports the evidence is based on thermogravimetric studies, as is the thermal decomposition of the products to $ThO(NCS)_2$ at a variety of higher temperatures. The existence of the anhydrous thiocyanate under the reported conditions is doubtful, and requires analytical or other confirmation, for the hydrate, $Th(NCS)_4 \cdot 4H_2O$ (see below) certainly loses 4 molecules of water at temperatures below 200°C and does not exhibit a plateau ascribable to $Th(NCS)_4$ owing to decomposition [2].

A study of the conductance of $Cs_4Th(NCS)_8 \cdot 2H_2O$ (p. 23) in water indicates dissociation to $Th(NCS)_4 \cdot 4H_2O$ and subsequent hydrolysis of the latter [1], and this hydrate has been prepared by melting hydrated $Th(NO_3)_4$ with the stoichiometric quantity of NaSCN, followed by extraction of the product into ethanol, in which $NaNO_3$ is virtually insoluble. The ethanol extracts were finally evaporated to an oily liquid which became crystalline on standing in a desiccator over NaOH or KOH [4]. It has also been prepared by treating a boiling aqueous solution of $Th(SO_4)_2$ with a hot solution of $Ba(SCN)_2$ in the presence of sufficient dilute H_2SO_4 to ensure precipitation of all of the barium; after standing for 1 day, the filtrate was evaporated to an oily syrup from which the crystalline product separated [4]. The hydrate decomposes in bright sunlight and is very hygroscopic and deliquescent; it is a non-electrolyte in methanol [4]. It is very soluble in water, in which extensive dissociation occurs, and is readily soluble in acetone, ethanol and methanol, but insoluble in benzene, carbon tetrachloride and toluene [4].

The IR spectrum of the hydrate indicates that no ionic or bridging thiocyanate groups are present [5].

References for 15.3.2:

[1] Molodkin, A. K.; Skotnikova, G. A. (Zh. Neorgan. Khim. **7** [1962] 1548/51; Russ. J. Inorg. Chem. **7** [1962] 800/2).
[2] Molodkin, A. K.; Skotnikova, G. A. (Zh. Neorgan. Khim. **8** [1963] 2080/7; Russ. J. Inorg. Chem. **8** [1963] 1086/90).
[3] Molodkin, A. K.; Skotnikova, G. A. (Zh. Neorgan. Khim. **8** [1963] 2240/7; Russ. J. Inorg. Chem. **8** [1963] 1173/7).
[4] Molodkin, A. K.; Skotnikova, G. A. (Zh. Neorgan. Khim. **9** [1964] 60/9; Russ. J. Inorg. Chem. **9** [1964] 32/7).

[5] Kharitonov, Yu. Ya.; Molodkin, A. K.; Babaeva, A. V. (Izv. Akad. Nauk SSSR Ser. Khim. **1964** 618/22; Bull. Acad. Sci. USSR Div. Chem. Sci. **1964** 578/81).
[6] Belova, V. I.; Syrkin, Ya. K.; Molodkin, A. K.; Ivanova, O. M.; Shiporina, L. M. (Zh. Neorgan. Khim. **13** [1968] 1458/60; Russ. J. Inorg. Chem. **13** [1968] 766/7).
[7] Mohanta, H.; Dash, K. C. (J. Indian Chem. Soc. **54** [1977] 166/8).
[8] Agarwal, R. K.; Jain, P. C.; Kapur, V.; Sharma, S.; Srivastava, A. K. (Transition Metal Chem. [Weinheim] **5** [1980] 237/9).
[9] Agarwal, R. K.; Srivastava, A. K.; Srivastava, T. N. (J. Inorg. Nucl. Chem. **42** [1980] 1366/8).
[10] Agarwal, R. K.; Srivastava, M.; Srivastava, A. K. (J. Inorg. Nucl. Chem. **43** [1981] 203/4).

[11] Srivastava, A. K.; Agarwal, R. K. (Thermochim. Acta **56** [1982] 247/52).

15.3.3 N-Thiocyanatothorates

Information on these compounds is summarised in Table 10.

Table 10
N-Thiocyanatothorates.

$RbTh(NCS)_5 \cdot 3H_2O$	[5]; decomposes at 30 to 40°C [2], thermal decomposition curve illustrated in [2] IR spectrum: ν(CN), ca. 2066 cm^{-1}, ν(CS), ca. 820, ca. 808, ca. 786 cm^{-1}, δ(NCS), 491, 483 cm^{-1} [7] molar refraction, R_D =105.6 cm^3 [3]
$(NH_4)_3Th(NCS)_7 \cdot 4H_2O$	decomposes at 275 to 380°C [2]
$(NH_4)_3Th(NCS)_7 \cdot 5H_2O$	[5], loses 1 H_2O at 120 to 156°C [2], thermal decomposition curve illustrated in [2] IR spectrum: ν(CN), ca. 2060 cm^{-1}, ν(CS), 791 cm^{-1}, δ(NCS), 475 cm^{-1} [7] molar refraction, R_D=159.5 cm^3 [3]
$K_4[Th(NCS)_8] \cdot 1.5H_2O$	[2]
$K_4[Th(NCS)_8] \cdot 3.5H_2O$	[5], loses ca. 2 H_2O from 40 to 160°C [2], thermal decomposition curve illustrated in [2] IR spectrum: ν(CN), ca. 2060 cm^{-1}, ν(CS), 792 cm^{-1}, δ(NCS), 475 cm^{-1} [7] molar refraction, R_D=166.9 cm^3 [3]
$Rb_4[Th(NCS)_8]$	decomposes above 200°C, vigorously above 400°C [2], density, d_4^{25}= 2.513 g/cm^3 [3] crystal optical properties: refractive indices, n_g=1.740, n_m=1.692, n_p=1.670, mean refractive index=1.701 molar refraction, R_D=159.9 cm^3 [3]
$Rb_4[Th(NCS)_8] \cdot 2H_2O$	[1, 5]; loses 2 H_2O at 60 to 100°C [5] or up to ca. 145°C [2], thermal decomposition curve illustrated in [2] IR spectrum: ν(CN), 2087, 2077, 2062, 2051 cm^{-1} [7], 2097, 2086, 2073, 2060 (Nujol mull) cm^{-1} [9], 2088, 2055 cm^{-1} (acetone solution) [9], ν(CS), 797 cm^{-1} [9], 794 cm^{-1} [7], δ(NCS), 478 cm^{-1} [9], 477 cm^{-1} [7], density, d_4^{25}=2.484 g/cm^3 [3] crystal optical properties: refractive indices, n_g=1.754, n_m= 1.685, n_p=1.666, mean refractive index=1.701, molar refraction,

References for 15.3.3 on p. 25

Table 10 (continued)

	R_D = 167.4 cm^3 [3]; binding energies, Th $5d_{5/2}$ = 88.0 eV and N1s = 398.5 eV [11] diamagnetic susceptibility, $\chi_m = -394 \times 10^{-6}$ [8]
$Rb_4[Th(NCS)_8] \cdot 3H_2O$	diamagnetic susceptibility, $\chi_m = -413 \times 10^{-6}$ [8]
$Cs_4[Th(NCS)_8]$	decomposes above 100°C [1], above 200°C [2], vigorously at 430°C [2]; density, d_4^{25} = 2.804 g/cm^3 [3] crystal optical properties: refractive indices, n_g = 1.740, n_m = 1.699, n_p = 1.678, mean refractive index = 1.704, molar refraction, R_D = 170.0 cm^3 [3]
$Cs_4[Th(NCS)_8] \cdot 2H_2O$	loses $2H_2O$ at 60 to 100°C [5], at 100°C [1], at 130 to 155°C [2], thermal decomposition curve illustrated in [2] crystallographic data: monoclinic, space group $P2_1/n$-C_{2h}^5 (No. 14), lattice parameters, a = 13.5, b = 14.0, c = 16.0 Å [1], a = 13.520, b = 13.696, c = 16.22 Å [3, 4], β = 90°, Z = 4 [1, 3, 4] density, d_4^{25} = 2.755 g/cm^3 [3] crystal optical properties: refractive indices, n_g = 1.754, n_m = 1.688, n_p = 1.668, mean refractive index = 1.702, molar refraction, R_D = 177.70 cm^3 [3] IR spectrum: ν(CN), 2084, 2073, 2057, 2050 cm^{-1} [7], 2092, 2083, 2062, 2055 (Nujol mull) cm^{-1} [9], 2087, 2048 cm^{-1} (acetone solution) [9], ν(CS), 794 cm^{-1} [9], 793 cm^{-1} [7], δ(NCS) 480 cm^{-1} [7, 9] diamagnetic susceptibility, $\chi_m = -468 \times 10^{-6}$ [8]
$(NH_4)_4[Th(NCS)_8]$	decomposes above 150°C [2]
$(NH_4)_4[Th(NCS)_8] \cdot 2H_2O$	[1], loses $2H_2O$ at 100°C [5], up to 150°C [2], thermal decomposition curve illustrated in [2] density, d_4^{25} = 1.871 g/cm^3 [3] crystal optical properties: refractive indices, n_g = 1.740, n_m = 1.735, n_p = 1.700, mean refractive index = 1.725, molar refraction, R_D = 170.9 cm^3 [3] IR spectrum: ν(CN), ca. 2060 cm^{-1}, ν(CS), 790 cm^{-1}, δ(NCS) 477 cm^{-1} [7] diamagnetic susceptibility, $\chi_m = -311$ to -351×10^{-6} [8]
$[N(C_2H_5)_4]_4[Th(NCS)_8]$	crystallographic data; cubic, space group, $Pm3m$-O_h^1 (No. 221) lattice parameter, a = 11.589(5) Å; Z = 1, density, d_{calc} = 1.300(2) g/cm^3 [12] IR spectrum: ν(CN), 2045 cm^{-1} [10]

Oxo- and hydroxothiocyanatothorates

$Na_2[Th(NCS)_5(OH)(H_2O)_{2\ to\ 3}]$	thermogram for dihydrate illustrated in [6]
$K_4[ThO(NCS)_6]$	see "Thorium" 1955, p. 334
$3Hg(CN)_2 \cdot Th(OH)(NCS)_3 \cdot 12H_2O$	see "Thorium" 1955, p. 350
$Hg(CN)_2 \cdot Th(OH)_3(NCS) \cdot (?)H_2O$	see "Thorium" 1955, p. 350

The only recorded pentathiocyanato complex salt, $RbTh(NCS)_5 \cdot 3H_2O$, is obtained by adding a hot aqueous solution containing the stoichiometric quantity of $Ba(SCN)_2 \cdot 2H_2O$ to a boiling solution containing $Th(SO_4)_2 \cdot 8H_2O$ and the calculated quantity of Rb_2SO_4. The mixture was allowed to stand for 1 day, after which the filtrate was evaporated at 40 to 60°C and then left in air; the solid which separated was recrystallised from water. The water molecules are probably bonded to the thorium atom, perhaps as $[Th(NCS)_5(H_2O)_3]^-$ [5], but this requires verification. The salt is very hygroscopic and deliquesces [5]. A thermogravimetric study of this salt indicates that decomposition begins at 30 to 40°C, and an endotherm at 75 to 90°C corresponds to the loss of 1 H_2O; two merging endotherms were observed at 115 to 140°C and 140 to 160°C, and both loss of water and decomposition occurred at 170°C, so that the anhydrous salt is evidently very unstable [2].

Only one heptathiocyanatothorate(IV) salt has been reported. $(NH_4)_3Th(NCS)_7 \cdot 5H_2O$ has been prepared by addition of the calculated quantity of NH_4SCN to an aqueous solution of hydrated $Th(NCS)_4$ obtained by reaction of aqueous $Th(SO_4)_2 \cdot 8H_2O$ with an aqueous solution containing the stoichiometric quantity of $Ba(SCN)_2 \cdot 2H_2O$, followed by evaporation until crystals of the product separated [5]. A thermogravimetric study revealed endotherms at 120 to 156°C (loss of H_2O) and 210 to 240°C, and an exotherm at 275 to 380°C (decomposition) [2]. Conductance measurements in aqueous solution indicate displacement of the NCS groups by H_2O molecules from this salt and from $RbTh(NCS)_5 \cdot 3H_2O$ (see above), but the salts behave as 4-ion and 2-ion electrolytes respectively in methanol [5].

Several salts of the octathiocyanatothorate(IV) anion have been recorded. The hydrated potassium salt, $K_4[Th(NCS)_8] \cdot 3.5H_2O$, has been prepared by heating an aqueous solution containing $Th(SO_4)_2 \cdot 8H_2O$ and the calculated quantity of K_2SO_4 almost to boiling, followed by addition of the stoichiometric quantity of $Ba(SCN)_2 \cdot 2H_2O$. After standing overnight, the solution was filtered and the filtrate was evaporated on a water bath and then in air until the product crystallised. It is readily soluble in water and is very hygroscopic (deliquescent) [5]. A thermogravimetric study indicates that the salt loses ca. 2 H_2O up to 160°C (endotherm at 110 to 135°C) and endotherms were observed at 160 to 163°C, 166 to 170°C, and 235 to 255°C (fusion), as well as an exotherm at 340°C [2].

The rubidium salt, $Rb_4[Th(NCS)_8] \cdot 2H_2O$, has been prepared in the same way as the potassium salt (see above) [5]. Thermogravimetry indicates loss of water at 85 to 115°C and 130 to 145°C (endotherms), followed by decomposition of the anhydrous salt, beginning at above ca. 200°C. The anhydrous salt rapidly absorbs water in air, ultimately reforming the dihydrate [2]. The binding energies for the Th $5d_{5/2}$ and N1s electrons in the dihydrate (Table 10, p. 22) were derived from the X-ray electronic spectra obtained under high vacuum [11], so that some or all of the water of crystallisation may have been removed.

The caesium salt, $Cs_4[Th(NCS)_8] \cdot 2H_2O$, has also been prepared in the same way as the potassium and rubidium salts [1]. Thermogravimetry indicates loss of water at 105 to 115°C and 130 to 155°C (endotherms), followed by decomposition of the anhydrous salt, beginning at above ca. 200°C. The anhydrous salt rehydrates in air like its rubidium analogue [2]. The dihydrate is not isostructural with $Cs_4[U(NCS)_8] \cdot 2H_2O$ ("Uranium" Suppl. Vol. C 13, 1983, p. 37) [1, 4].

The dihydrated ammonium salt has been prepared in the same way as the alkali metal salts and is also formed when the stoichiometric quantity of NH_4SCN is added to an aqueous solution of $Th(NCS)_4 \cdot xH_2O$ obtained by reaction of aqueous $Th(SO_4)_2 \cdot 8H_2O$ with the calculated amount of $Ba(SCN)_2 \cdot 2H_2O$. The product crystallises from the syrupy liquid formed by extensive evaporation of the filtrate [5].

The hydrated salts $M_4[Th(NCS)_8] \cdot 2H_2O$ (M = Rb [5], Cs [1, 5] and NH_4 [5]) are readily soluble in water, methanol, ethanol, and acetone, but are insoluble in toluene and carbon tetrachlo-

References for 15.3.3 on p. 25

ride; the caesium salt is also insoluble in benzene and diethyl ether [5]. The molar conductivity of these salts in methanol corresponds to that expected for a 5-ion electrolyte whereas the conductance in water indicates decomposition to $[Th(NCS)_4(H_2O)_4]$ (p. 20) and subsequent hydrolysis. However, the caesium salt can apparently be recrystallised from aqueous solution [1]. All three salts are hydrolysed by alkali, yielding white precipitates which may consist of a hydrated basic thiocyanate or hydroxide [5].

The three dihydrates lose water slowly in a desiccator over concentrated H_2SO_4 and partially decompose in bright sunlight, becoming yellow or red [5]. The anhydrous rubidium and caesium salts, obtained from the dihydrates at 60 to 100°C, are stable at 100°C, whereas the ammonium salt decomposes partially at 100°C [5].

The IR spectra of the dihydrated rubidium and caesium salts suggest that the $[Th(NCS)_8]^{4-}$ anion is square antiprismatic in acetone solution, whereas the symmetry is lowered to dodecahedral in the solid state [9]. The IR spectra of these, and other complex salts, indicate that ionic or bridging NCS groups are absent [7].

The tetraethylammonium salt $[N(C_2H_5)_4]_4[Th(NCS)_8]$, has been obtained by reaction of $ThCl_4$ with the stoichiometric quantities of KSCN and $[N(C_2H_5)_4]Cl$ in methyl cyanide. The product crystallises from the filtrate after heating for a few minutes; it can be recrystallised from methyl cyanide [10]. The coordination geometry of the $[Th(NCS)_8]^{4-}$ anion in this salt is cubic (**Fig. 3**) [12, 13], as is the $[U(NCS)_8]^{4-}$ anion in the analogous salt ("Uranium" Suppl. Vol. C 13, 1983, p. 38). Interatomic distances and angles are given in Table 11.

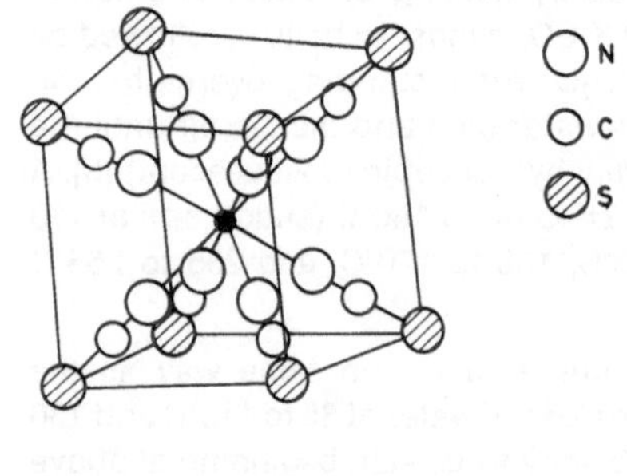

Fig. 3. View of the cubic environment of Th^{IV} in the anion of $[N(C_2H_5)_4]_4[Th(NCS)_8]$ [12, 13].

Table 11
Interatomic Distances (in Å) and Angles (in °) of $[Th(NCS)_8]^{4-}$ in $[N(C_2H_5)_4]_4[Th(NCS)_8]$ [12].

Th–N(1)	2.47(2)	N(3)–C(32)	1.52(3)	C(22)–N(2)–C(22)	110(1)
N(1)–C(1)	1.15(2)	C(32)–C(31)	1.55(2)	N(2)–C(22)–C(21)	111(2)
C(1)–S(1)	1.59(2)	N(1)–N(1)	2.85(2)	C(32)–N(3)–C(32)	113(3)
N(2)–C(22)	1.53(2)	C(1)–C(1)	4.17(3)	N(3)–C(32)–C(31)	114(3)
C(22)–C(21)	1.55(2)	S(1)–S(1)	6.02(1)		

A basic compound of composition $Na_2[Th(NCS)_5(OH)(H_2O)_{2\ to\ 3}]$ is reported to be formed when an aqueous solution containing $Th(NCS)_4 \cdot xH_2O$ (p. 20) and NaSCN, in the stoichiometric proportions required for the formation of $Na_2Th(NCS)_6$, is left to crystallise in a desiccator over NaOH. The water content of this product is variable, and it is very hygroscopic, deliquescing in air. It loses water and decomposes when heated, and there is no evidence for the formation of any stable intermediate. The salt behaves as a 3-ion electrolyte in methanol [6]. Salts of other basic complex anions, recorded in "Thorium" 1955, are included in Table 10.

References for 15.3.3:

[1] Molodkin, A. K.; Skotnikova, G. A. (Zh. Neorgan. Khim. **7** [1962] 1548/51; Russ. J. Inorg. Chem. **7** [1962] 800/2).
[2] Molodkin, A. K.; Skotnikova, G. A. (Zh. Neorgan. Khim. **8** [1963] 2080/7; Russ. J. Inorg. Chem. **8** [1963] 1086/90).
[3] Molodkin, A. K.; Skotnikova, G. A. (Zh. Neorgan. Khim. **8** [1963] 2240/7; Russ. J. Inorg. Chem. **8** [1963] 1173/7).
[4] Arutyunyan, E. G.; Porai-Koshits, M. A. (Zh. Strukt. Khim. **4** [1963] 110/1; J. Struct. Chem. [USSR] **4** [1963] 96/7).
[5] Molodkin, A. K.; Skotnikova, G. A. (Zh. Neorgan. Khim. **9** [1964] 60/9; Russ. J. Inorg. Chem. **9** [1964] 32/7).
[6] Molodkin, A. K.; Skotnikova, G. A. (Zh. Neorgan. Khim. **9** [1964] 1493/4; Russ. J. Inorg. Chem. **9** [1964] 811/2).
[7] Kharitonov, Yu. Ya.; Molodkin, A. K.; Babaeva, A. V. (Izv. Akad. Nauk SSSR Ser. Khim. **1964** 618/22; Bull. Acad. Sci. USSR Div. Chem. Sci. **1964** 578/81).
[8] Belova, V. I.; Syrkin, Ya. K.; Molodkin, A. K.; Ivanova, O. M.; Shipkina, L. M. (Zh. Neorgan. Khim. **13** [1968] 1458/60; Russ. J. Inorg. Chem. **13** [1968] 766/7).
[9] Grey, I. E.; Smith, P. W. (Australian J. Chem. **22** [1969] 311/5).
[10] Al-Kazzaz, Z. M. S.; Bagnall, K. W.; Brown, D.; Whittaker, B. (J. Chem. Soc. Dalton Trans. **1972** 2273/7).

[11] Nefedov, V. I.; Molodkin, A. K.; Salyn', Ya. V.; Ivanova, O. M.; Porai-Koshits, M. A.; Balakaeva, T. A.; Belyakova, Z. V. (Zh. Neorgan. Khim. **19** [1974] 2628/31; Russ. J. Inorg. Chem. **19** [1974] 1435/7).
[12] Charpin, P.; Lance, M.; Navaza, A. (Acta Cryst. C **39** [1983] 190/2).
[13] Johnson, C. K. (ORNL-3794 [1965], from [12]).

15.4 Thorium Selenocyanates

The only known compounds are the N,N-dimethylformamide (dmf) complexes, $Th(NCSe)_4 \cdot 4dmf$, $M_2Th(NCSe)_6 \cdot xdmf$ (M = Na, x = 3 and M = K, x = 4.5) and $K_4Th(NCSe)_8 \cdot 2dmf$, described in "Thorium" Suppl. Vol. E, 1985, p. 45.

15.5 Thorium Alkyl (Aryl) Oxides and Alkane (Arene) Di- and Polyolates

15.5.1 Introduction

The compounds described in this section are formally derived from alcohols, phenols, diols and polyols by replacement of the hydrogen of the OH group, or groups, by Th. The products are commonly referred to as thorium alkoxides (= alkyl oxides) even when they are not derived from alkanols. Similarly, the term "phenoxide" is often used to include all aryl oxides.

Compounds of thorium which include coordinated neutral alcohols and phenols are covered in "Thorium" Suppl. Vol. E, 1985, pp. 28/30.

Reviews which deal specifically with the thorium alkoxides and related compounds are not available. A short survey of actinide alkoxides has been published [1] and these compounds are briefly mentioned in other reviews [2 to 9]; reference [3] includes data on thorium compounds which are not available elsewhere.

References for 15.5.1 on p. 26

References for 15.5.1:

[1] Bagnall, K. W. (in: Bailar, J. C.; Emeléus, H. J.; Nyholm, R.; Trotman-Dickenson, A. F., Comprehensive Inorganic Chemistry, Vol. 5, Pergamon, Oxford 1973, pp. 412/6).
[2] Wardlaw, W.; Bradley, D. C. (Endeavour **14** [1955] 140/5).
[3] Mehrotra, R. C.; Mehrotra, A. (Inorg. Chim. Acta Rev. **5** [1971] 127/36).
[4] Bradley, D. C. (Advan. Chem. Ser. No. 23 [1959] 10/36).
[5] Bradley, D. C. (Progr. Inorg. Chem. **2** [1960] 303/61).
[6] Bradley, D. C. (Coord. Chem. Rev. **2** [1967] 299/318).
[7] Bradley, D. C.; Fisher, K. J. (MTP [Med. Tech. Publ. Co.] Intern. Rev. Sci. Inorg. Chem. Ser. One **5** [1972] 65/91).
[8] Bradley, D. C. (Advan. Inorg. Chem. Radiochem. **15** [1972] 259/322).
[9] Bradley, D. C.; Mehrotra, R. C.; Gaur, D. P. (Metal Alkoxides, Academic, London 1978, pp. 1/399).

15.5.2 Thorium Compounds with Monohydroxo Alcohols and Phenols

15.5.2.1 Thorium Alkyl Oxides

15.5.2.1.1 Introduction

The known compounds are listed in Table 12. Apart from the preparative details and the degrees of complexity of these compounds in non-aqueous media, little is known about them. The methoxide and ethoxide, $Th(OCH_3)_4$ and $Th(OC_2H_5)_4$, are involatile and are probably polymeric [1], consistent with their low solubilities in organic solvents, whereas most of the compounds derived from higher alcohols can be sublimed under reduced pressure (see Table 12). The volatilities of these compounds increase with increasing bulk of the alkyl group; for example, $Th\{OC(C_2H_5)_3\}_4$ and $Th\{OC(CH_3)(C_2H_5)(i\text{-}C_3H_7)\}_4$ are more volatile than the isomeric $Th\{OC(CH_3)(C_2H_5)(n\text{-}C_3H_7)\}_4$. The first two are monomers in boiling benzene whereas the last has a "molecular complexity" of 1.7 in that solvent [2].

Table 12
Thorium Alkyl Oxides.

$Th(OCH_3)_4$	decomposes at 320°C/0.05 Torr [1]
$Th(OC_2H_5)Cl_3\cdot 2phen$	phen = 1,10-phenanthroline = $C_{12}H_8N_2$, see Th E, pp. 17, 21
$Th(OC_2H_5)Br_3\cdot x(C_6H_5)_3PO$	x = 2 or 3; see Th E, pp. 90/1
$Th(OC_2H_5)_4$	[3]; involatile [1, 2]
$Th(O\text{-}i\text{-}C_3H_7)Cl_3\cdot C_5H_{10}O_2$	$C_5H_{10}O_2 = CH_3COO\text{-}i\text{-}C_3H_7$, see Th E, p. 64
$Th(O\text{-}i\text{-}C_3H_7)_2Cl_2\cdot 0.5C_5H_{10}O_2$	see Th E, p. 64
$Th(O\text{-}i\text{-}C_3H_7)_3Cl$	[11]
$Th(O\text{-}i\text{-}C_3H_7)_4$	m.p. 110°C (under reduced pressure) [1]; sublimed at 200 to 210°C/0.05 to 0.1 Torr [1]; molecular complexity, 3.8 in C_6H_6 [1, 5], 1.8 in $i\text{-}C_3H_7OH$ [1] IR spectrum (285 to 5000 cm^{-1}) [9]
$Th(O\text{-}n\text{-}C_4H_9)_4$	molecular complexity, 6.44 in C_6H_6 [4]
$Th\{OC(CH_3)_3\}Cl(CH_3COO)_2$	[11]

Table 12 (continued)

Th{OC$(CH_3)_3$}$Cl_{1.3}(CH_3COO)_{1.7}$	[11]
Th{OC$(CH_3)_3$}$_{1.6}Cl_{1.4}(CH_3COO)$	[11]
Th{OC$(CH_3)_3$}$_2Cl_{1.4}(CH_3COO)_{0.6}$	[11]
Th{OC$(CH_3)_3$}$_{3.1}Cl_{0.9}$	[11]
Th{OC$(CH_3)_3$}$_4$	sublimed at 160°C/0.1Torr [2, 11]; molecular complexity, 3.4 in C_6H_6 [2]; IR spectrum illustrated in [9]
Th{OCH$(CH_3)(C_2H_5)$}$_4$	molecular complexity, 4.2 in C_6H_6 [5]
Th(O-*n*-$C_5H_{11})_4$	molecular complexity, 6.20 in C_6H_6 [4]
Th{$OCH_2C(CH_3)_3$}$_4$	molecular complexity, 4.01 in C_6H_6 [4]
Th{OC$(CH_3)_2(C_2H_5)$}$_4$	b.p. 208°C/0.3 Torr; molecular complexity, 2.8 in C_6H_6 [2]
Th{OCH$(C_2H_5)_2$}$_4$	molecular complexity, 4.1 in C_6H_6 [5]
Th{OC$(CH_3)(C_2H_5)_2$}$_4$	b.p. 148°C/0.1Torr; molecular complexity, 1.8 in C_6H_6 [2]
Th{OC$(CH_3)_2$(*n*-C_3H_7)}$_4$	decomposes at >120°C/0.1Torr; molecular complexity, 2.6 in C_6H_6 [2]
Th{OC$(CH_3)_2$(*i*-C_3H_7)}$_4$	decomposes at >120°C/0.1Torr; molecular complexity, 2.3 in C_6H_6 [2]
Th{OC$(C_2H_5)_3$}$_4$	b.p. 148°C/0.05 Torr [2]; monomer in C_6H_6 [2], parachor, 1231.5 (mean value from calculated values at 25, 30, 35, 40°C); temperature dependance of density, $d_t=1.2505-0.00110\,t$ g/cm^3; temperature dependance of surface tension, $\gamma_t=22.61-0.027\,t$ [6]; viscosity in poise, 0.492 (25°C), 0.387 (30°C), 0.322 (35°C), 0.269 (40°C) [8]; activation energy for viscosity, 31.9 kJ/mol (=7.6 kcal/mol), free energy of activation of viscous flow, 27.7 kJ/mol (=6.6 kcal/mol); entropy ΔS, +14.7 $J\cdot mol^{-1}\cdot K^{-1}$ (=3.5 cal $\cdot mol^{-1}\cdot K^{-1}$) [8]
Th{OC$(CH_3)(C_2H_5)$(*n*-C_3H_7)}$_4$	b.p. 153°C/0.1Torr; molecular complexity, 1.7 in C_6H_6 [2]
Th{OC$(CH_3)(C_2H_5)$(*i*-C_3H_7)}$_4$	b.p. 139°C/0.05 Torr; monomer in C_6H_6 [2]
Th{OC$(CH_3)_2(CCl_3)$}$_4\cdot$2py	see Th E, pp. 12, 14
Th[Al(O-*i*-$C_3H_7)_4]_4$	[10]
$LiTh_2$(O-*i*-$C_3H_7)_9$	[10]
$NaTh_2$(O-*i*-$C_3H_7)_9$	[10]

All of the alkoxides are moisture-sensitive and are readily hydrolysed. Both $Th(OC_2H_5)_4$ and Th(O-*i*-$C_3H_7)_4$ are more basic than NH_3 in ethanol [1].

References for 15.5.2.1 on pp. 28/9

15.5.2.1.2 Preparation

$ThCl(O\text{-}i\text{-}C_3H_7)_3$ is reported to be formed when $Th(O\text{-}i\text{-}C_3H_7)_4$ (see below) is heated with the stoichiometric quantity of CH_3COCl in benzene under reflux for 1 h [11]. The corresponding *t*-butoxide, $ThCl_{0.9}\{OC(CH_3)_3\}_{3.1}$, has been obtained in the same way; with larger quantities of CH_3COCl, products of composition $ThCl_{1.4}(CH_3COO)_{0.6}\{OC(CH_3)_3\}_2$, $ThCl_{1.4}(CH_3COO)\{OC(CH_3)_3\}_{1.6}$, $ThCl_{1.3}(CH_3COO)_{1.7}\{OC(CH_3)_3\}$ and $ThCl(CH_3COO)_2\{OC(CH_3)_3\}$ are formed; some of these may be mixtures [11].

$Th(O\text{-}i\text{-}C_3H_7)_4$, which is commonly used as the starting material for the preparation of the other alkoxides, has been prepared by treating a suspension of $ThCl_4 \cdot 4\,i\text{-}C_3H_7OH$ ("Thorium" Suppl. Vol. E, 1985, pp. 28/9) in $i\text{-}C_3H_7OH$ and benzene with an excess of dry NH_3 gas. After removal of free NH_3 under reduced pressure, the filtrate was evaporated to leave a viscous solid which was dried at 80°C under reduced pressure. However, this product contained some chloride ion even after vacuum sublimation at 200 to 210°C [1]. A better route is the same procedure, but using $(C_5H_5NH)_2ThCl_6$ in place of $ThCl_4 \cdot 4\,i\text{-}C_3H_7OH$. After evaporation of the filtrate, the product was treated with $NaO\text{-}i\text{-}C_3H_7$ in boiling $i\text{-}C_3H_7OH$, yielding a chloride-free product [1]. The attempted preparation of $Th(OC_2H_5)_4$ from $ThCl_4 \cdot 4C_2H_5OH$ ("Thorium" Suppl. Vol. E, 1985, pp. 28/9) or $(C_5H_5NH)_2ThCl_6$ in ethanol and benzene yielded impure products contaminated with chloride [1]. A similar route to $Th(OC_2H_5)_4$ is by reaction of a suspension of $ThCl_4$ in anhydrous ethanol with dry NH_3 gas; the solid was subsequently extracted into C_6H_6 [3].

Another preparative route leading to $Th(O\text{-}i\text{-}C_3H_7)_4$ is by heating a suspension of $(C_5H_5NH)_2ThCl_6$ in $i\text{-}C_3H_7OH$ with $NaO\text{-}i\text{-}C_3H_7$, then evaporating the resulting yellow solution to leave a brownish residue. The best preparative method is by reaction of $ThCl_4 \cdot 4\,i\text{-}C_3H_7OH$ in $i\text{-}C_3H_7OH$ with $NaO\text{-}i\text{-}C_3H_7$. Evaporation of the filtrate leaves a white, analytically pure product [1].

The best preparative route for the other alkoxides is from $Th(O\text{-}i\text{-}C_3H_7)_4$ by alcohol exchange. $Th(OCH_3)_4$ is obtained as a white gelatinous precipitate when methanol is added to a solution of the isopropoxide in benzene; after boiling for 1 h, the mixture was evaporated to dryness, after which the residue was treated with methanol and the solution was again evaporated to dryness [1]. The ethoxide, $Th(OC_2H_5)_4$, remains as a solid residue when $Th(O\text{-}i\text{-}C_3H_7)_4$ is heated in boiling ethanol for 4 h. It is only slightly soluble in ethanol or benzene [1]. The alkoxides $Th(OR)_4$, where $R = n\text{-}C_4H_9$ [4], $CH(CH_3)(C_2H_5)$ [5], $C(CH_3)_3$ [2], $n\text{-}C_5H_{11}$, $(CH_3)_3CCH_2$(neopentyl) [4], $C(CH_3)_2(C_2H_5)$ [2], $CH(C_2H_5)_2$ [5], $C(CH_3)(C_2H_5)_2$, $C(CH_3)_2(n\text{-}C_3H_7)$, $C(CH_3)_2(i\text{-}C_3H_7)$, $C(CH_3)(C_2H_5)(n\text{-}C_3H_7)$, $C(CH_3)(C_2H_5)(i\text{-}C_3H_7)$, and $C(C_2H_5)_3$ [2] have all been prepared from $Th(O\text{-}i\text{-}C_3H_7)_4$ by the exchange reaction with the appropriate alcohol in benzene. However, $Th\{OC(CH_3)_2(CCl_3)\}_4$ could not be obtained by this method, but the pyridine adduct, $Th\{OC(CH_3)_2(CCl_3)\}_4 \cdot 2C_5H_5N$ ("Thorium" Suppl. Vol. E, 1985, pp. 12, 14) has been prepared by treatment of $(C_5H_5NH)_2ThCl_6$ in the alcohol with gaseous NH_3 [7].

An alternative method of preparation for $Th\{OC(CH_3)_3\}_4$ is by transesterification, in which $Th(O\text{-}i\text{-}C_3H_7)_4$ is heated with an excess of $CH_3COO\{C(CH_3)_3\}$; $CH_3COO\text{-}i\text{-}C_3H_7$ formed in the reaction is removed by fractionation [11].

$Th[Al(O\text{-}i\text{-}C_3H_7)_4]_4$ is reported to be formed by reaction of $ThCl_4$ with the stoichiometric quantity of $K[Al(O\text{-}i\text{-}C_3H_7)_4]$ in $i\text{-}C_3H_7OH$. The compounds $MTh_2(O\text{-}i\text{-}C_3H_7)_9$ (M = Li, Na) were obtained similarly from $Th(O\text{-}i\text{-}C_3H_7)_4$ and $M(O\text{-}i\text{-}C_3H_7)$ [10].

References for 15.5.2.1:

[1] Bradley, D. C.; Saad, M. A.; Wardlaw, W. (J. Chem. Soc. **1954** 1091/4).
[2] Bradley, D. C.; Saad, M. A.; Wardlaw, W. (J. Chem. Soc. **1954** 3488/90).

[3] Müller, R.; Sliwinski, S. (Ger. [East] 10596 [1955]; C.A. **1958** 19948).
[4] Bradley, D. C.; Chatterjee, A. K.; Wardlaw, W. (J. Chem. Soc. **1956** 2260/4).
[5] Bradley, D. C.; Chatterjee, A. K.; Wardlaw, W. (J. Chem. Soc. **1956** 3469/72).
[6] Bradley, D. C.; Prevedorou, C. C. A.; Swanwick, J. D.; Wardlaw, W. (J. Chem. Soc. **1958** 1010/4).
[7] Bradley, D. C.; Sinha, R. N. P.; Wardlaw, W. (J. Chem. Soc. **1958** 4651/4).
[8] Bradley, D. C.; Prevedorou, C. C. A.; Wardlaw, W. (Can. J. Chem. **39** [1961] 1619/24).
[9] Lynch, C. T.; Mazdiyasni, K. S.; Smith, J. S.; Crawford, W. J. (Anal. Chem. **36** [1964] 2332/7).
[10] Agarwal, M. M. (Diss. Univ. Rajasthan, India, 1968 from Mehrotra, R. C.; Mehrotra, A., Inorg. Chim. Acta Rev. **5** [1971] 127/36).

[11] Mehrotra, R. C.; Misra, R. A. (Proc. 3rd. Nucl. Radiat. Chem. Symp.; Poona, India, 1967, pp. 473/82; C.A. **70** [1969] No. 43618).

15.5.2.2 Thorium Aryl Oxides

The available information on the known compounds is summarised in Table 13.

Table 13
Thorium Aryl Oxides.

$Th(OC_6H_5)_4 \cdot 2L$	$L = (CH_3)_2PCH_2CH_2P(CH_3)_2$, see Th E, p. 26
$ThCl_3(OC_6H_4\text{-}2\text{-}CH_3)$	pinkish white [4]
$ThCl_3(OC_6H_4\text{-}3\text{-}CH_3)$	dirty white [4]
$ThCl_2(OC_6H_4\text{-}2\text{-}NH_2)_2$	IR spectrum: ν(NH), 3230, 3150 cm^{-1}, ν(ThO), 390 cm^{-1}, ν(ThN), 325 cm^{-1}, ν(ThCl), 220 cm^{-1} [12] molar conductivity = 45.5 $\Omega^{-1} \cdot cm^2 \cdot mol^{-1}$ in $(CH_3)_2SO$ [12]
$Th(OO)(OC_6H_4\text{-}2\text{-}NH_2)_2$	IR spectrum: ν(NH), 3200, 3110 cm^{-1} Raman spectrum, ν_1(OO), 825 cm^{-1} [12] molar conductivity = 3.04 $\Omega^{-1} \cdot cm^2 \cdot mol^{-1}$ in $(CH_3)_2SO$ [12]
$ThCl_2(OC_6H_4\text{-}2\text{-}NO_2)_2$	dirty brown [4]
$ThCl_2(OC_6H_4\text{-}4\text{-}NO_2)_2$	greyish white [4]
$Th(C_9H_6NO)_4 \cdot 2HOC_6H_3\text{-}2,4\text{-}(NO_2)_2$	C_9H_7NO = 8-quinolinol, see Th E, p. 29
$Th(C_9H_6NO)_4 \cdot 2HOC_6H_2\text{-}2,4,6\text{-}(NO_2)_3$	see Th E, pp. 29/30
$[Th(acac)(CH_3COO)_2(H_2O)_5]\text{-}(OC_6H_2\text{-}1,3,5\text{-}(NO_2)_3)$	see Th E, pp. 134/5
$Th(OC_6H_2\text{-}4,6\text{-}(NO_2)_2\text{-}2\text{-}NH_2)_4 \cdot 2H_2O$	explodes at 628.93 K (explosion delay = 10 s), activation energy of process controlling the explosion delay = 56.0656(!) kJ/mol [9] IR spectrum: ν(NH), 3250 cm^{-1} (Na salt, ν(NH), 3400 cm^{-1}) [9]

References for 15.5.2.2 on pp. 31/2

Table 13 (continued)

$ThCl_3(1\text{-}OC_{10}H_7)$	pinkish grey [4]
$ThCl_2(1\text{-}OC_{10}H_7)_2$	brownish black [4]
$ThCl_3(2\text{-}OC_{10}H_7)$	pinkish brown [4]
$ThCl_2(2\text{-}OC_{10}H_7)_2$ (1-, 2-naphtholates)	brownish black [4]
$Th(OH)_2(C_{13}H_8N_3OS)_2$ ($C_{13}H_9N_3OS$ = 1-(2-thiazolylazo)-2-naphthol)	decomposition at 200°C [6]
$Th(C_{13}H_7O_3)_4$(?) ($C_{13}H_8O_3$ = 1-hydroxyxanthone)	yellow [2]; no analytical data available
$Th(C_{14}H_9O_4)_4$(?) ($C_{14}H_{10}O_4$ = 1-hydroxy-3-methoxyxanthone)	[1]; no analytical data available
$Th(C_{10}H_7N_4O_5)_4$ ($C_{10}H_8N_4O_5$ = 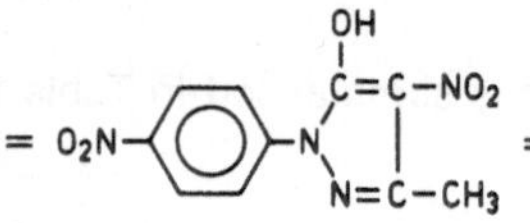 = picrolonic acid)	[5]; explodes at 268°C [10] thermal decomposition curve illustrated in [10]
$Th(C_{15}H_9N_2O_3)_4$ ($C_{15}H_{10}N_2O_3$ = 3-phenylazo-4-hydroxycoumarin) 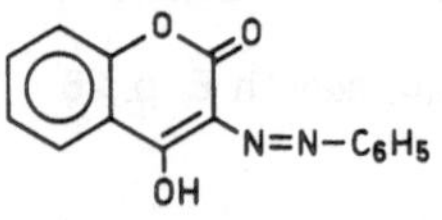	m.p. 182°C [8] IR spectrum: ν(CO), 1650 cm^{-1} (free ligand, 1740 cm^{-1}) absorption spectrum, λ_{max} = 256, 425 nm [8].

The only recorded unsubstituted phenoxide is the diphosphane complex, $Th(OC_6H_5)_4 \cdot 2(CH_3)_2PCH_2CH_2P(CH_3)_2$ (see "Thorium" Suppl. Vol. E, 1985, p. 26).

The 2- and 3-methylphenoxides (*o*-and *m*-cresolates), $ThCl_3(OC_6H_4\text{-}2(\text{or } 3)\text{-}CH_3)$, have been prepared by heating the 1:1 adducts, $ThCl_4 \cdot (HOC_6H_4CH_3)$ (see "Thorium" Suppl. Vol. E, 1985, p. 29) with an excess of the cresol [4]. The 2-aminophenoxide, $ThCl_2(OC_6H_4\text{-}2\text{-}NH_2)_2$, has been prepared by treating a suspension of $ThCl_4$ in methyl cyanide with the stoichiometric quantity of the phenol. It can be recrystallised from hot acetone by partial evaporation and addition of ether. The corresponding peroxide, $Th(OO)(OC_6H_4\text{-}2\text{-}NH_2)_2$, is obtained by dissolving $ThCl_2(OC_6H_4\text{-}2\text{-}NH_2)_2$ in a 1:1 mixture of acetone and methanol containing a little pyridine and added H_2O_2. After boiling the mixture for 5 min and cooling, the precipitate was washed with acetone. The relatively high molar conductivity of $ThCl_2(OC_6H_4\text{-}2\text{-}NH_2)_2$ in $(CH_3)_2SO$ is presumably due to partial displacement of Cl^- by the solvent [12]. The 2- and 4-nitrophenoxides, $ThCl_2(OC_6H_4\text{-}2(\text{or } 4)\text{-}NO_2)_2$, are obtained by evaporating an ethereal solution of the respective phenol and $ThCl_4$ at 60 to 65°C to remove the solvent, followed by heating at 10 to 15°C above the m.p. of the respective phenol until the evolution of HCl gas ceases [4]. The same procedure with naphthalen-1(or 2)-ol yielded $ThCl_3(1\text{-}(\text{or } 2)\text{-}OC_{10}H_7)$ whereas $ThCl_2(1\text{-}(\text{or } 2)\text{-}OC_{10}H_7)_2$ was obtained when the residue left on evaporation of the ether, which contained an excess of the naphthol, was heated at 160 to 170°C (naphthalen-1-ol) or 180 to 200°C (naphthalen-2-ol), temperatures which are about 60 to 70°C above the m.p. of the naphthol concerned [4]. These compounds are insoluble in benzene, toluene, and carbon tetrachloride, but are sparingly soluble in ethanol, propanol, and acetone [4].

Thorium picramate, $Th(OC_6H_2\text{-}4,6\text{-}(NO_2)_2\text{-}2\text{-}NH_2)_4 \cdot 2H_2O$, is precipitated when an aqueous solution of sodium picramate is added to aqueous $Th(NO_3)_4$. The compound is explosive, and the activation energy of the process involving the explosion delay may be related to the dissociation energy of the N–O bond in the nitro group, releasing oxygen for the oxidation of the C and H of the ring. The shift in ν(NH) in the IR spectrum of the thorium compound ($150\ cm^{-1}$) indicates that the amino group N atom is coordinated to the metal atom [9].

The addition of an alcoholic solution of 1-hydroxyxanthone to an aqueous solution of a Th^{IV} salt yields a deep yellow solution which gives a yellow precipitate on dilution with water. This is probably $Th(C_{13}H_7O_3)_4$, but analytical data are lacking. The ligand can be used to separate Th from La^{III} and Ce^{III} and for the determination of Th following ignition to ThO_2 [2]; see also [7]. The corresponding 1-hydroxy-3-methoxyxanthone derivative, probably $Th(C_{14}H_9O_4)_4$ (analytical data are lacking), is obtained by adding an excess of the xanthone in ethanol to a boiling solution of $Th(NO_3)_4$ in aqueous ethanol (30 mL C_2H_5OH, ca. 100 mL H_2O). After heating on a water bath to remove most of the ethanol, 100 mL H_2O was added to precipitate the complex [1]. The optimum pH for formation of this compound is 2.6 to 3.2, but precipitation is satisfactory up to pH 4.0. The ligand can be used for the gravimetric determination of Th (as ThO_2 following ignition) and for the separation of Th from U, Ce, and the cerite earths [1].

Conductometric studies in the picrolonic acid ($C_{10}H_8N_4O_5$)–Th^{IV} aqueous system confirm that the complex $Th(C_{10}H_7N_4O_5)_4$ is formed [5]. The compound is precipitated from aqueous solutions of Th salts and this precipitation can be used for the gravimetric estimation of Th; in this procedure the picrolonate is weighed as $Th(C_{10}H_7N_4O_5)_4$ after drying at 60 to 200°C [10, 11].

The 3-phenylazo-4-hydroxycoumarin ($=C_{15}H_{10}N_2O_3$) compound, $Th(C_{15}H_9N_2O_3)_4$, results when equimolar quantities of aqueous Th^{IV} and the ligand in methanol are mixed. The product can be recrystallised from dioxane and is a non-electrolyte in dioxane and in methanol. It is also soluble in non-polar solvents [8].

Orange II (the sodium salt of 1-(2-hydroxynaphthyl)azo-4-benzene sulfonic acid)

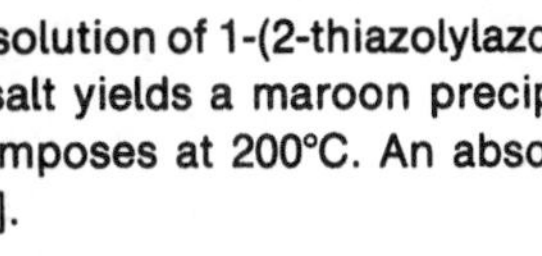

has been used for the gravimetric determination of Th. Addition of a 1% aqueous solution of the reagent to a near-boiling solution of $Th(NO_3)_4$ precipitates Th, the product being calcined to ThO_2 [3]. The composition of the precipitate is unknown.

The addition of an ethanolic solution of 1-(2-thiazolylazo)-naphthalen-2-ol ($=C_{13}H_9N_3OS$) to an aqueous solution of a Th^{IV} salt yields a maroon precipitate of approximate composition $Th(C_{13}H_8N_3OS)_2(OH)_2$; this decomposes at 200°C. An absorption at 540 nm in the UV-visible spectrum has been reported [6].

References for 15.5.2.2:

[1] Saxena, G. M.; Seshadri, T. R. (Proc. Indian Acad. Sci. A **47** [1958] 238/43).
[2] Dev, B.; Jain, B. D. (Proc. Indian Acad. Sci. A **54** [1961] 341/4).
[3] Popa, G.; Baiulescu, Gh.; Iliescu, V. (Acad. Rep. Populare Romine Studii Cercetari Chim. **10** [1962] 367/70).
[4] Prasad, S.; Kumar, S. (J. Indian Chem. Soc. **40** [1963] 531/3).
[5] Joshi, D. P.; Jain, D. V. (J. Indian Chem. Soc. **41** [1964] 711/4).
[6] Nickless, G.; Pollard, F. H.; Samuelson, T. J. (Anal. Chim. Acta **39** [1967] 37/46).

[7] Murata, A.; Nakamura, M. (Bunseki Kagaku **21** [1972] 487/91; C.A. **77** [1972] No. 69748).
[8] Kumar, B. B.; Rao, K. S. R. K. M.; Ganorkar, M. C. (Current Sci. [India] **42** [1973] 461/3).
[9] Srivastava, R. S.; Agrawal, S. P.; Bhargava, H. N. (Propellants Explosives **1** [1976] 101/3).
[10] Dupuis, T.; Duval, C. (Anal. Chim. Acta **3** [1949] 589/98).
[11] Dupuis, T.; Duval, C. (Compt. Rend. **228** [1949] 401/2).
[12] Westland, A. D.; Tarafder, M. T. H. (Inorg. Chem. **21** [1982] 3228/32).

15.5.3 Thorium Compounds with Aliphatic and Aromatic Diols

The glycoloxide which is obtained by refluxing $Th(CH_3COO)_4$ (p. 47) with ethylene glycol is presumably $Th(C_2H_4O_2)_2$. It can be used as a catalyst for the production of high molecular weight polymers from dihydric alcohols and dicarboxylic acids [3, 4].

The majority of the recorded diolates (Table 14) are derivatives of pyrocatechol (=1,2-$C_6H_4(OH)_2$). $ThCl_2(1,2\text{-}C_6H_4O_2)$ has been prepared in the same way as the corresponding 2- and 4-nitrophenoxides (p. 30) except that the final heating was at a temperature well below the m.p. of the diol in order to avoid decomposition [1]. The analogous resorcinol (H_2L = 1,3-$C_6H_4(OH)_2$), hydroquinone (H_2L = 1,4-$C_6H_4(OH)_2$) and orcinol (H_2L = 2,5-$CH_3C_6H_3(OH)_2$) compounds, $ThCl_2L$, have been prepared in the same way [1]. These compounds are insoluble in benzene, toluene, and carbon tetrachloride, but are sparingly soluble in acetone, ethanol, and propanol [1]. However, a product of composition $H_2[Th(1,2\text{-}C_6H_4O_2)_3]$ is obtained when $ThCl_4$ is heated with molten pyrocatechol at ca. 125°C until the evolution of HCl ceases; the excess of the diol was removed from the cooled melt by washing with diethyl ether [5].

Table 14
Thorium Diolates.

$Th(1,2\text{-}C_6H_4O_2)_2 \cdot 2C_5H_5N$ (1,2-$C_6H_4(OH)_2$ = pyrocatechol)	yellow [2]
$Th(1,2\text{-}C_6H_4O_2H)_4 \cdot 2C_5H_5N$	yellow [2]; possibly $(C_5H_5NH)_2H_2[Th(1,2\text{-}C_6H_4O_2)_4]$
$ThCl_2(1,2\text{-}C_6H_4O_2)$	grey [1]
$H_2[Th(1,2\text{-}C_6H_4O_2)_3]$	[5]
$Na_4[Th(1,2\text{-}C_6H_4O_2)_4] \cdot 21H_2O$	crystallographic data: tetragonal, space group $I\bar{4}\text{-}S_4^2$ (No. 82), lattice parameters, a = 14.709(4), c = 9.978(3) Å, Z = 2, density, d_{calc} = 1.75, d_{obs} = 1.74 g/cm³ [7] infrared spectrum: 1570, 1025, 910, 864 cm^{-1} [7]
$K_2[Th_3(1,2\text{-}C_6H_4O_2)_7] \cdot 20H_2O$	see "Thorium" 1955, p. 336
$(NH_4)_2[Th(1,2\text{-}C_6H_4O_2)_3] \cdot 5H_2O$	see "Thorium" 1955, p. 342
$(NH_4)_2H_2[Th(1,2\text{-}C_6H_4O_2)_4]$	[2]
$(NH_4)_2[Th_3(1,2\text{-}C_6H_4O_2)_6(OH)_2] \cdot 10H_2O$	(2); see "Thorium" 1955, p. 342
$(C_5H_5NH)_2H_2[Th(1,2\text{-}C_6H_4O_2)_4]$	see "Thorium" 1955, p. 343
$(C_5H_5NH)_2[Th_2(1,2\text{-}C_6H_4O_2)_3(OH)_4] \cdot 10H_2O$	see "Thorium" 1955, p. 343

Table 14 (continued)

$(CN_3H_6)_2[Th_2(1,2\text{-}C_6H_4O_2)_3(OH)_4]\cdot 10H_2O$ (CN_3H_6 = guanidinium)	see "Thorium" 1955, p. 343
$ThCl_2(1,3\text{-}C_6H_4O_2)$ ($1,3\text{-}C_6H_4(OH)_2$ = resorcinol)	brick red [1]
$ThCl_2(1,4\text{-}C_6H_4O_2)$ ($1,4\text{-}C_6H_4(OH)_2$ = hydroquinone)	black [1]
$ThCl_2(2,5\text{-}CH_3C_6H_3O_2)$ ($2,5\text{-}CH_3C_6H_3(OH)_2$ = 2,5-dihydroxytoluene = orcinol)	yellowish brown [1]
$Th(1,8\text{-}C_{10}H_6O_2)_2$ ($1,8\text{-}C_{10}H_6(OH)_2$ = 1,8-dihydroxynaphthalene)	[5]
$Th(C_{12}H_8O_2)_2$ (2,2'-dihydroxybiphenyl = $2\text{-}HOC_6H_4C_6H_4OH\text{-}2'$)	[5]
$Th(C_{20}H_{12}O_2)_2$ (2,2'-dihydroxy-1,1'-binaphthyl = $2\text{-}HOC_{10}H_6C_{10}H_6OH\text{-}2'$)	[5]
$Th(4\text{-}CH_3\text{-}6,7\text{-}(O)_2C_9H_3O_2)(OH)_2$ ($4\text{-}CH_3\text{-}6,7\text{-}(HO)_2C_9H_3O_2$ = 4-methyl-6,7-dihydroxycoumarin = 4-methylesculetin)	IR spectrum: ν(CO), 1660 cm^{-1}, ν(C=C), 1600, 1560 cm^{-1}, ϱ(Th–OH), 1220, 1180 cm^{-1}, ν(Th–O)(?), 450 cm^{-1} [6]
$Th(4\text{-}CH_3\text{-}7,8\text{-}(O)_2C_9H_3O_2)(OH)_2\cdot 4H_2O$ ($4\text{-}CH_3\text{-}7,8\text{-}(HO)_2C_9H_3O_2$ = 4-methyl-7,8-dihydroxycoumarin = 4-methyldaphnetin)	loses $4H_2O$ at 180°C, decomposes at >180°C [6] IR spectrum: ν(CO), 1680 cm^{-1}, ν(C=C), 1590, 1560 cm^{-1}, ϱ(Th–OH), 1250 cm^{-1}, ν(Th–O)(?), 430 cm^{-1} [6]
$Th(C_6H_5N_2O_2)_2(C_{16}H_{15}N_3O_4)$	$C_6H_5N_2O_2$ = cupferronate anion, $C_{16}H_{17}N_3O_4$ = 5-isopropyl-4-methyl-4'-nitro-1,1'-azobenzene-2,2'-diol, see Th E, pp. 166/7

Treatment of aqueous $Th(NO_3)_4$ with 3.8 to 4.0 equivalents of KOH in the presence of varying amounts of pyrocatechol yields a greenish white precipitate in which the molar ratio Th : catechol is approximately unity; this product may contain the $[Th(1,2\text{-}C_6H_4O_2)]^{2+}$ cation [2]. A similar product is obtained when aqueous NH_3 is used, but when an aqueous mixture of $Th(NO_3)_4$ (0.01 mol) and pyrocatechol (0.08 mol) is treated with aqueous NH_3 until the solution is faintly alkaline, and the filtrate is heated on a water bath for 5 min, then cooled, a white precipitate of composition $(NH_4)_2H_2[Th(1,2\text{-}C_6H_4O_2)_4]$ is obtained. This could also be written as $(NH_4)_2[Th(1,2\text{-}C_6H_4O_2)_3]\cdot 1,2\text{-}(HO)_2C_6H_4$ [2]. Under similar conditions, but with more NH_3 present and with boiling for ca. 2 min, the previously reported (see "Thorium" 1955, p. 342) compound $(NH_4)_2[Th_3(C_6H_4O_2)_6(OH)_2]\cdot 10H_2O$ is obtained [2]. When the reaction mixture is treated with pyridine, the composition of the product is $Th(1,2\text{-}C_6H_4O_2H)_4\cdot 2C_5H_5N$ [2] (possibly $(C_5H_5NH)_2H_2[Th(1,2\text{-}C_6H_4O_2)_4]$) or, with a larger quantity of pyridine, $Th(1,2\text{-}C_6H_4O_2)_2\cdot 2C_5H_5N$ [2].

References for 15.5.3 on p. 35

The hydrated salt, $Na_4[Th(1,2\text{-}C_6H_4O_2)_4]\cdot 21H_2O$, separates when an oxygen-free solution of $ThCl_4$ in H_2O is added to a solution of the diol in oxygen-free aqueous NaOH [7]. It is isomorphous with the uranium analogue (see "Uranium" Suppl. Vol. C 13, 1983, pp. 89/90). The 8-coordination geometry about the Th atom (**Fig. 4**) is a trigonal faced dodecahedron [7, 8]. All of the water O atoms form a hydrogen-bonded network throughout the crystal [7]. Bond distances and angles in $[Th(1,2\text{-}C_6H_4O_2)_4]^{4-}$ are shown in Table 15.

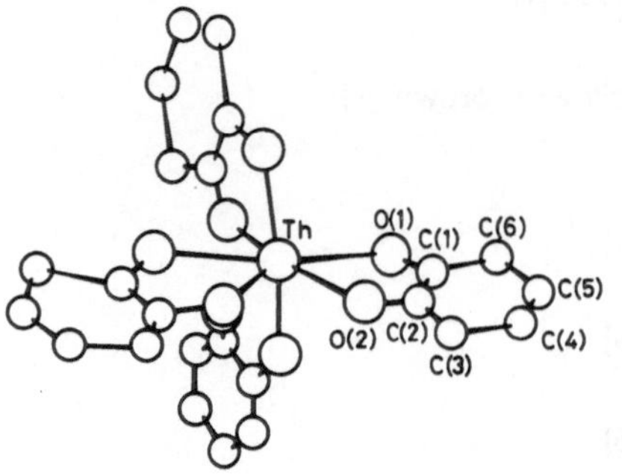

Fig. 4. The idealised view of the coordination arrangement about the thorium atom in the $[Th(1,2\text{-}C_6H_4O_2)_4]^{4-}$ anion in $Na_4[Th(1,2\text{-}C_6H_4O_2)_4]\cdot 21H_2O$ [7].

Table 15

Bond Distances (in Å) and Angles in (°) of $[Th(1,2\text{-}C_6H_4O_2)_4]^{4-}$ in $Na_4[Th(1,2\text{-}C_6H_4O_2)_4]\cdot 21H_2O$. The angle between catechols containing the two oxygens is given in parentheses [7].

Th–O(1) = 2.418(3), Th–O(2) = 2.421(3)

O(1)–Th–O(2)	66.8(1)	O(1)–Th–O(1) (90°)	93.64(4)	O(2)–Th–O(2) (90°)	128.6(1)
O(1)–Th–O(2) (90°)	80.7(1)	O(1)–Th–O(1) (180°)	150.8(2)	O(2)–Th–O(2) (180°)	75.7(2)
O(1)–Th–O(2) (180°)	142.3(1)				

Ultracentrifugation experiments involving Th salts and sodium pyrocatechol-3,5-disulfonate have provided evidence for the formation of a complex salt formulated as $Na_2[Th(C_6H_2O_2(SO_3)_2)_{1.5}]$[10], but this product was not isolated.

The thorium salt of 1,8-dihydroxynaphthalene, $Th(1,8\text{-}C_{10}H_6O_2)_2$, precipitates when a solution of the diol in a mixture of methanol and triethylamine is added dropwise to a solution of $ThCl_4$ in methanol [5]. The 2,2′-biphenyldiolate, $Th[2\text{-}OC_6H_4C_6H_4O\text{-}2']_2$, is precipitated when tri- or diethylamine is added dropwise to a methanol solution containing $ThCl_4$ and the diol [5]. The corresponding binaphthyl derivative, $Th(2\text{-}OC_{10}H_6C_{10}H_6O\text{-}2')_2$, is prepared in the same way using triethylamine as the precipitating base [5].

The basic 4-methylesculetin and 4-methyldaphnetin derivatives, $Th(4\text{-}CH_3\text{-}6,7\text{-}(O)_2\text{-}C_9H_3O_2)(OH)_2$ and $Th(4\text{-}CH_3\text{-}7,8\text{-}(O)_2C_9H_3O_2)(OH)_2\cdot 4H_2O$, are precipitated when an aqueous solution of $Th(NO_3)_4$ is mixed with an ethanol solution of the ligand (1:2 molar ratio), followed by adjustment to pH 5.0. Excess ligand was removed by washing the precipitate with ethanol and then water. The products were vacuum-dried at 100°C or over P_2O_5 in vacuum at room temperature [6].

A 1:1 Th^{IV} complex with 3,4-dihydroxy-4′-nitroazobenzene is reported to be formed in aqueous media [9], but this compound does not appear to have been isolated.

References for 15.5.3:

[1] Prasad, S.; Kumar, S. (J. Indian Chem. Soc. **40** [1963] 531/3).
[2] Agarwal, R. P.; Mehrotra, R. C. (J. Inorg. Nucl. Chem. **24** [1963] 821/7).
[3] Monsanto Company (Neth. 66-09020 [1966]; C.A. **67** [1967] No. 12439).
[4] Monsanto Company (Brit. 1132749 [1968]; C.A. **70** [1969] No. 20469).
[5] Andrä, K. (Z. Anorg. Allgem. Chem. **361** [1968] 254/8).
[6] Singh, D.; Singh, H. B. (Z. Naturforsch. **32b** [1977] 438/42).
[7] Sofen, S. R.; Abu-Dari, K.; Freyberg, D. P.; Raymond, K. N. (J. Am. Chem. Soc. **100** [1978] 7882/7).
[8] Raymond, K. N.; Weitl, F. L.; Sofen, S. R.; Cooper, S. R.; Smith, W. L.; Abu-Dari, K. (Abstr. Papers 178th Ann. Meeting Am. Chem. Soc., Washington, D.C., 1979, p. 189).
[9] Arkhangel'skaya, A. S.; Khomik, L. I. (Izv. Vysshikh Uchebn. Zavedenii Khim. Khim. Tekhnol. **23** [1980] 48/50; C.A. **93** [1980] No. 18245).
[10] Gustafson, R. L.; Martell, A. E. (J. Am. Chem. Soc. **82** [1960] 5610/6).

15.5.4 Thorium Compounds with Polyhydric Alcohols

The few known compounds are listed in Table 16.

Table 16
Thorium Compounds with Polyhydric Alcohols.

$Th(OH)_2(1,2,3\text{-}C_6H_3O_3)_2$(?) ($1,2,3\text{-}C_6H_3(OH)_3$ = pyrogallol)	decomposition to ThO_2 at 675°C; thermolysis curve illustrated in [7]
$Na_2[Th(1,2,3\text{-}C_6H_3O_3)_2]\cdot7H_2O$	[2]
$K_2[Th(1,2,3\text{-}C_6H_3O_3)_2]\cdot7H_2O$	[2]
$ThCl_2(HOC_6H_3O_2)$ ($1,3,5\text{-}C_6H_3(OH)_3$ = phloroglucinol)	dark brown [1]

Pyrogallic acid (=$1,2,3\text{-}C_6H_3(OH)_3$) precipitates Th from aqueous solution in the pH range 3.5 to 6.0 [6], but the composition of the precipitate was not reported. The thermal decomposition of a pyrogallate formulated as $Th(OH)_2(C_6H_3O_3)_2$ [7] obtained in this way has been studied; this product could be $Th(OH)_2(C_6H_3O(OH)_2)_2$.

The pyrogallol compounds $M^I_2[Th(1,2,3\text{-}C_6H_3O_3)_2]\cdot7H_2O$ (M^I = Na or K) are obtained when an aqueous 1:3 mixture of $Th(NO_3)_4\cdot6H_2O$ and $1,2,3\text{-}C_6H_3(OH)_3$ is treated with 6 equivalents of aqueous molar M^IOH under nitrogen, followed by addition of anhydrous ethanol to precipitate the complex [2]. The phloroglucinol (=$1,3,5\text{-}C_6H_3(OH)_3$) compound, $ThCl_2(HOC_6H_3O_2)$, is prepared in the same way as the pyrocatechol complex, $ThCl_2(1,2\text{-}C_6H_4O_2)$ and has the same solubility characteristics [1] (see p. 32).

An aqueous solution of polyacrolein, which can be regarded as a polyhydric alcohol, gives a

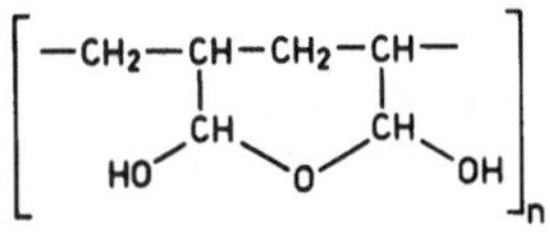

precipitate with aqueous solutions of thorium salts. Precipitation is quantitative at pH > 4 [3] or pH < 5 [4]. The reported compositions of the precipitates range from $Th(C_6H_{10}O_3)_{28}$ to

References for 15.5.4 on p. 36

$Th(C_6H_{10}O_3)_{8\ to\ 9}$ as the pH is increased from 2 to 8 [3]. $Th(C_6H_{10}O_3)_{28}$ and $Th(C_6H_{10}O_3)_{24}$ are reported to be formed at pH 2.36 (molar ratio, polymer:Th=1:3) and 2.40 (ratio 1:1.66), respectively, while $Th(C_6H_{10}O_3)_8$ and $Th(C_6H_{10}O_3)_9$ appear to be formed at pH 4.73 (polymer:Th=1:3) and 8.32 (polymer:Th=1:3), respectively [4].

The product of the polycondensation of 4-(2-pyridylazo)resorcinol with resorcinol and HCHO (see formula below) forms a chelate with Th^{4+} and behaves as a cation exchanger [5].

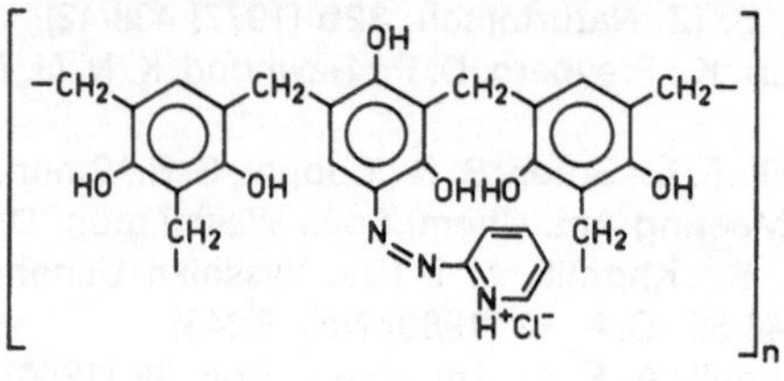

References for 15.5.4:

[1] Prasad, S.; Kumar, S. (J. Indian Chem. Soc. **40** [1963] 531/3).
[2] Agarwal, R. P.; Mehrotra, R. C. (J. Indian Chem. Soc. **42** [1965] 61/6).
[3] Koton, M. M.; Andreeva, I. V.; Andreev, P. F.; Danilov, L. G.; Rogozina, E. M. (Dokl. Akad. Nauk SSSR **146** [1962] 608/10; Dokl. Chem. Proc. Acad. Sci. USSR **142/147** [1962] 835/6).
[4] Andreeva, I. V.; Andreev, P. F.; Danilov, L. T.; Rogozina, E. M. (Radiokhimiya **6** [1964] 86/93; Soviet Radiochem. **6** [1964] 76/82).
[5] Szczepaniak, W.; Siepak, J. (Polimery **14** [1969] 538/40; C.A. **73** [1970] No. 4457).
[6] Deshmukh, G. S.; Xavier, J. (J. Indian Chem. Soc. **29** [1952] 911/4).
[7] Wendlandt, W. W. (Anal. Chem. **29** [1957] 800/2).

15.6 Thorium Carboxylates

15.6.1 Introduction

Reviews of lanthanide and actinide [1] and actinide [2] carboxylates which have appeared since the publication of "Thorium" 1955 include much useful information. Less comprehensive treatments are also available in books on the chemistry of the actinides [3 to 5] and in a review of actinide chemistry [6]. Thorium compounds with Schiff bases derived from salicylaldehyde and aminocarboxylic acids are described in "Thorium" Suppl. Vol. E, 1985, pp. 185/6 and 194/5.

References for 15.6.1:

[1] Bagnall, K. W. (MTP [Med. Tech. Publ. Co.] Intern. Rev. Sci. Inorg. Chem. Ser. Two **7** [1975] 41/63).
[2] Casellato, U.; Vigato, P. A.; Vidali, M. (Coord. Chem. Rev. **26** [1978] 85/159).
[3] Katz, J. J.; Seaborg, G. T. (The Chemistry of the Actinide Elements, Methuen, London 1957, pp. 1/508).
[4] Bagnall, K. W. (The Actinide Elements, Elsevier, Amsterdam 1972, pp. 1/272).
[5] Bailar, J. C.; Emeléus, H. J.; Nyholm, R.; Trotman-Dickenson, A. F. (Comprehensive Inorganic Chemistry, Vol. 5: The Chemistry of the Actinides, Pergamon, Oxford 1975, pp. 1/715).
[6] Comyns, A. E. (AERE-C-R-2320 [1957] 1/16; N.S.A. **12** [1958] No. 4776; Chem. Rev. **60** [1960] 115/46 [584 references]).

15.6.2 Aliphatic Monocarboxylates of Thorium

15.6.2.1 Thorium Formates and Formato Complexes

15.6.2.1.1 Thorium Formates

The known formates and their properties are listed in Table 17.

Table 17
Thorium Formates and Related Compounds.

thorium formates	
$Th(HCOO)_4$	decomposition at ca. 210 to 345°C [9], 220 to ca. 260°C [1], >230°C [14], 250 to 300°C [4], 289°C [6], activation energy of decarboxylation 36 kcal/mol (151 kJ/mol) at 225 to 280°C [14]; density, d_{obs} = 3.415 g/cm^3 [9] crystallographic data: orthorhombic, see λ-$Th(HCOO)_4$ below; tetragonal [19, 22, 24], space group $I\bar{4}$-S_4^2 (No. 82) [20, 22, 24], lattice parameters, a = 7.973(2), c = 6.588(2) Å [22, 24]; Z = 2 [17, 22, 24]; density, d_{calc} = 3.267, d_{obs} = 3.22(2) g/cm^3 [22, 24]; X-ray powder diffraction data given in [6] crystal modifications: λ-form, orthorhombic [15, 17], space group $Pmn2_1$-C_{2v}^7 (No. 31) or Pmmn-D_{2h}^{13} (No. 59), lattice parameters, a = 6.78(2), b = 7.32(4), c = 16.92(6) Å; Z = 4; density, d_{calc} = 3.26, d_{obs} = 3.1 g/cm^3 [12, 15] α-form (130°C), monoclinic [15, 17], space group B2-C_2^3 (No. 5), Bm-C_s^3 (No. 8), or B2/m-C_{2h}^3 (No. 12) [17], lattice parameters, a = 7.465(10), b = 6.672(10), c = 8.515(10) Å, β = 91.35°; Z = 2; density, d_{calc} = 3.228 g/cm^3 [15] γ-form (165°C), tetragonal [15, 17], lattice parameters, a = 7.309(5), c = 8.175(5) Å; Z = 2; density, d_{calc} = 3.134 g/cm^3 [15] enthalpy of transformation, α to γ, +1.7 kcal/mol (7.1 kJ/mol) [11]; transition temperatures, μ→δ (reversible), 95°C [11]; δ→γ, ca. 122°C, γ→α on cooling [11], at 95°C [17], λ→α 64.5°C [11], 65°C [17], reversible IR spectrum: ν(CH), 2949, 2919, 2863 cm^{-1}, ν_{as}(CO), 1582 cm^{-1}, ν_s(CO), 1385 cm^{-1}, π(CH), 1165 cm^{-1}, δ(OCO), 808, 784, 777 cm^{-1}, ν(Th–O), 291 cm^{-1}, δ(OThO), 173 cm^{-1} [24] α-form: ν_s(CO) + ν_{as}(CO), 3030 cm^{-1}, 2ν_s(CO), 2750 cm^{-1}, ν_{as}(CO), 1570 cm^{-1}, ϱ_r(COO), 1383 cm^{-1}, ν_s(CO), 1374 cm^{-1}, δ(OCO), 780, 771, 765 cm^{-1} [16] γ-form: ν_s(CO) + ν_{as}(CO), 2960 cm^{-1}, 2ν_s(CO), 2740 cm^{-1}, ν_{as}(CO), 1570 cm^{-1}, ϱ_r(COO) + ν_s(CO), 1370 cm^{-1}, δ(OCO), 783, 764 cm^{-1} [16] δ-form: ν_s(CO) + ν_{as}(CO), 3030, 2960 cm^{-1}, 2ν_s(CO), 2750 cm^{-1}, ν_{as}(CO), 1570 cm^{-1}, ϱ_r(COO), 1386, 1374 cm^{-1}, ν_s(CO), 1362, 1343 cm^{-1}, δ(OCO), 791, 788, 770 cm^{-1} [16] μ-form: ν_s(CO) + ν_{as}(CO), 2922 cm^{-1}, 2ν_s(CO), 2755, 2730, 2675 cm^{-1}, ν_{as}(CO), 1560 cm^{-1}, ϱ_r(COO), 1392, 1384 cm^{-1}, ν_s(CO), 1363, 1349, 1340 cm^{-1}, δ(OCO), 791, 783, 770 cm^{-1}, ν(Th–O), 307, 285, 266, 232, 214 cm^{-1} [16]

References for 15.6.2.1.1 to 15.6.2.1.3 on pp. 41/2

Table 17 (continued)

	crystal optical properties, refractive indices, $n_g=1.686$, $n_p=1.620$, molar refraction, $R_D=44.18\ cm^3$ [9]; ^1H NMR spectrum [13]; diamagnetic susceptibility, $\chi_m=-78\times10^{-6}$ [10]
$Th(HCOO)_4\cdot$ca. $0.25\,H_2O$	[2]
$Th(HCOO)_4\cdot{}^2\!/\!_3\,H_2O$	dehydrated at 80 to109°C [6] crystallographic data: L-form, orthorhombic [15, 18], space group $P2_12_12_1\text{-}D_2^4$ (No. 19) [18], lattice parameters, a=6.78(2), b=7.32(4), c=16.92(6) Å; Z=4; density, $d_{calc}=3.26$, $d_{obs}=3.1\ g/cm^3$ [15]; lattice parameters, a=6.761(3), b=10.491(5), c= 13.323 Å; Z=4; density, $d_{calc}=3.212$, $d_{obs}=3.21\ g/cm^3$ [18] M-form, cubic, lattice parameter, a=6.80 Å [15]; X-ray powder diffraction data given in [6, 12] IR spectrum: L-form, $2\nu_{as}(CO)$, 3200 cm^{-1}, $\nu_s(CO)+\nu_{as}(CO)$, 2965 cm^{-1}, $2\nu_s(CO)$, 2740 cm^{-1}, $\nu_{as}(CO)+\delta(OCO)$, 2305 cm^{-1}, $\nu_{as}(CO)$, 1575 cm^{-1}, $\varrho_r(COO)$, 1398, 1390, 1384 cm^{-1}, $\nu_s(CO)$, 1374, 1363, 1345 cm^{-1}, $\delta(OCO)$, 789, 778, 775, 770, 767, 759 cm^{-1} [16], 791, 783, 770 cm^{-1} [18], $\nu(Th\text{–}OH_2)$, 385 cm^{-1}, $\nu_{as}(Th\text{–}O)$, 268 cm^{-1}, $\nu_s(Th\text{–}O)$, 259 cm^{-1} [16] M (or μ)-form, the same as μ-$Th(HCOO)_4$ (see above) [16] ^1H NMR spectrum [13]
$Th(HCOO)_4\cdot2H_2O$	diamagnetic susceptibility, $\chi_m=-112\times10^{-6}$ [10]
$Th(HCOO)_4\cdot2.5H_2O$	diamagnetic susceptibility, $\chi_m=-128\times10^{-6}$ [10]
$[Th(HCOO)_4(H_2O)_2]\cdot0.7H_2O$(?)	diamagnetic susceptibility, $\chi_m=-124\times10^{-6}$ [10]
$Th(HCOO)_4\cdot3H_2O$	dehydrated at 62.4 to 67.0°C [11], 80 to 100°C [9], 100 to 120°C [4], 171°C [6], 180°C/50 min [14], thermogravimetric and differential analysis curves illustrated in [6], see also [9, 14]; enthalpy of combustion, $\Delta H=929\pm13$ kJ/mol; enthalpy of formation, $\Delta H_{form}=-3300\pm13$ kJ/mol [26] crystallographic data: monoclinic [7, 12, 15], space group $P2_1\text{-}C_2^2$ (No. 4); lattice parameters, a=6.66, b=8.72, c=10.0 Å, β=109°; Z=2; density, $d_{calc}=2.88$ [7], $d_{obs}=2.865\ g/cm^3$ [7, 9]; space group $P2_1\text{-}C_2^2$ (No. 4) or $P2_1/m\text{-}C_{2h}^2$ (No. 11), lattice parameters, a=6.77(1), b=8.81(2), c=9.61(2) Å, β=109.25(8)°; Z=2; density, $d_{obs}=2.86\ g/cm^3$ [12, 15] (also indexable as pseudoorthorhombic, lattice parameters, a=6.77, b=8.81, c=18.15 Å, β=91.4° [12, 15]; Z=4; density, $d_{calc}=2.871$, $d_{obs}=2.862\ g/cm^3$ [15]); X-ray powder diffraction data given in [6]; crystal optical properties; refractive indices, $n_g=1.702$, $n_p=1.602$, $n_m=1.626$ [4, 9], molar refraction, $R_D=58.77\ cm^3$ [9] IR spectrum: $\nu_s(CO)+\nu_{as}(CO)$, 3035, 2940 cm^{-1}, $2\nu_s(CO)$, 2760 cm^{-1} [16], $\nu_{as}(CO)$, 1590 [8], 1575 cm^{-1}, $\varrho_r(COO)$, 1398, 1382 cm^{-1} [16], $\nu_s(CO)$, 1403, 1382, ca. 1355 cm^{-1} [8],

Table 17 (continued)

	1372, 1367 cm^{-1}, δ(OCO), 798, 793, 781 cm^{-1} [16], 782(?) cm^{-1} [8], ν(ThO), 309, 284, 262, 247 cm^{-1} [16], IR spectrum (3800 to 500 cm^{-1}) illustrated in [8]; ^1H NMR spectrum reported in [13]; diamagnetic [4]
$Th_3(HCOO)_{12}\cdot 9H_2O$	presumably $Th(HCOO)_4\cdot 3H_2O$, see "Thorium" 1955, p. 302
$Th_3(HCOO)_{12}\cdot 3.5H_2O$	see "Thorium" 1955, p. 302
$Th(HCOO)_4\cdot 1.5CO(NH_2)_2$	see Th E, pp. 39, 42
$Th(HCOO)_4\cdot (CH_3)_2SO$	see Th E, pp. 100, 103
$ThCl_2(HCOO)_2$	[3]
$ThCl_3(HCOO)$	light brown [3]
hydroxide formates	
$Th(OH)(HCOO)_3$	[6]
$Th(OH)_2(HCOO)_2$	decomposes at 210°C [21]
$Th(OH)_2(HCOO)_2\cdot 2H_2O$	loses $2H_2O$ at 140°C; differential thermal analysis and thermogravimetric analysis curves illustrated in [21] density, d_{obs}, 3.30 g/cm^3 [21]
$Th(OH)_3(HCOO)$	[21]; decomposition: OH group at 260°C, HCOO at 210°C [27]
$Th(OH)_3(HCOO)\cdot nH_2O$	dehydrated at 140°C [27]
$[Th_3(HCOO)_6(OH)_5](HCOO)\cdot xH_2O$	x=2 or 4, "Thorium" 1955, p. 302
$[Th_3(HCOO)_6(OH)_5X]\cdot yH_2O$	$X=NO_3$ (y=10), ClO_3 (y=16), ClO_4 (y=12), NCS (y=7); see "Thorium" 1955, pp. 302/3
$Th_3(HCOO)_6(OH)_6\cdot 4H_2O$	see "Thorium" 1955, pp. 302/3
salt of (ethoxy-hydroxy-phosphoryl)-formic acid ethyl ester (diethylphosphono formate)	
$Th\{(C_2H_5O)P(:O)(O)COOC_2H_5\}_4$	IR spectrum: ν(CO), 1702, 1180 cm^{-1}, ν_{as}(POO), 1225, 1160 cm^{-1}, $\nu(C_2H_5OP)+\nu(C_2H_5OC)$, 1143 cm^{-1}, ν_s(POO), 1087, 1070 cm^{-1}, ν(ThO(POO)), 442, 407 cm^{-1} [25]

The mixed formate chlorides, $Th(HCOO)Cl_3$ and $Th(HCOO)_2Cl_2$, are reported to be formed when $ThCl_4$ is heated at 160°C under reflux with the calculated quantities of HCOOH [3]. It is not certain whether these are genuine compounds or mixtures of $Th(HCOO)_4$ and $ThCl_4$.

Anhydrous $Th(HCOO)_4$ is obtained when $ThCl_4$ is heated with HCOOH at 60 to 70°C [19, 20, 24]. Crystals of this product possess tetragonal symmetry and the compound is isomorphous with $Pa(HCOO)_4$ and $Np(HCOO)_4$, but not with $U(HCOO)_4$ prepared in this manner [19, 22, 24]. The anhydrous formate is also reported to be formed by heating hydrated $Th(NO_3)_4$ with a large excess of HCOOH [1]. This reaction is vigorous, so that it is preferable to add the $Th(NO_3)_4$ in small portions to the acid, and then reflux the precipitate for 2 h with HCOOH [2]. However, even after heating the solid so obtained at 80 to 90°C in a vacuum, the product appeared to be slightly hydrated, $Th(HCOO)_4\cdot$ ca. $0.25H_2O$ [2] (see $Th(HCOO)_4\cdot ⅔H_2O$, p. 40).

References for 15.6.2.1.1 to 15.6.2.1.3 on pp. 41/2

The anhydrous compound is formed when $Th(HCOO)_4 \cdot 3H_2O$ (see below) is washed with acetone or diethyl ether [4], and when the trihydrate is heated at 80 to 100°C [9], 100 to 120°C [4], 171°C [6], or at 62.4 to 67.0°C when the heating rate is slow (0.5°C/day) [11]. It is also obtained when $Th(HCOO)_4 \cdot 2/3 H_2O$ (see below) is heated at 88 to 109°C [6].

Anhydrous $Th(HCOO)_4$ is polymorphic [14 to 18], but see also [24]. The λ-form results when the trihydrate is dehydrated at 171°C [6] and this form, when slowly heated to constant weight at 0.5°C/day to 65°C [16, 17], 64.5°C and above [11], or at 130°C [15], yields the α-form which is isostructural with α-$U(HCOO)_4$ ("Uranium" Suppl. Vol. C 13, 1983, pp. 97/8). This change is reversed on cooling, and α-$Th(HCOO)_4$ hydrates readily [15]. The α-form transforms to γ-$Th(HCOO)_4$, isomorphous with γ-$U(HCOO)_4$ ("Uranium" Suppl. Vol. C 13, 1983, pp. 97/8) on heating [11] (e.g. at 125°C [16, 17]) and this transformation is rapid and reversible [11] unless residual water is present [17]; it occurs at 95°C if the sample has been pre-heated to 153°C in order to eliminate water [17]. μ-$Th(HCOO)_4$ is obtained when M-$Th(HCOO)_4 \cdot 2/3 H_2O$ (see bewlow) is dehydrated [11] (e.g., at 93°C [16]) and this form has the same crystal structure as the parent hydrate, whereas the L-form of $Th(HCOO)_4 \cdot 2/3 H_2O$ yields λ-$Th(HCOO)_4$ on dehydration [11] (e.g. at 62°C [16]). μ-$Th(HCOO)_4$ undergoes a reversible transformation to δ-$Th(HCOO)_4$ at 95°C [11, 16] or 130°C [15] and this form can also be obtained by rapidly heating $Th(HCOO)_4 \cdot 3H_2O$ (see below) or by slowly (0.5°C/day) heating M-$Th(HCOO)_4 \cdot 2/3 H_2O$ [11]. The δ-form transforms to γ-$Th(HCOO)_4$ above 122°C [11], 125°C [16], or 160°C [15]. X-ray powder diffraction data for the λ-, α-, and γ-forms are given in Table 17 (p. 37). The shapes of the split features assigned to the δ(OCO) modes in the IR spectra of these crystal modifications are apparently characteristic for the phases concerned [17].

The kinetics of decarboxylation of $Th(HCOO)_4$ have been studied; the primary gaseous products are HCHO and CO_2, and the secondary products noted were H_2, CO, H_2O, CH_3OH, $HCOOCH_3$, HCOOH, and $(CH_3)_2O$ [14]. Although decomposition to ThO_2 is reported to take place at ca. 310°C, some decomposition was observed at 230°C [14], consistent with the observations of other authors [1, 4, 6, 9]; see Table 12, p. 37.

$Th(HCOO)_4 \cdot 2/3 H_2O$ separates when small portions of solid hydrated $Th(NO_3)_4$ are added to pure HCOOH [6]. Formation of this hydrate has been observed in the thermogravimetric curve for the trihydrate at 50 to 70°C when the rate of heating is slow, or at 70 to ca. 90°C under a controlled pressure of H_2O vapour at a heating rate of 1°C/min [6], and also at 52.0 to 60°C [11]. The product obtained when the rate of heating is slow differs crystallographically (form L) from that obtained by direct synthesis (form M). X-ray powder diffraction data for the L- and M-forms are given in Table 17 (p. 38). $Th(HCOO)_4 \cdot 2/3 H_2O$ dehydrates when heated at 180°C for 50 min [14].

The trihydrate, $Th(HCOO)_4 \cdot 3H_2O$, has been prepared by dissolving freshly precipitated thorium hydroxide in 84, 40, or 20% aqueous HCOOH at 60°C or at room temperature; the best conditions are 40% HCOOH at 50 to 60°C. Crystals of the product separate on cooling [4, 6], see also [26].

$Th(HCOO)_4 \cdot 3H_2O$ is not appreciably soluble in organic solvents, and its solubility in water is 80.7 g/L [6]. A study of the ternary system H_2O–HCOOH–$Th(HCOO)_4$ is reported in [6]. Thermogravimetric [9], differential thermal analysis [6, 14], and emanating power [14] experiments have also been recorded. The trihydrate begins to lose water at 92°C when heated at 4°C/min in a current of dry N_2 [6].

The IR spectrum has been interpreted as supporting the formulation $[Th(HCOO)_4(H_2O)_2] \cdot H_2O$ [16], but the structure has been described as a 3-dimensional network of Th atoms bound by HCOO bridges. Each Th atom is surrounded by 8 oxygen atoms of 8 HCOO groups, with 6 oxygen atoms at the apices of a trigonal prism and the remaining two capping rectangular faces. The coordination polyhedron has been described as a distorted square antiprism formed by the oxygen atoms of the HCOO groups with the H_2O molecules located on the square faces [7]. An electron density map is given in [7]. X-ray diffraction data are summarised in Table 17 (p. 38).

The ^1H NMR spectrum of the trihydrate appears to indicate 2 possible environments for the H_2O molecules [13].

15.6.2.1.2 Thorium Hydroxide Formates

$Th(HCOO)_3(OH)$ is reported to be formed by the action of water vapour on $Th(HCOO)_4$ at ca. 25°C and the kinetics of this slow reaction have been studied [6].

The addition of aqueous NH_3 to solutions of $Th(NO_3)_4$ containing HCOOH and HNO_3 leads to a product of composition $Th(HCOO)_2(OH)_2 \cdot 2H_2O$, the yield of which is reduced at pH 4 to 5.3, possibly owing to the formation of a soluble complex, such as $(NH_4)_2Th(HCOO)_6$ (p. 43). As the relative amount of added NH_3 increases up to pH 3.85, crystals of $Th(HCOO)_2(OH)_2 \cdot 2H_2O$ predominate in the solid phase, while with further addition of NH_3 to pH 5.3, the solubility increases. Then, with further addition of aqueous NH_3, the solubility increases and the solid phase is a mixture of hydrated $Th(HCOO)_2(OH)_2$ and $Th(HCOO)(OH)_3$ [21]. The tendency to hydrolyse to species of the type $Th(HCOO)_{4-n}(OH)_n \cdot mH_2O$ has been discussed in [23]. In another paper, the precipitation of thorium from aqueous media in the presence of HCOOH, presumably as a basic formate, is said to begin at pH 6 [5].

The precipitation of Th as a basic formate (composition unspecified), by generating NH_3 from the hydrolysis of urea, has been used for the separation of Th from the lanthanides [28].

15.6.2.1.3 Thorium Salt of (Ethoxy-hydroxy-phosphoryl)-formic Acid Ethyl Ester

$Th\{(C_2H_5O)P(:O)(O)COOC_2H_5\}_4$ has been obtained by heating a suspension of $ThCl_4$ in diethoxyphosphorylformic acid ethyl ester ($=(C_2H_5O)_2P(:O)COOC_2H_5$) at 50 to 200°C until dissolution is followed by precipitation of the complex. The compound is a linear chain-like polymeric species involving bidentate bridging anions which are coordinated through two PO oxygen atoms to adjacent metal cations; the carboxylate O atoms are not bonded to the Th atom [25]. IR data are included in Table 17 (p. 39).

References for 15.6.2.1.1 to 15.6.2.1.3:

[1] Okubo, M.; Goto, R. (Nippon Kagaku Zasshi **81** [1960] 1132/6; C.A. **56** [1962] 3342).
[2] Sahoo, B.; Panda, S.; Patnaik, D. (J. Indian Chem. Soc. **37** [1960] 594).
[3] Jaura, K. L.; Bajwa, P. S. (J. Sci. Ind. Res. [India] **20 B** [1961] 391/4).
[4] Golovnya, V. A.; Ivanova, O. M. (Zh. Neorgan. Khim. **8** [1963] 2462/9; Russ. J. Inorg. Chem. **8** [1963] 1290/4).
[5] Tserkovnitskaya, I. A.; Charykov, A. K. (Izv. Vysshikh Uchebn. Zavedenii Khim. Khim. Tekhnol. **7** [1964] 544/50; C.A. **62** [1965] 3385).
[6] Claudel, B.; Mentzen, B. (Bull. Soc. Chim. France **1966** 1547/52).
[7] Arutyunyan, E. G.; Antsyshkina, A. S.; Balta, E. Ya. (Zh. Strukt. Khim. **7** [1966] 471/2; J. Struct. Chem. [USSR] **7** [1966] 448/9).
[8] Evstaf'eva, O. N.; Ivanova, O. M.; Molodkin, A. K.; Dvoryantseva, G. A.; Struchkova, M. I. (Dokl. Akad. Nauk SSSR **172** [1967] 860/2; Dokl. Chem. Proc. Acad. Sci. USSR **172/177** [1967] 118/20).

[9] Molodkin, A. K.; Ivanova, O. M.; Kozina, L. E.; Petrov, K. I. (Zh. Neorgan. Khim. **13** [1968] 1327/36; Russ. J. Inorg. Chem. **13** [1968] 694/9).
[10] Belova, V. I.; Syrkin, Ya. K.; Molodkin, A. K.; Ivanova, O. M.; Shiporina, L. M. (Zh. Neorgan. Khim. **13** [1968] 1458/60; Russ. J. Inorg. Chem. **13** [1968] 766/7).

[11] Breysse, M.; Mentzen, B.; Navarro, A. (Compt. Rend. C **267** [1968] 1091/2).
[12] Chevreton, M.; Claudel, B.; Mentzen, B. (J. Chim. Phys. **65** [1968] 890/4).
[13] Demarquay, J.; Tho, Pham Quang; Mentzen, B.; Claudel, B. (J. Chim. Phys. **65** [1968] 1380/5).
[14] Mentzen, B. (Ann. Chim. [Paris] [14] **3** [1968] 367/84).
[15] Mentzen, B. (Rev. Chim. Minérale **6** [1969] 713/25).
[16] Mentzen, B. F. (J. Solid State Chem. **3** [1971] 12/9).
[17] Mentzen, B. F. (J. Solid State Chem. **3** [1971] 20/5).
[18] Mentzen, B.; Prost, M. (Compt. Rend. C **276** [1973] 229/32).
[19] Bohres, E. W. (JUEL-1080-NC [1974] 1/66; C.A. **82** [1975] No. 67534).
[20] Bohres, E. W.; Hauck, J.; Schenk, H. J.; Schwochau, K. (Proc. 16th Intern. Conf. Coord. Chem., Dublin 1974, Paper 2.16b; C.A. **85** [1976] No. 38949).

[21] Andryushin, V. G.; Kozhnevnikov, P. B.; Pozharskaya, M. E.; Shmidt, V. S. (Radiokhimiya **17** [1975] 555/9; Soviet Radiochem. **17** [1975] 534/7).
[22] Hauck, J. (Inorg. Nucl. Chem. Letters **12** [1976] 617/22).
[23] Shmidt, V. S.; Andryushin, V. G. (Radiokhimiya **18** [1976] 506/11; Soviet Radiochem. **18** [1976] 439/43).
[24] Greis, O.; Bohres, E. W.; Schwochau, K. (Z. Anorg. Allgem. Chem. **433** [1977] 111/8).
[25] Mikulski, C. M.; Sanford, P.; Harris, N.; Rabin, R.; Karayannis, N. M. (J. Coord. Chem. **12** [1983] 187/95).
[26] Thakur, L.; Thakur, A. K.; Ahmad, M. F. (Indian J. Chem. A **19** [1980] 793/5).
[27] Shmidt, V. S.; Andryushin, V. G. (Radiokhimiya **24** [1982] 601/6; Soviet Radiochem. **24** [1982] 498/503).
[28] Willard, H. H.; Gordon, L. (Anal. Chem. **20** [1948] 165/9).

15.6.2.1.4 Thorium Formato Complexes

The known compounds are listed in Table 18.

Table 18
Thorium Formato Complexes.

$K[Th(HCOO)_5]$[a)]	[1], density, $d_{obs}=3.047\ g/cm^3$ [3] IR spectrum: $\nu_{as}(CO)$, (1670?), 1630 cm^{-1}, $\nu_s(CO)+\delta(CH)$, 1385, 1298 cm^{-1}, $\delta(OCO)$, 795, 782, 778, 768 cm^{-1} [5]; crystal optical properties: refractive indices, $n_g=1.658$, $n_p=1.600$, molar refraction, $R_D=57.87\ cm^3$ [3]; see also "Thorium" 1955, p. 334
$Rb[Th(HCOO)_5]$[a)]	density, $d_{obs}=3.235\ g/cm^3$ [3] IR spectrum: $\nu_{as}(CO)$, (1690?), 1590 cm^{-1}, $\nu_s(CO)+\delta(CH)$, 1400, 1355 cm^{-1}, $\delta(OCO)$, 798, 785, 778, 770 cm^{-1} [5]

Table 18 (continued)

$Cs[Th(HCOO)_5]$[a)]	density, $d_{obs}=3.586$ g/cm^3 [3]; decomposes at >80°C [1] IR spectrum: $\nu_{as}(CO)$, 1610 cm^{-1}, $\nu_s(CO)+\delta(CH)$, 1400, 1350, (1330?) cm^{-1}, $\delta(OCO)$, 800, 765 cm^{-1} [5]; crystal optical properties: refractive indices, $n_g=1.608$, $n_p=1.518$, molar refraction, $R_D=53.4$ cm^3 [3]
$(C_5H_5NH)[Th(HCOO)_5]$	see "Thorium" 1955, p. 343
$Rb_2[Th(HCOO)_6]$	[3]
$Rb_2[Th(HCOO)_6]\cdot 2H_2O$	loses $2H_2O$ at 80 to 100°C in air [3]
$Rb_2[Th(HCOO)_6]\cdot 3H_2O$[a)]	crystal optical properties: refractive indices, $n_g=1.577$, $n_m=1.562$, $n_p=1.546$ [3]
$Cs_2[Th(HCOO)_6]$[a)]	density, $d_{obs}=3.548$ g/cm^3 [3]; decomposes at >80°C [1] IR spectrum: $\nu_{as}(CO)$, 1620 to 1560 cm^{-1}, $\nu_s(CO)+\delta(CH)$, 1410, 1350, 1315, 1300 cm^{-1}, $\delta(OCO)$, 806, 785, 780 cm^{-1} [5]; crystal optical properties: refractive indices, $n_g=1.616$, $n_p=1.522$, $n_m=1.575$ [1, 3], molar refraction, $R_D=71.12$ cm^3 [3]; diamagnetic susceptibility, $\chi_m=-192\times10^{-6}$ [4]
$(NH_4)_2[Th(HCOO)_6]$	see "Thorium" 1955, p. 339
$Sr[Th(HCOO)_6]\cdot 2H_2O$	see "Thorium" 1955, p. 348
$Ba[Th(HCOO)_6]$[a)]	density, $d_{obs}=3.554$ g/cm^3 [3]; decomposes at 250 to 300°C [1] IR spectrum: $\nu_{as}(CO)$, 1625, 1610, 1590 cm^{-1}, $\nu_s(CO)+\delta(CH)$, 1405, 1330, 1320, 1310 cm^{-1}, $\delta(OCO)$, 795, 788, 778 cm^{-1} [5]; crystal optical properties: refractive indices, $n_g=1.674$, $n_p=1.573$ [1, 3], molar refraction, $R_D=63.45$ cm^3 [3]; diamagnetic susceptibility, $\chi_m=-150\times10^{-6}$ [4]
$Ba[Th(HCOO)_6]\cdot 2H_2O$[a)]	density, $d_{obs}=3.413$ g/cm^3 [3]; loses $2H_2O$ at 80 to 100°C [3], 100 to 120°C [1] IR spectrum: $\nu_{as}(CO)$, 1627(?) cm^{-1}, $\nu_s(CO)$, 1422, 1408, ca. 1326, 1318 cm^{-1}, $\delta(OCO)$, 792 cm^{-1} [2] IR spectrum (3800 to 500 cm^{-1}) illustrated in [2]; crystal optical properties: refractive indices, $n_g=1.674$, $n_m=1.655$, $n_p=1.644$; molar refraction, $R_D=72.89$ cm^3 [3], see also "Thorium" 1955, p. 349
$Cs_3[Th(HCOO)_7]$[a)]	density, $d_{obs}=3.458$ g/cm^3 [3]; decomposes at >80°C [1] IR spectrum: $\nu_{as}(CO)$, 1590 cm^{-1}, $\nu_s(CO)$, 1340, 1323, ca. 1310 cm^{-1}, $\delta(OCO)$, 795 cm^{-1} IR spectrum (3800 to 500 cm^{-1}) illustrated in [2]; crystal optical properties: refractive indices, $n_g=1.610$, $n_p=1.520$ [1, 3]; molar refraction, $R_D=88.97$ cm^3 [3]; diamagnetic susceptibility, $\chi_m=-265\times10^{-6}$ [4]

References for 15.6.2.1.4 on p. 45

Table 18 (continued)

$Na_4[Th(HCOO)_8]$ [a)]	density, $d_{obs}=2.592$ g/cm³ [3] IR spectrum: $\nu_{as}(CO)$, 1610 cm^{-1}, $\nu_s(CO)+\delta(CH)$, 1405, 1365 cm^{-1}, $\delta(OCO)$, 798 cm^{-1} [5]; crystal optical properties: refractive indices, $n_g=1.645$, $n_m=1.615$, $n_p=1.569$, molar refraction, $R_D=91.39$ cm³ [3]
$Rb_4[Th(HCOO)_8]$ [a)]	density, $d_{obs}=3.208$ g/cm³ [3] IR spectrum: $\nu_{as}(CO)$, 1600 cm^{-1}, $\nu_s(CO)+\delta(CH)$, 1400, 1350 cm^{-1}, $\delta(OCO)$, 795, 780, 775 cm^{-1} [5] crystal optical properties: refractive indices, $n_g=1.658$, $n_m=1.633$, $n_p=1.609$, molar refraction, $R_D=101.00$ cm³ [3]
$Cs_4[Th(HCOO)_8]$ [a)]	density, $d_{obs}=3.672$ g/cm³ [3]; decomposition at >80°C [1] IR spectrum: $\nu_{as}(CO)$, ca. 1620 cm^{-1}, $\nu_s(CO)$, ca. 1350, 1326 cm^{-1}, $\delta(OCO)$, 776 cm^{-1} [2] IR spectrum (3800 to 500 cm^{-1}) illustrated in [2] crystal optical properties: refractive indices, $n_g=1.620$, $n_m=1.561$, $n_p=1.521$ [1, 3], molar refraction, $R_D=99.97$ cm³ [3]; diamagnetic [1], susceptibility, $\chi_m=-303\times10^{-6}$ [4]
$Na[Th_3(HCOO)_6O(OH)_4X]\cdot yH_2O$	$x=NO_3$, $y=10.5$; $x=ClO_3$, $y=13$; $x=ClO_4$, $y=9$ see "Thorium" 1955, p. 326/7
$K[Th_3(HCOO)_6O(OH)_4(NCS)]\cdot 7H_2O$	see "Thorium" 1955, pp. 326, 335
$K_2[Th_3(HCOO)_6O_2(OH)_2(NCS)_2]$	see "Thorium" 1955, pp. 326, 335
$Na[Th_3(HCOO)_6O(OH)_4(NO_3)]\cdot[Th_3(HCOO)_6(OH)_5(NO_3)]\cdot 21H_2O$	see "Thorium" 1955, p. 326
$2K[Th_3(HCOO)_6O(OH)_4(NO_3)]\cdot[Th_3(HCOO)_6(OH)_5(NO_3)]\cdot 30H_2O$	see "Thorium" 1955, pp. 326, 334

a) Thermogravimetric heating curve illustrated in [3].

The IR spectra of these compounds do not help in the assignment of the bonding mode of the HCOO group, which is probably not unidentate in the majority of the salts [5], although some or all of the HCOO groups are likely to be unidentate in the octaformatocomplex anion, $[Th(HCOO)_8]^{4-}$.

Alkali metal salts of the pentaformatothorate(IV) anion are quite well known. $K[Th(HCOO)_5]$ is obtained when a tenfold excess of K(HCOO) in water is added to an aqueous solution of $Th(HCOO)_4\cdot 3H_2O$ (p. 40); the salt precipitates on standing for a few hours in air [1]. $Cs[Th(HCOO)_5]$ has been prepared in a similar manner [1], and $Rb[Th(HCOO)_5]$ is prepared by adding the stoichiometric quantity of $Ba(HCOO)_2$ in water to a hot (70 to 80°C) aqueous solution containing $Th(SO_4)_2\cdot 8H_2O$ and Rb_2SO_4; the salt was isolated from the filtrate by leaving it to evaporate in air [3]. Aqueous solutions of the potassium and caesium salts have abnormally high molar conductances, indicating hydrolysis [1].

$Rb_2[Th(HCOO)_6]\cdot 2H_2O$ has been prepared in the same way as $Rb[Th(HCOO)_5]$ (see above). The salt loses water readily on heating at 80 to 100°C in air and when washed with acetone, ethanol, or ether. The anhydrous product forms the trihydrate on standing in air [3]. Anhydrous $Cs_2[Th(HCOO)_6]$ is also obtained in the same way as $Rb[Th(HCOO)_5]$. Its conductance in water corresponds to that expected for a 3-ion electrolyte [1].

Anhydrous $Ba[Th(HCOO)_6]$ is obtained when the dihydrate is heated at 80 to 100°C [3] or 100 to 120°C [1] and on washing the dihydrate with acetone, ethanol, or ether [1]. The dihydrate is precipitated when a 6-fold excess of $Ba(HCOO)_2$ in water is added to a hot (50°C) aqueous solution of $Th(HCOO)_4 \cdot 3H_2O$. The precipitate becomes contaminated with $Ba(HCOO)_2$ if left in contact with the mother liquor for more than 2 to 3 h. The molar conductance of the salt in water is abnormally high, indicating hydrolysis [1].

The only recorded heptaformato complex salt, $Cs_3[Th(HCOO)_7]$, and the octaformato complex, $Cs_4[Th(HCOO)_8]$, have been prepared in the same way as $Cs_2[Th(HCOO)_6]$ (see above). They are very hygroscopic and decompose on storage. Their molar conductivities in water correspond to those expected for a 4-ion and 5-ion electrolyte, respectively [1]. $Na_4[Th(HCOO)_8]$ and $Rb_4[Th(HCOO)_8]$ have also been prepared in the same manner [3].

The caesium salts, $Cs_n[Th(HCOO)_{4+n}]$ (n=2, 3 or 4) are more soluble in water than $K[Th(HCOO)_5]$, $Cs[Th(HCOO)_5]$, or $Th(HCOO)_4 \cdot 3H_2O$ [1].

References for 15.6.2.1.4:

[1] Golovnya, V. A.; Ivanova, O. M. (Zh. Neorgan. Khim. **8** [1963] 2462/9; Russ. J. Inorg. Chem. **8** [1963] 1290/4).
[2] Evstaf'eva, O. N.; Ivanova, O. M.; Molodkin, A. K.; Dvoryantseva, G. A.; Struchkova, M. I. (Dokl. Akad. Nauk SSSR **172** [1967] 860/2; Dokl. Chem. Proc. Acad. Sci. USSR **172/177** [1967] 118/20).
[3] Molodkin, A. K.; Ivanova, O. M.; Kozina, L. E.; Petrov, K. I. (Zh. Neorgan. Khim. **13** [1968] 1327/36; Russ. J. Inorg. Chem. **13** [1968] 694/9).
[4] Belova, V. I.; Syrkin, Ya. K.; Molodkin, A. K.; Ivanova, O. M.; Shiporina, L. M. (Zh. Neorgan. Khim. **13** [1968] 1458/60; Russ. J. Inorg. Chem. **13** [1968] 766/7).
[5] Petrov, K. I.; Molodkin, A. K.; Saralidze, O. D.; Ivanova, O. M. (Zh. Neorgan. Khim. **13** [1968] 1716/9; Russ. J. Inorg. Chem. **13** [1968] 895/7).

15.6.2.2 Thorium Acetates and Acetato Complexes

15.6.2.2.1 Thorium Acetates

The known compounds and their properties are listed in Table 19. Mixed acetate/carboxylate compounds are described on pp. 138 (aminobenzoates), 142 (4-methoxybenzoate), 149 (hydroxynaphthoates), 153 (pyridine-3-carboxylate), 155/6 (pyridine-2,6-dicarboxylates), and 157 (thiophene-2-carboxylate).

Table 19
Thorium Acetates.

$Th(CH_3COO)_4$	[2, 5]; decompostion: >180°C (max. 242 to 260°C) [20], 235 to 300°C [36], 240°C [6], >270°C [1], 289°C [10], >300°C [14], 310°C [37], >370°C [23] thermogram [10, 14, 23], differential thermal analysis, thermogravimetric analysis and differential thermogravimetric analysis (under N_2) curves [31] illus-

References for 15.6.2.2.1 on pp. 49/50

Table 19 (continued)

	trated; activation energy for decomposition, E_{act} = 26 ± 3 kcal/mol (109 ± 13 kJ/mol) [32], 111 ± 9 kJ/mol [36], 26.5 ± 2.5 kcal/mol (111.3 ± 10.5 kJ/mol) [31, 40], 151 ± 8 kJ/mol [36]; enthalpy of decomposition to ThO_2, $\Delta H = 36 \pm 2$ kcal/mol (151 ± 4 kJ/mol) [32]
	crystallographic data: monoclinic, space group C2/c-C_{2h}^6 (No. 15), lattice parameters a = 17.829(6), b = 8.285(3), c = 8.388(3) Å, β = 105.41(8); Z = 4 [17]
	density, d_{calc} = 2.604, d_{obs} = 2.52(12) [17], 2.464 g/cm³ [10, 14], X-ray powder diffraction pattern illustrated in [23]
	crystal optical properties: refractive indices n_1 = 1.520, n_2 = 1.666 [14]
	IR spectrum: ν_{as}(CO), 1720, 1564, 1518 cm^{-1} [22], 1632, 1567 cm^{-1} [25], 1560, 1520 cm^{-1} [38, 41], $\nu_{as}(CO) + \nu_{as}(CH_3)$, 1449 cm^{-1} [22], ν_s(CO) = 1410, 1370 cm^{-1} [38, 41], $\nu(CO) + \delta(CH_3)$, 1453, 1411, 1385, 1348 cm^{-1} [25], ν(CC), 985, 960, 950 cm^{-1} [38, 41], 960, 945 cm^{-1} [14], δ(OCO), 695 cm^{-1} [38, 41], 675, 640, 618, 610 cm^{-1} [14], 643 cm^{-1} [25], δ_s(OCO), 676 cm^{-1} [22], ϱ_w(OCO), 640 cm^{-1} [22]
	IR spectrum (4000 to 400 cm^{-1}) illustrated in [14]
	enthalpy of formation, $\Delta H_f = -2674 \pm 1$ kJ/mol [34], -2663 ± 5 kJ/mol [35]
	Th–O bond strength, 299 kJ/mol [34]
	diamagnetic susceptibility, $\chi_m = -95$ to -119×10^{-6} [18]
	binding energy, $Th5d_{5/2}$ = 87.9 eV [24]; see also "Thorium" 1955, p. 303
$ThCl_n(CH_3COO)_{4-n}$	n = 0.3, 1.8, 2.2, 2.3, 3.0, and 3.4 [15]
$Th(OH)(CH_3COO)_3$	[14]; decomposes at 289°C [9]; thermogram illustrated in [9]; density, d_{obs} = 2.499 g/cm³ [9]
	solubility product, $S = [Th^{4+}][CH_3COO^-]^3[OH^-] = 5.4 \times 10^{-23}$ [9]
$Th(OH)(CH_3COO)_3 \cdot 1.5H_2O$	decomposition 250 to 370°C [14]
$Th(OH)_{1.5}(CH_3COO)_{2.5} \cdot 2H_2O$	decomposition 250 to 370°C [14]
$Th(OH)_2(CH_3COO)_2$	decomposition to $ThO(CH_3COO)_2$ at ca. 235 [26, 37] to 240°C [26], enthalpy of formation (combustion), $\Delta H_f = -275$ kcal/mol (−1165 kJ/mol) [21], enthalpy of sublimation, ΔH_{subl} = 27.3 kcal/mol (114.7 kJ/mol), entropy of sublimation, ΔS_{subl} = 52.5 cal · $mol^{-1} \cdot K^{-1}$, (220.5 J · $mol^{-1} \cdot K^{-1}$) [21], Th–O bond energy = 73.5 kcal/mol (309kJ/mol) [21]
$Th(OH)_2(CH_3COO)_2 \cdot H_2O$	[8]; density, d_{obs} = 2.89 g/cm³ [25]; loses H_2O at ca. 170 to 180°C [26]; differential thermal and thermogravimetric curves illustrated in [26]

Table 19 (continued)

	IR spectrum: ν(CO), 1576 cm^{-1}, δ(CH_3)+ν(COO), 1429, 1378, 1352, 1306 cm^{-1}, δ(OCO), 650 cm^{-1}, π(COO), 555 cm^{-1} [25]; see also "Thorium" 1955, p.305
$Th(OH)_2(CH_3COO)_2 \cdot 2H_2O$	IR spectrum: ν(CO), 1594 cm^{-1}, $ν_s$(CO), 1423 cm^{-1} [11, 39], δ(OCO), 680 cm^{-1} [39], $δ_s$(OCO), 647 cm^{-1} [11, 39] IR spectrum illustrated in [12] ^1H NMR spectrum [11, 39]
$Th(OH)_2(CH_3COO)_2 \cdot 2.5H_2O$	decomposes at 250 to 370°C [14]
$Th(OH)_2(CH_3COO)_2 \cdot 3H_2O$	loses $2H_2O$ at ca. 150°C [26]; density, d_{obs}=2.75 g/cm^3 [25] IR spectrum: ν(CO), 1576 cm^{-1}, ν(COO)+δ(CH_3), 1439, 1378, 1352, 1306 cm^{-1}, δ(OCO), 650 cm^{-1}, π(COO), 555 cm^{-1} [25]; differential thermal and thermogravimetric analysis curves illustrated in [26]
$ThO(CH_3COO)_2$	decomposes at 300 to 328°C [26], 320°C [37]
$ThO(CH_3COO)_2 \cdot 4H_2O$	[13] ($Th(OH)_2(CH_3COO)_2 \cdot 3H_2O$?)
$Th(OH)_{2.5}(CH_3COO)_{1.5} \cdot H_2O$	decomposes at 250 to 370°C [14]
$Th(OH)_3(CH_3COO)$	loses $1.5H_2O$ at ca. 200°C [26, 37] and decomposition at 300 to 360°C [26], 360°C [37]
$Th(OH)_3(CH_3COO) \cdot H_2O$	loses H_2O at ca. 120°C, differential thermal and thermogravimetric analysis curves illustrated in [26], see also [27] density, d_{obs}=3.75 g/cm^3 [25] IR spectrum: ν(CO), 1576 cm^{-1}, ν(COO)+δ(CH_3), 1439, 1378, 1352, 1306 cm^{-1}, δ(OCO), 650 cm^{-1}, π(COO), 555 cm^{-1} [25]
$ThCl(OC(CH_3)_3)_{3-n}(CH_3COO)_n$	n≧2 [15]
$[Th(acac)(CH_3COO)_2(H_2O)_5](C_6H_2N_3O_7)$	$C_6H_3N_3O_7$=1,3,5-trinitrophenol, see Th E, pp. 134/5
$Th(acac)(CH_3COO)Cl_2 \cdot 4H_2O$	see Th E, pp. 134/5
$Th(C_{14}H_7O_4)_x(CH_3COO)_{4-x} \cdot 4H_2O$	x=1 or 2; $C_{14}H_8O_4$=alizarin, see Th E, p. 140
$Th(C_{14}H_7O_5)_x(CH_3COO)_{4-x} \cdot yH_2O$	x=1, y=2 and x=2, y=1; $C_{14}H_8O_5$=purpurin, see Th E, p. 140
$Th(C_{14}H_7O_6)_x(CH_3COO)_{4-x} \cdot yH_2O$	x=1, y=4 and x=2, y=3; $C_{14}H_8O_6$=quinalizarin, see Th E, p. 141
$Th(C_{14}H_7O_4)_x(CH_3COO)_{4-x} \cdot H_2O$	x=1 or 2; $C_{14}H_8O_4$=quinizarin, see Th E, p. 141
$Th(C_{17}H_{11}NO_7SNa)_2(CH_3COO)_2 \cdot 2H_2O$(?)	see Th E, p. 141

$Th(CH_3COO)_4$ has been prepared by heating hydrated $Th(NO_3)_4$ with $(CH_3CO)_2O$ [2, 10] or with a 2:1 mixture of glacial CH_3COOH and $(CH_3CO)_2O$ [34, 35] until the evolution of oxides of nitrogen ceases, and by heating hydrated $ThCl_4$ either with a mixture of CH_3COOH and $(CH_3CO)_2O$ under reflux (130°C) for 2 h [5] or with a large excess of CH_3COOH [6].

References for 15.6.2.2.1 on pp. 49/50

$Th(CH_3COO)_4$ precipitates immediately when freshly precipitated Th^{IV} hydroxide is treated with 50% aqueous CH_3COOH at 40 to 60°C [14] or with glacial CH_3COOH [17]. Direct precipitation of $Th(CH_3COO)_4$ by treating $Th(NO_3)_4$ with the stoichiometric quantity of CH_3COONa in concentrated aqueous solution gives only a poor yield [6]. When $Th(NO_3)_4 \cdot 2.75\,H_2O$ is warmed with the stoichiometric quantity of glacial CH_3COOH a viscous solution is obtained from which $Th(NO_3)_4 \cdot 5\,H_2O$ crystallises on standing in air [10].

$Th(CH_3COO)_4$ is isomorphous with $U(CH_3COO)_4$ ("Uranium" Suppl. Vol. C 13, 1983, pp. 112/3) and $Ce(CH_3COO)_4$ [22] and the thorium atom is presumably 10-coordinate [16, 17, 22]. The Th salt is readily soluble in 40% HCOOH [14] and is also soluble in monoethanolamine (70.81 g/100 g solvent) [3], 1,2-diaminoethane (4.30 g/100 g solvent) [3, 4], and 1,2-dihydroxyethane (0.41 g/100 g solvent) [3]. These data refer to 30°C. The conductivities of the solutions in 1,2-dihydroxy- [3] and 1,2-diaminoethane [3, 4] at 30°C are low and suggest weak electrolyte behaviour [4]. The acetate is also slightly soluble in water (1.5 g/100 g H_2O) and the conductivity of the aqueous solution suggests extensive (84.8%) hydrolysis [10]. It is almost insoluble in acetone, chloroform, dioxane, ethanol, and ether [10].

Acetone, CO_2 and ThO_2 are formed in the thermal decomposition of $Th(CH_3COO)_4$ [1, 7]; acetic acid is the primary volatile product [31] and this is then catalytically decomposed on ThO_2 [32]. The rate of decarboxylation has been studied by thermogravimetry [7] and the mechanism of the thermal decomposition has been investigated using the catalytic decomposition of $CH_3COOC_2H_5$ on ThO_2 at 300 to 400°C as a model [33]. The nature of the decomposition differs from that of $U(CH_3COO)_4$, possibly because of a difference in the state of the surfaces of the crystals of these compounds, but the activation energies for decomposition are the same for the two compounds [31, 32, 40]. The kinetic and thermodynamic parameters for the thermal decomposition of $Th(CH_3COO)_4$ have been reviewed [36] and the kinetics of the decomposition at 100 to 300°C under isothermal conditions have been studied [40]. A review of published data on the thermal decomposition of thorium acetates together with thoriumhalogeno acetates (see p. 55) is given in [42].

Satellite [28] and shakeup satellites [29] in the X-ray photoelectron (ESCA) spectrum of $Th(CH_3COO)_4$ have been studied and configuration interaction satellites in the ESCA spectrum have been discussed [30].

$Th(CH_3COO)_4$ has been proposed as a catalyst for the production of high molecular weight polyesters from dihydric alcohols and dicarboxylic acids or their derivatives [19].

$Th(OH)(CH_3COO)_3$ is precipitated when glacial CH_3COOH is added to an acetone solution of $Th(NO_3)_4 \cdot 2.75\,H_2O$ (molar ratio, CH_3COOH:Th = 4:1); the yield is increased by adding a large excess of CH_3COOH. This product is almost insoluble in acetone, carbon tetrachloride, chloroform, dioxane, dichloroethane, diethyl ether, ethanol, or water [9]. The hydrate, $Th(OH)(CH_3COO)_3 \cdot 1.5\,H_2O$, is obtained by treating solid $Th(SO_4)_2 \cdot 8\,H_2O$ with CH_3COOH (molar ratio, CH_3COOH:Th = 5:1), whereas a product of composition $Th(OH)_{1.5}(CH_3COO)_{2.5} \cdot 2\,H_2O$ is obtained when the molar ratio is 6:1 [14]. The same reaction, but with aqueous CH_3COONH_4 in place of CH_3COOH, yields $Th(OH)_2(CH_3COO)_2 \cdot 2.5\,H_2O$ at a molar ratio of the reactants of 7:1 and $Th(OH)_{2.5}(CH_3COO)_{1.5} \cdot H_2O$ at a molar ratio of 4:1 [14]. These products probably involve OH and CH_3COO bridges in a polymeric structure [14].

$Th(OH)_2(CH_3COO)_2 \cdot 2\,H_2O$ is precipitated when methanol is added to an aqueous mixture of CH_3COONa (1 to 2 M) and $Th(NO_3)_4$ (0.1 M) at pH 4.5 to 5.0 [11, 12] and when aqueous 2 M CH_3COONa is mixed with 0.5 M $Th(NO_3)_4$ in methanol; the optimum pH for precipitation is again 4.5 to 5.0 [39]. It has also been obtained by adding 20% aqueous NH_3 to a mixture of Th^{IV} in 0.4 M HNO_3 and 1.4 M CH_3COOH [37]. $Th(OH)_3(CH_3COO) \cdot n\,H_2O$ was apparently obtained in a similar manner [37]. The IR spectrum of $Th(OH)_2(CH_3COO)_2 \cdot 2\,H_2O$ has been discussed in terms of the intermolecular hydrogen bonding between the OH groups and the carbonyl O atoms of the CH_3COO groups [12]. Molecular weight determinations (cryoscopy) suggest that the compound is a dimer [39].

$Th(OH)_2(CH_3COO)_2 \cdot 3H_2O$ is reported to be the predominant product which is precipitated from an aqueous solution containing $Th(NO_3)_4$, 0.4 M HNO_3, and 1.4 M CH_3COOH on addition of concentrated aqueous NH_3 to pH 3.17 [25] whereas the monohydrate is obtained at pH 6.12 [25] and $Th(OH)_3(CH_3COO) \cdot H_2O$ at pH 8 to 9.7 [25]. The IR spectrum of $Th(OH)_2(CH_3COO)_2 \cdot H_2O$ suggests that the compound is a polymer [8]. The tendency of basic acetates of the form $Th(OH)_n(CH_3COO)_{4-n} \cdot mH_2O$ to hydrolyse has been discussed [27].

$ThO(CH_3COO)_2$ may be formed as an intermediate in the decarboxylation of CH_3COOH on ThO_2 as catalyst [1, 7]. A hydrate, $ThO(CH_3COO)_2 \cdot 4H_2O$, is reported to be precipitated from an aqueous mixture of $Th(NO_3)_4$ and CH_3COONa [13]; this product is presumably the same as $Th(OH)_2(CH_3COO)_2 \cdot 3H_2O$.

The products of composition $ThCl_x(CH_3COO)_y(OC(CH_3)_3)_{4-x-y}$ (x=1.4, y=0.6 or 1; x=1.3, y=1.7) obtained by the reaction of $Th(OC(CH_3)_3)_4$ (p. 28) with CH_3COCl [15], are probably mixtures (p. 28). Similarly, the chloride acetates, $ThCl_x(CH_3COO)_{4-x}$ (x=0.3, 1.8, 2.2, 2.3, 3 and 3.4), obtained by treating $ThCl_4$ with $CH_3COOC(CH_3)_3$ (large excess and molar ratio Th : $CH_3COOC(CH_3)_3$ = 1:4, 1:3, 1:2, or 1:1 under reflux for 1 h and 1:1 at room temperature overnight, respectively [15]) are also likely to be mixtures.

References for 15.6.2.2.1:

[1] Kuriacose, J. C.; Jungers, J. C. (Bull. Soc. Chim. Belges **64** [1955] 502/34).
[2] Panda, S.; Patnaik, D. (J. Indian Chem. Soc. **33** [1956] 877/8).
[3] Muniyappan, T.; Anjaneyalu, B. (Current Sci. [India] **26** [1957] 319/20).
[4] Muniyappan, T.; Anjaneyalu, B. (Proc. Indian Acad. Sci. A **45** [1957] 412/7).
[5] Kapoor, R. N.; Pande, K. C.; Mehrotra, R. C. (J. Indian Chem. Soc. **35** [1958] 157/60).
[6] Okubo, M.; Goto, R. (Nippon Kagaku Zasshi **81** [1960] 1132/6; C.A. **56** [1962] 3342).
[7] Okubo, M.; Goto, R. (Nippon Kagaku Zasshi **82** [1961] 620/3; C.A. **56** [1962] 11439).
[8] Geleceanu, I.; Lapitskii, A. V. (Dokl. Akad. Nauk SSSR **144** [1962] 573/5; Dokl. Proc. Acad. Sci. USSR **142/147** [1962] 460/2).
[9] Kovalenko, K. N.; Kazachenko, D. V.; Samsonova, O. N. (Zh. Neorgan. Khim. **8** [1963] 797/801; Russ. J. Inorg. Chem. **8** [1963] 407/9).
[10] Kovalenko, K. N.; Kazachenko, D. V.; Samsonova, O. N. (Zh. Neorgan. Khim. **8** [1963] 2222/5; Russ. J. Inorg. Chem. **8** [1963] 1163/5).

[11] Geleceanu, I.; Lapitskii, A. V.; Veiner, M.; Salimov, M. A.; Artamonova, E. P. (Radiokhimiya **6** [1964] 93/101; Soviet Radiochem. **6** [1964] 83/90).
[12] Galateanu, I. (Oesterr. Chemiker-Ztg. **66** [1965] 275/85).
[13] Kazachenko, D. V.; Kovalenko, K. N. (Zh. Neorgan. Khim. **11** [1966] 1631/6; Russ. J. Inorg. Chem. **11** [1966] 871/4).
[14] Molodkin, A. K.; Ivanova, O. M.; Skotnikova, G. A. (Zh. Neorgan. Khim. **12** [1967] 116/25; Russ. J. Inorg. Chem. **12** [1967] 57/62).
[15] Mehrotra, R. C.; Misra, R. A. (Proc. 3rd Nucl. Radiat. Chem. Symp., Poona, India, 1967, pp. 473/82).
[16] Eliseev, A. A.; Molodkin, A. K.; Ivanova, O. M. (Zh. Neorgan. Khim. **12** [1967] 2854/5; Russ. J. Inorg. Chem. **12** [1967] 1507/8).
[17] Bressat, R.; Claudel, B.; Giorgio, G.; Mentzen, B. (J. Chim. Phys. **65** [1968] 1615/7).
[18] Belova, V. I.; Syrkin, Ya. K.; Molodkin, A. K.; Ivanova, O. M.; Shiporina, L. M. (Zh. Neorgan. Khim. **13** [1968] 1458/60; Russ. J. Inorg. Chem. **13** [1968] 766/7).

[19] Monsanto Co. (Brit. 1132749 [1968]; C.A. **70** [1969] No. 20469).
[20] Paul, R. C.; Saran, M. S.; Bains, M. S. (Indian J. Chem. **7** [1969] 384/6).

[21] Athavale, V. T.; Kalyanaraman, R.; Sundaresan, M. (Indian J. Chem. **7** [1969] 386/91).
[22] Giorgio, G. (Ann. Chim. [Paris] [14] **6** [1971] 53/66).
[23] Schmidt, V. S.; Andryushin, V. G.; Rybakov, K. A.; Pozharskaya, M. E.; Kozhevnikov, P. B. (Radiokhimiya **15** [1973] 130/2; Soviet Radiochem. **15** [1973] 131/2).
[24] Nefedov, V. I.; Molodkin, A. K.; Salyn, Ya. V.; Ivanova, O. M.; Porai-Koshits, M. A.; Balakaeva, T. A.; Belyakova, Z. V. (Zh. Neorgan. Khim. **19** [1974] 2628/31; Russ. J. Inorg. Chem. **19** [1974] 1435/7).
[25] Schmidt, V. S.; Andryushin, V. G.; Rybakov, K. A.; Teterin, E. G. (Radiokhimiya **16** [1974] 391/7; Soviet Radiochem. **16** [1974] 390/5).
[26] Andryushin, V. G.; Kozhevnikov, P. B.; Pozharskaya, M. E.; Rybakov, K. A.; Shmidt, V. S. (Radiokhimiya **16** [1974] 413/5; Soviet Radiochem. **16** [1974] 411/3).
[27] Shmidt, V. S.; Andryushin, V. G. (Radiokhimiya **18** [1976] 506/11; Soviet Radiochem. **18** [1976] 439/43).
[28] Allen, G. C.; Tucker, P. M. (Chem. Phys. Letters **43** [1976] 254/7).
[29] Bancroft, G. M.; Sham, T. K.; Esquivel, J. L.; Larsson, S. (Chem. Phys. Letters **51** [1977] 105/10).
[30] Bancroft, G. M.; Sham, T. K.; Larsson, S. (Chem. Phys. Letters **46** [1977] 551/7).

[31] Dubrovin, A. V.; Skripitsyna, O. V.; Zakharova, T. V.; Dunaeva, K. M.; Spitsyn, V. I. (Zh. Neorgan. Khim. **23** [1978] 1495/1500; Russ. J. Inorg. Chem. **23** [1978] 823/7).
[32] Dubrovin, A. V.; Dunaeva, K. M.; Agafonov, V. N. (Radiokhimiya **21** [1979] 584/91; Soviet Radiochem. **21** [1979] 509/15).
[33] Dubrovin, A. V.; Volod'kina, L. V.; Dunaeva, K. M. (Radiokhimiya **21** [1979] 592/7; Soviet Radiochem. **21** [1979] 515/20).
[34] Thakur, L.; Thakur, A. K.; Ahmad, M. F. (Indian J. Chem. A **19** [1980] 793/5).
[35] Thakur, L.; Thakur, A. K.; Prasad, R.; Ahmad, M. F. (J. Indian Chem. Soc. **57** [1980] 339/40).
[36] Spitsyn, V. I.; Dunaeva, K. M.; Dubrovin, A. V.; Zhirov, A. I.; Santalova, N. A.; Mazo, G. N. (Zh. Neorgan. Khim. **25** [1980] 29/44; Russ. J. Inorg. Chem. **25** [1980] 15/22).
[37] Shmidt, V. S.; Andryushin, V. G. (Radiokhimiya **24** [1982] 601/6; Soviet Radiochem. **24** [1982] 498/503).
[38] Matveev, Yu. S.; Dunaeva, K. M. (Koord. Khim. **9** [1983] 361/5).
[39] Galateanu, I. (Acad. Rep. Populare Romine Studii Cercetari Fiz. **14** [1963] 557/70).
[40] Dubrovin, A. V.; Dunaeva, K. M. (Tezisy Dokl. 7th Soveshch. Kinet. Mekh. Khim. Reacts. Tverd. Tele, Novosibirsk 1977, Vol. 1, pp. 144/5; C.A. **88** [1978] No. 198422).

[41] Matveev, Yu. S.; Dunaeva, K. M. (Koord. Khim. **11** [1985] 621/7; Soviet J. Coord. Chem. **11** [1985] 352/8).
[42] Matveev, Yu. S.; Dunaeva, K. M. (Vestn. Mosk. Univ. Khim. **27** [1986] 389/97; C.A. **105** [1986] No. 197811).

15.6.2.2.2 Acetatothorates

The available information on the few known compounds is summarised in Table 20.

Table 20
Acetatothorates.

$(NH_4)_2[Th(CH_3COO)_6]$	loses $2CH_3COONH_4$ at about 150°C [4], thermogram and X-ray powder diffraction pattern illustrated in [4], density, $d_{obs}=2.041$ g/cm^3 [4] IR spectrum: ν(CO), 1579 cm^{-1}, δ(CH_3)+ν(CO)+δ(NH), 1461, 1444, 1381, 1355, 1326 cm^{-1}, δ(OCO), 656 cm^{-1}, π(COO), 550 cm^{-1} [5] IR spectrum illustrated in [5]
$(CN_3H_6)_2[Th(CH_3COO)_6]$ (CN_3H_6 = guanidinium)	m.p. 175°C, decomposition at 200 to 280°C [2]; crystallographic data: cubic, space group Pa3-T_h^6 (No. 205) [1, 2], lattice parameter, a=13.5 Å; Z=4 [1, 2]; density, $d_{obs}=2$ [1], 2.001 g/cm^3 [2]; crystals isotropic, refractive index, n=1.567 [2]; see also "Thorium" 1955, p. 343 IR spectrum: ν(CC), 943 cm^{-1}, δ(OCO), 697, 675, 665, 618 cm^{-1} [2] IR spectrum (4000 to 400 cm^{-1}) and thermogram illustrated in [2]; diamagnetic susceptibility, $\chi_m = -257\times10^{-6}$ [3]
$(CN_3H_6)_2[Th(CH_3COO)_6]\cdot3$ to $3.5H_2O$	[2]
$(CN_3H_6)_2[Th(CH_3COO)_6]\cdot8H_2O$	[2]
Al$[Th(CH_3COO)_6]$(?)	[6]; (thorium (III)?)

$(NH_4)_2[Th(CH_3COO)_6]$ is precipitated when $Th(CH_3COO)_4$ is treated with a saturated aqueous solution of CH_3COONH_4 for 3 h. It is also obtained when $Th(OH)_2(CH_3COO)_2$ (p. 48) is added to a saturated aqueous solution of CH_3COONH_4 at room temperature and the filtrate is then left in a desiccator over concentrated H_2SO_4 to crystallise [4]. It decomposes to $Th(CH_3COO)_4$ at about 150°C and is hydrolysed to basic thorium acetates on dissolution in water. The salt is insoluble in benzene, ethanol, or ether [4].

Crystals of the guanidinium salt, $(CN_3H_6)_2[Th(CH_3COO)_6]$, precipitate from aqueous acetic acid solutions containing $Th(CH_3COO)_4$ and a large excess of $CH_3COO(CN_3H_6)$ [1]. The coordination geometry of the complex anion is reported to be eicosahedral [1]. The salt can be recrystallised from cold water, but hydrolysis to a basic acetate occurs if the solution is heated [2]. The molar conductance of aqueous solutions of the salt indicates ionic dissociation. The salt is insoluble in organic solvents such as methanol, ethanol, or hydrocarbons [2].

The hydrated salt, $(CN_3H_6)_2[Th(CH_3COO)_6]\cdot8\,H_2O$, has been obtained by dissolving freshly precipitated thorium hydroxide in aqueous $CH_3COO(CN_3H_6)$ acidified with CH_3COOH, with heating at 50 to 60°C to increase the rate of dissolution. The hydrate crystallises when the cooled solution (pH 5) is left in air [2]. It is also obtained when an aqueous solution containing $Th(CH_3COO)_4$ and $CH_3COO(CN_3H_6)$ (molar ratio = 1:12) is left to crystallise [2]. The octahydrate loses H_2O to form $(CN_3H_6)_2[Th(CH_3COO)_6]\cdot3$ to $3.5H_2O$ on drying in air and is completely dehydrated on standing over concentrated H_2SO_4 [2].

References for 15.6.2.2.2 on p. 52

An aluminium salt of composition $AlTh(CH_3COO)_6$ is reported to be formed when a Th^{IV} compound in a mixture of CH_3COOH and $(CH_3CO)_2O$ is heated with a Cu–Al alloy under reflux [6]. This product is unlikely to be a Th^{III} compound and may be either a basic salt, $AlTh(CH_3COO)_6(OH)$, or the simple salt $Al_2[Th(CH_3COO)_6]_3$.

References for 15.6.2.2.2:

[1] Arutyunyan, E. G.; Porai-Koshits, M. A.; Molodkin, A. K.; Ivanova, O. M. (Zh. Strukt. Khim. **7** [1966] 813/4; J. Struct. Chem. [USSR] **7** [1966] 760/1).
[2] Molodkin, A. K.; Ivanova, O. M.; Skotnikova, G. A. (Zh. Neorgan. Khim. **12** [1967] 116/25; Russ. J. Inorg. Chem. **12** [1967] 57/62).
[3] Belova, V. I.; Syrkin, Ya. K.; Molodkin, A. K.; Ivanova, O. M.; Shiporina, L. M. (Zh. Neorgan. Khim. **13** [1968] 1458/60; Russ. J. Inorg. Chem. **13** [1968] 766/7).
[4] Shmidt, V. S.; Andryushin, V. G.; Rybakov, K. A.; Pozharskaya, M. E.; Kozhevnikov, P. B. (Radiokhimiya **15** [1973] 130/2; Soviet Radiochem. **15** [1973] 131/2).
[5] Shmidt, V. S.; Andryushin, V. G.; Rybakov, K. A.; Teterin, E. G. (Radiokhimiya **16** [1974] 391/7; Soviet Radiochem. **16** [1974] 390/5).
[6] Bhaskare, C. K.; Ganage, K. N. (Proc. Symp. Chem. React. Non-Aqueous Media Molten Salts, Hyderabad, India, 1978 [1980], pp. 212/6; C.A. **96** [1982] No. 114843).

15.6.2.3 Thorium Compounds with Substituted Acetic Acids

15.6.2.3.1 Thorium Hydroxyacetates (Glycolates)

The few known compounds and their properties are listed in Table 21.

Table 21
Thorium Hydroxyacetates (Glycolates).

$Th(HOCH_2COO)_4 \cdot 2H_2O$	dehydrated(?) at 165°C, decomposes >319°C, thermogram illustrated in [1] IR spectrum: $\nu_{as}(CO)$, 1575 cm^{-1}, $\nu_s(CO)$, 1391 cm^{-1} [2]
$Th(OH)(HOCH_2COO)_3 \cdot H_2O$	[1]
$Th(OH)(HOCH_2COO)_2((NO_2)OCH_2COO) \cdot H_2O$	[1]
$ThO(HOCH_2COO)_2 \cdot 2H_2O$	[1]

The hydrated glycolate, $Th(HOCH_2COO)_4 \cdot 2H_2O$, crystallises when an aqueous solution containing a Th^{IV} salt and $HOCH_2COONa$ (molar ratio = 1:4.5) is allowed to stand [1] and when freshly precipitated thorium(IV) hydroxide is treated with an excess of $HOCH_2COOH$ [2]. It is insoluble in most organic solvents, but is slightly soluble in methanol. Its solubility in water at 22°C is low and corresponds to a 0.00035 M solution; the salt is ca. 40% hydrolysed in saturated aqueous solution [1]. The endotherm observed in the thermogram at 165°C may be due to dehydration [1].

The basic glycolate, $ThO(HOCH_2COO)_2 \cdot 2H_2O$, is precipitated from aqueous solutions of Th^{IV} salts in the presence of $HOCH_2COONa$ and NaOH [1]. The hydroxide compound, $Th(OH)(HOCH_2COO)_3 \cdot H_2O$, is precipitated when an acetone solution of $Th(NO_3)_4 \cdot 5H_2O$ is treated with an excess of $HOCH_2COOH$ (molar ratio = 1:10). The precipitates obtained after standing for 1 to 5 days appear to contain some $Th(OH)(HOCH_2COO)_2((NO_2)OCH_2COO) \cdot H_2O$; the nitration appears to be due to reaction of liberated HNO_3 with the OH group of the glycolate ion [1].

References for 15.6.2.3.1:

[1] Kazachenko, D. V.; Kovalenko, K. N. (Zh. Neorgan. Khim. **11** [1966] 1631/6; Russ. J. Inorg. Chem. **11** [1966] 871/4).
[2] Sbrignadello, G.; Tomat, G.; Battiston, G.; Vigato, P. A. (J. Inorg. Nucl. Chem. **40** [1978] 1647/52).

15.6.2.3.2 Thorium Phenoxyacetates

Information on the known compounds is summarised in Table 22.

Table 22
Thorium Phenoxyacetates and Related Compounds.

$Th(C_6H_5OCH_2COO)_4$	[10, 13]
$Th(OH)(C_6H_5OCH_2COO)_3 \cdot xH_2O$	[8, 9]
$Th(OH)_2(C_6H_5OCH_2COO)_2$	rapid decomposition above 180°C; thermolysis curve illustrated in [12]
$Th(OH)(2\text{-}CH_3C_6H_4OCH_2COO)_3 \cdot xH_2O$	[8, 9] (*o*-cresoxyacetate)
$Th(OH)(3\text{-}CH_3C_6H_4OCH_2COO)_3 \cdot 3H_2O$	[3] (*m*-cresoxyacetate)
$Th(OH)_2(3\text{-}CH_3C_6H_4OCH_2COO)_2$	decomposes >210°C, to ThO_2 at 500°C; thermolysis curve illustrated in [7]
$Th(OH)(4\text{-}CH_3C_6H_4OCH_2COO)_3 \cdot xH_2O$	[8, 9] (*p*-cresoxyacetate)
$Th(OH)(2\text{-}CH_3OC_6H_4OCH_2COO)_3 \cdot 3H_2O$	[4] (guaiacoxyacetate)
$Th(OH)(2\text{-}CH_3OC_6H_4OCH_2COO)_3 \cdot xH_2O$	[8, 9]
$Th(OH)(5\text{-}CH_3\text{-}2\text{-}i\text{-}C_3H_7C_6H_3OCH_2COO)_3 \cdot xH_2O$	[8, 9] (thymoxyacetate)
$Th(OH)(2,4\text{-}Cl_2C_6H_3OCH_2COO)_3$	decomposes >140°C, to ThO_2 at 475°C; thermolysis curve illustrated in [7]
$Th(OH)(2,4\text{-}Cl_2C_6H_3OCH_2COO)_3 \cdot 2H_2O$	decomposes at 280 to 285°C [4]
$Th(OH)(2,4\text{-}Cl_2C_6H_3OCH_2COO)_3 \cdot 4H_2O$	loses $2H_2O$ at 105 to 110°C [4]
$Th(OH)(2,4\text{-}Cl_2C_6H_3OCH_2COO)_3 \cdot xH_2O$	[8, 9]
$Th(OH)(2,4,5\text{-}Cl_3C_6H_2OCH_2COO)_3$	rapid decomposition above 205°C; thermolysis curve illustrated in [12]
$Th(OH)(2,4,5\text{-}Cl_3C_6H_2OCH_2COO)_3 \cdot H_2O$	[6]
$Th(OH)(2,4,5\text{-}Cl_3C_6H_2OCH_2COO)_3 \cdot xH_2O$	[8, 9]
$Th(OH)(4\text{-}O_2NC_6H_4OCH_2COO)_3 \cdot xH_2O$	yellow [8, 9]
$Th(OH)_2(2\text{-}(CHO)C_6H_4OCH_2COO)_2 \cdot 2H_2O$	[8, 9]
$Th(OH)(2\text{-}C_{10}H_7OCH_2COO)_3 \cdot xH_2O$	pinkish white [8, 9] (β-naphthoxyacetate)

References for 15.6.2.3.2 on p. 54

The phenoxyacetate, $Th(C_6H_5OCH_2COO)_4$, has been prepared by adding an excess of thorium hydroxide or nitrate to a hot aqueous solution of $C_6H_5OCH_2COOH$ [10]. The hydrated basic salt, $Th(OH)(C_6H_5OCH_2COO)_3 \cdot xH_2O$, is precipitated from a hot, but not boiling, solution of a Th^{IV} salt on addition of a hot, 2% aqueous solution of the acid (pH 4.2) [8, 9]. The precipitation of an unspecified thorium phenoxyacetate by addition of an excess of hot 2% aqueous $C_6H_5OCH_2COOH$ to a boiling solution of a Th^{IV} salt (neutral to Congo red), followed by cooling, redissolution of the precipitate in hot dilute HCl and reprecipitation, has been utilised for the separation of Th from the cerite earths and for the estimation of Th [1].

The hydrated basic salts, $Th(OH)(ROCH_2COO)_3 \cdot xH_2O$ (R = 2-$CH_3C_6H_4$ (pH = 4.2), 4-$CH_3C_6H_4$ (pH = 4.2), 2-$CH_3OC_6H_4$ (pH = 4.4), 5-CH_3, 2-*i*-$C_3H_7C_6H_3$ (pH = 4.2), 2,4-$Cl_2C_6H_3$ (pH = 2.8), 2,4,5-$Cl_3C_6H_2$ (pH = 2.2), 4-$O_2NC_6H_4$ (pH = 4.5), and 2-$C_{10}H_7$ (pH = 4.2)) are precipitated at the indicated values of the solution pH in the same way as the basic phenoxyacetate (see above) [8, 9]. The same procedure with 2-(CHO)$C_6H_4OCH_2COOH$ at pH = 4.2 yields the dihydrated compound, $Th(OH)_2$(2-(CHO)$C_6H_4OCH_2COO)_2 \cdot 2H_2O$ [8, 9]. Th(OH)(3-$CH_3C_6H_4OCH_2COO)_3 \cdot 3H_2O$ is precipitated in a similar manner, but with NH_4NO_3 present in the solution [3], the precipitation has been used for the separation of Th^{IV} from U^{VI} and for the determination of Th (as ThO_2 after ignition) in monazite [3]; see also [2]. The basic guaiacoxyacetate, Th(OH)(2-$CH_3OC_6H_4OCH_2$-$COO)_3 \cdot 3H_2O$, is obtained in the same way by precipitation from a hot solution which is neutral to Congo red [4] and the 2,4-dichlorophenoxyacetate, Th(OH)(2,4-$Cl_2C_6H_3OCH_2COO)_3 \cdot 4H_2O$ is precipitated from a hot solution which is neutral to thymol blue. The latter loses 2 molecules of H_2O at 105 to 110°C [4]. The precipitation of Th with 2,4-$Cl_2C_6H_3OCH_2COOH$ at pH 2.8 to 3.0, followed by ignition of the precipitate to ThO_2, has been used as a method of estimating Th [5]. The precipitation has also been used to separate Th from Zr, Ti, and Fe [11]. The precipitation of the 2,4,5-trichlorophenoxyacetate, Th(OH)(2,4,5-$Cl_3C_6H_2OCH_2COO)_3 \cdot H_2O$, using a hot 0.5% aqueous solution of the acid or with a cold 1% aqueous solution of its sodium salt, followed by ignition of the precipitate to ThO_2, has also been used for the determination of Th [6].

The thermolysis of the *m*-cresoxyacetate, $Th(OH)_2$(*m*-$CH_3C_6H_4OCH_2COO)_2$ and 2,4-dichlorophenoxyacetate, Th(OH)(2,4-$Cl_2C_6H_3OCH_2COO)_3$ has been described; these compounds were precipitated on addition of an aqueous solution of the acid to a boiling dilute aqueous solution of $Th(NO_3)_4$ at a final pH of 3 to 5 [7]. It is not clear whether the initial precipitates were actually anhydrous.

References for 15.6.2.3.2:

[1] Venkataramaniah, M.; Satyanarayanamurthy, T. K.; Raghava Rao, B. S. V. (J. Indian Chem. Soc. **27** [1950] 81/6).
[2] Venkataramaniah, M.; Raghava Rao, B. S. V.; Rao, C. L. (Anal. Chem. **23** [1951] 539/40).
[3] Venkataramaniah, M.; Raghava Rao, B. S. V.; Rao, C. L. (Anal. Chem. **24** [1952] 747/9).
[4] Datta, S. K.; Banerjee, G. (J. Indian Chem. Soc. **31** [1954] 397/401).
[5] Datta, S. K.; Banerjee, G. (Anal. Chim. Acta **12** [1955] 323/8).
[6] Datta, S. K. (Anal. Chim. Acta **14** [1956] 39/44).
[7] Wendlandt, W. W. (Anal. Chem. **29** [1957] 800/2).
[8] Datta, S. K. (Z. Anal. Chem. **174** [1960] 104/8).
[9] Datta, S. K. (Z. Anal. Chem. **174** [1960] 109/18).
[10] Kumok, V. N.; Kaleeva, V. A.; Serebrennikov, V. V. (Izv. Vysshikh Uchebn. Zavedenii Khim. Khim. Tekhnol. **11** [1968] 633/6; C.A. **69** [1968] No. 92459).

[11] Datta, S. K.; Banerjee, G. (J. Indian Chem. Soc. **31** [1954] 773/8).
[12] Wendlandt, W. W. (Anal. Chim. Acta **18** [1958] 316/20).
[13] Tserkovnitskaya, I. A.; Charykov, A. K. (Izv. Vysshikh Uchebn. Zavedenii Khim. Khim. Tekhnol. **7** [1964] 544/50; C.A. **62** [1965] 3385).

15.6.2.3.3 Thorium Aminoacetate and Related Compounds

The available data for the few known compounds are summarised in Table 23.

Table 23
Thorium Aminoacetate and Related Compound.

aminoacetate (aminoacetic acid = glycine)	
$[Th(OH)(NH_2CH_2COO)(H_2O)_5](NO_3)_2$	IR spectrum: $\nu_{as}(CO)$, 1645 to 1605 cm^{-1} [3]; spectrum illustrated in [3]
$Th(C_{16}H_{16}N_2O_4)_2 \cdot 5H_2O$(?)	$C_{16}H_{18}N_2O_4$ = 2-$HOC_6H_4N(CH_2COOH)(CH_2)_2NH$(2-$HOC_6H_4$); formula written in [2] as $[Th(C_{16}H_{16}N_2O_4)] \cdot 5H_2O$

Very little is known about thorium compounds with these acids. The basic aminoacetate (glycinate), $[Th(OH)(NH_2CH_2COO)(H_2O)_5](NO_3)_2$, is obtained from aqueous solutions of Th^{IV} salts and the acid (molar ratio = 1:1) at pH 1 to 3 [3]. N-Phenylglycine precipitates Th^{IV} completely in the pH range 3.8 to 4.0; the mixture was boiled for 1 to 2 min after the addition of the acid. The composition of the precipitate was not recorded, but the precipitation was utilised for the determination of Th as ThO_2 after ignition [1].

Thorium is also precipitated when a hot 2% aqueous solution of the potassium salt of N-benzoylglycine (hippuric acid) is added to a boiling aqueous solution of a Th^{IV} salt; after adjustment of the pH to ca. 4.6, the mixture was heated on a water bath for 15 min [4]. The composition of the precipitate was not recorded.

A salt of N,N'-di(2-hydroxyphenyl)ethylenediamine-N-acetic acid, (2-$OHC_6H_4)N(CH_2COOH)$-$(CH_2)_2NH$(2-OHC_6H_4), reported as $[Th(C_{16}H_{16}N_2O_4)] \cdot 5H_2O$ (possibly $[Th(C_{16}H_{16}N_2O_4)_2] \cdot 5H_2O$), has been obtained by addition of a 4:1 mixture of methanol and ether to an equimolar mixture of aqueous $Th(NO_3)_4$ and the acid at pH 9 [2].

References for 15.6.2.3.3:

[1] Jain, B. D.; Singhal, S. P. (Current Sci. [India] **32** [1963] 66/7).
[2] Ryabchikov, D. I.; Volynets, M. P. (Zh. Neorgan. Khim. **10** [1965] 619/27; Russ. J. Inorg. Chem. **10** [1965] 334/9).
[3] Sergeev, G. M.; Korshunov, I. A. (Radiokhimiya **16** [1974] 787/90; Soviet Radiochem. **16** [1974] 771/4).
[4] Bajpai, S. K.; Shukla, V. K. (Z. Anal. Chem. **274** [1975] 303).

15.6.2.3.4 Thorium Halogenoacetates and Related Compounds

The individual compounds are listed in Table 24 (pp. 56/61) together with physical data, where available. 1,3-Diketonatothorium(IV) halogenoacetates, $Th(acac)_x(RCOO)_{4-x}$ (R = CF_3, $CHCl_2$ or CCl_3), and 8-quinolinatothorium(IV) chloroacetates are discussed in "Thorium" Suppl. Vol. E, 1985, pp. 134/5 and 158/61, respectively.

$Th(CF_3COO)_4$ has been prepared by evaporating the solution obtained by dissolving freshly precipitated thorium carbonate in CF_3COOH until a glassy residue remains. This yielded a crystalline solid after addition of $(CF_3CO)_2O$ followed by evaporation [1]. $Th(CF_3COO)_4$ is also obtained in 100% yield by treating $ThCl_4$ with CF_3COOH at 70°C/24 h [4] and by heating $Th(CH_3COO)_4$ with an excess of CF_3COOH under reflux, either alone or in C_6H_6 or $CH_3C_6H_5$ [6].

References for 15.6.2.3.4 on pp. 63/4

Table 24
Thorium Halogenoacetates and Related Compounds.

trifluoroacetates	
$Th(CF_3COO)_4$	decomposes at 272°C [4]; activation energy for decomposition, $E_{act} = 250 \pm 20$ kJ/mol [23] IR spectrum: $\nu_{as}(CO)$, 1634 cm^{-1}, $\nu_s(CO)$, 1492 cm^{-1}, $\delta(OCO)$, 727 cm^{-1} [6]; solubility in anhydrous CF_3COOH (100 g) = 0.016 ± 0.002 g [1]
$Th(CF_3COO)_4 \cdot H_2O$	decomposes to ThF_4 in air or N_2 at 310°C [14]; density, $d_{obs}(25°C) = 2.37(2)$ g/cm³ [14] IR spectrum: $\nu_{as}(CO)$, 1680, 1634, 1630 cm^{-1}, $\nu_s(CO)$, 1492, 1470 cm^{-1}, $\nu(CC)$, 865 cm^{-1}, $\delta(OCO)$, 800, 730 cm^{-1}, $\nu(ThO(H_2O))$, 435 cm^{-1} [14] X-ray diffraction data reported in [14]
$Th(CF_3COO)_4 \cdot 4(CH_3NH)_2CO$	see [21]
$Th(CF_3COO)_4 \cdot 2((CH_3)_2N)_2CO$	see [21]
$Th(CF_3COO)_4 \cdot 3CH_3CON(CH_3)_2$	see Th E, pp. 49, 52/3
$[Th(CF_3COO)_4\{(CH_3)_3CCON(CH_3)_2\}]_2O$	see Th E, pp. 55, 58
$Th(CF_3COO)_4 \cdot 3(CH_3)_3PO$	see [22]
$Th(CF_3COO)_4 \cdot 2(C_6H_5)_3PO$	see [22]
$Th(CF_3COO)_4 \cdot 3C_6H_5(CH_3)_2PO$	see [21]
$Th(CF_3COO)_4 \cdot 3CH_3(C_6H_5)_2PO$	see [21]
$Th(CF_3COO)_4 \cdot 4(CH_3)_2SO$	see [22]
$Th(CF_3COO)_4 \cdot 2(C_6H_5)_2SO$	see [22]
$(NH_4)_2[Th(CF_3COO)_6]$(?)	[12]
$Th(OH)_{1.5}(CF_3COO)_{2.5}$	decomposes to ThO_2 at ca. 563 K [12]
$Th(OH)_{1.5}(CF_3COO)_{2.5} \cdot H_2O$	loses H_2O at 443 K [12]
$Th(OH)_{1.5}(CF_3COO)_{2.5} \cdot 2H_2O$	loses $1H_2O$ at ca. 373 K; differential thermal analysis and thermogravimetric curves illustrated in [12] density, $d_{obs} = 2.55$ g/cm³ [12]
$Th(OH)_2(CF_3COO)_2 \cdot H_2O$	density, $d_{obs} = 2.80$ g/cm³ [12]
$Th(OH)_3(CF_3COO)$	[12]
$Th(OH)_3(CF_3COO) \cdot H_2O$	loses H_2O at 373 K; differential thermal analysis and thermogravimetric curves illustrated in [12] density, $d_{obs} = 3.03$ g/cm³ [12]

Table 24 (continued)

$Th(OH)_3(CF_3COO)\cdot 2.5H_2O$	density, $d_{obs}=2.86 g/cm^3$ [12]
$Th(OH)_3(CF_3COO)\cdot 4H_2O$	density, $d_{obs}=2.80 g/cm^3$ [12]
$Th(OH)_3(CF_3COO)\cdot 6H_2O$	density, $d_{obs}=2.70 g/cm^3$ [12]
$ThO(OH)(CF_3COO)$(?)	decomposition at 543 K [12]
$Th(acac)_3(CF_3COO)\cdot 4H_2O$	see Th E, pp. 134/5
monochloroacetates	
$Th(CH_2ClCOO)_4$	m.p. 182°C [7]; decomposes >75°C (maximum at 253 to 268°C) [17], to $ThO(CH_2ClCOO)_2$ at 260°C (in N_2) [13], 280 to 320°C [11], 295°C [10]; activation energy for decomposition, $E_{act}=88\pm8$ kJ/mol [23] IR spectrum: $\nu_{as}(CO)$, 1595 cm^{-1} [10, 15], 1580, 1540 cm^{-1} [6], $\nu_s(CO)$, 1430, 1395 cm^{-1} [6], 1390 cm^{-1} [10, 15], $\nu(CC)$, 970, 940 cm^{-1} [10, 15], $\delta(OCO)$, 715, 700 cm^{-1} [6], 700, 680 cm^{-1} [10, 15], $\nu(ThO)$, 350 to 340 cm^{-1} [7] X-ray powder diffraction data are given in [10] molar conductivity in $C_6H_5NO_2$ (10^{-3}M solution at 25°C) = 21.03 $\Omega^{-1}\cdot cm^2\cdot mol^{-1}$ [7]
$Th(CH_2ClCOO)_4\cdot CH_3CON(CH_3)_2$	see Th E, pp. 49, 52/3
$ThO(CH_2ClCOO)_2$	decomposes to ThO_2 at 300°C in air, 310°C in N_2 [13], 320°C in air or N_2 [10], 320 to 480°C in air or N_2 [11]
$Th(C_9H_6NO)(CH_2ClCOO)_3$	see Th E, pp. 159/60 (C_9H_7NO = 8-quinolinol)
$Th(C_9H_6NO)_2(CH_2ClCOO)_2$	see Th E, pp. 159/60
$[Th(C_9H_6NO)_2(CH_2ClCOO)_2]_2\cdot CH_2ClCOOH$	see Th E, pp. 159/60
$Th(C_9H_6NO)_3(CH_2ClCOO)$	see Th E, pp. 159/60
dichloroacetates	
$Th(CHCl_2COO)_4$	decomposes >230°C (maximum at 282 to 315°C) [17]; decomposition to $ThOCl(CHCl_2COO)$ at 280 to 340°C in air or N_2 [11], 310°C in air, 320°C in N_2 [13], 345°C [10]; activation energy for decomposition to $ThOCl(CHCl_2COO)$, $E_{act}=117\pm10$ kJ/mol and for decomposition to $ThCl_4+0.5ThOCl_2$, $E_{act}=159\pm12$ kJ/mol [23] IR spectrum: $\nu_{as}(CO)$, 1590, 1570 cm^{-1}, $\nu_s(CO)$, 1425 cm^{-1}, $\delta(OCO)$, 738, 715 cm^{-1} [6] Raman spectrum: $\nu_{as}(CO)$, 1582 cm^{-1}, $\nu_s(CO)$, 1460 cm^{-1}, $\delta(OCO)$, 736, 721 cm^{-1} [6]

References for 15.6.2.3.4 on pp. 63/4

Table 24 (continued)

$Th(CHCl_2COO)_4 \cdot H_2O$	loses H_2O at 120 to 130°C [13], 130°C [10] IR spectrum: $\nu_{as}(CO)$, 1595 cm^{-1}, $\nu_s(CO)$, 1390 cm^{-1}, $\nu(CC)$, 980 cm^{-1}, $\delta(OCO)$, 665, 648 cm^{-1} [10] X-ray powder diffraction data are given in [10]
$Th(CHCl_2COO)_4 \cdot CH_3CON(CH_3)_2$	see Th E, pp. 49, 52/3
$Th(CHCl_2COO)_4 \cdot CH_3CON(C_6H_5)_2$	see Th E, pp. 53/4
$[Th(CHCl_2COO)_3\{(CH_3)_3CCON(CH_3)_2\}]_2O$	see Th E, pp. 55, 58
$Th(CHCl_2COO)_4 \cdot 4(CH_3)_3PO$	see [22]
$Th(CHCl_2COO)_4 \cdot 2(C_6H_5)_3PO$	see [22]
$Th(CHCl_2COO)_4 \cdot (CH_3)_2SO$	see [22]
$ThOCl(CHCl_2COO)$	decomposition to $ThOCl_2$ at 370 to 500°C [11], 410°C in air, 420°C in N_2 [13], ca. 500°C [10]
$Th(acac)_3(CHCl_2COO) \cdot 4H_2O$	see Th E, pp. 134/5
$[Th_6(CHCl_2COO)_{12}(OH)_{12}(OH_2)_2]$	crystallographic data: monoclinic, space group $P2_1/n$-C_{2h}^5 (No. 14); lattice parameters, a = 15.087(9), b = 15.111(11), c = 16.084(5) Å, β = 99.73(4)°; Z = 4; density, d_{calc} = 2.90, d_{obs} = 2.85 g/cm³ [16]
$Th(C_9H_6NO)(CHCl_2COO)_3$	see Th E, pp. 159/60 (C_9H_7NO = 8-quinolinol)
$Th(C_9H_6NO)(CHCl_2COO)_3 \cdot C_9H_7NO$	see Th E, pp. 159/60
$[Th(C_9H_6NO)(CHCl_2COO)_3(C_9H_7NO)$-$(CHCl_2COOH)]_2 \cdot CHCl_2COOH$	see Th E, pp. 159/60
$Th(C_9H_6NO)_2(CHCl_2COO)_2$	see Th E, pp. 159/60
$[Th(C_9H_6NO)_2(CHCl_2COO)_2] \cdot C_9H_7NO$	see Th E, pp. 159/60
$Th(C_9H_6NO)_3(CHCl_2COO)$	see Th E, pp. 159/60
trichloroacetates	
$Th(CCl_3COO)_4$	m.p. 210°C (decomposition) [8]; decomposition >175°C (maximum at 233 to 260°C) [17]; decomposition to $ThCl_4$ at 260 to 340°C [11], 340°C in N_2 [10], or to $ThOCl(CCl_3COO)$ in air at 290°C [10] or 295°C [13]; activation energy for decomposition to $ThCl_4 + 0.3\,ThOCl_2$, E_{act} = 146 ± 10 kJ/mol [23] IR spectrum: $\nu_{as}(CO)$, 1618 cm^{-1}, $\nu_s(CO)$, 1425, 1400 cm^{-1}, $\delta(OCO)$, 740, 692 cm^{-1} [6], $\nu(ThO)$, 335 cm^{-1} [8]

Table 24 (continued)

$Th(CCl_3COO)_4 \cdot H_2O$	loses H_2O at 120°C in air, 130°C in N_2 [13], 140°C [10] IR spectrum: $\nu_{as}(CO)$, 1615 cm^{-1}, $\nu_s(CO)$, 1405, 1390 cm^{-1}, $\nu(CC)$, 980 cm^{-1}, $\delta(OCO)$, 690 cm^{-1} [10] X-ray powder diffraction data are given in [10]
$Th(CCl_3COO)_4 \cdot 2H_2O$	IR spectrum (3274 to 447 cm^{-1}) illustrated in [2, 5], $\nu(CO)$, 1668 cm^{-1}, $\nu(CO)+\delta(OH)$ [2] or $\nu_s(CO)$ [3], 1379 cm^{-1}, $\delta(OCO)$, 841 cm^{-1}, $\nu(CCl)$, 770, 684 cm^{-1}, $\delta_{as}(CCO)$, 554 cm^{-1}, $\delta_s(CCO)$, 447 cm^{-1} [2, 3] ^1H NMR spectrum reported in [2, 3]
$Th(CCl_3COO)_4 \cdot 4(CH_3NH)_2CO$	see [21]
$Th(CCl_3COO)_4 \cdot 2.5((CH_3)_2N)_2CO$	see [21]
$Th(CCl_3COO)_4 \cdot 3CH_3CON(CH_3)_2$	see Th E, pp. 49, 52/3
$Th(CCl_3COO)_4 \cdot 2C_5H_5N$	see Th E, pp. 12, 14
$Th(CCl_3COO)_4 \cdot 2C_5H_5NO$	see Th E, pp. 69, 74
$Th(CCl_3COO)_4 \cdot 3C_6H_5(CH_3)_2PO$	see [21]
$Th(CCl_3COO)_4 \cdot 3CH_3(C_6H_5)_2PO$	see [21]
$Th(CCl_3COO)_4 \cdot 4(CH_3)_3PO$	see [22]
$Th(CCl_3COO)_4 \cdot n(C_6H_5)_3PO$	n = 3, see [22]; n = 2, see Th E, pp. 90, 92
$Th(CCl_3COO)_4 \cdot 3(CH_3)_2SO$	see [22]
$Th(CCl_3COO)_4 \cdot 2(C_6H_5)_2SO$	see [22]
$Na_2[Th(CCl_3COO)_6]$	m.p. 175°C [9] IR spectrum: $\nu_{as}(CO)$, 1590 cm^{-1}, $\nu_s(CO)$, 1362 cm^{-1}, $\delta(OCO)$, 770, 675 cm^{-1} [9] molar conductivity in $C_6H_5NO_2$, 24.80 $\Omega^{-1} \cdot cm^2 \cdot mol^{-1}$ [9]
$K_2[Th(CCl_3COO)_6]$	m.p. 155°C [9] IR spectrum: $\nu_{as}(CO)$, 1592 cm^{-1}, $\nu_s(CO)$, 1362 cm^{-1}, $\delta(OCO)$, 771, 689 cm^{-1} [9] molar conductivity in $C_6H_5NO_2$, 22.42 $\Omega^{-1} \cdot cm^2 \cdot mol^{-1}$ [9]
$Th(C_9H_6NO)(CCl_3COO)_2(OCH_3) \cdot CH_3OH$	see Th E, pp. 29/30 (C_9H_7NO = 8-quinolinol)
$Th(C_9H_6NO)(CCl_3COO)_3 \cdot C_9H_7NO$	see Th E, pp. 159/60
$Th(C_9H_6NO)(CCl_3COO)_3 \cdot C_9H_7NO \cdot CCl_3COOH$	see Th E, pp. 159/60
$Th(C_9H_6NO)_2(CCl_3COO)_2$	see Th E, pp. 159/60

References for 15.6.2.3.4 on pp. 63/4

Table 24 (continued)

$Th(C_9H_6NO)_2(CCl_3COO)_2 \cdot CCl_3COOH$	see Th E, pp. 159/60
$Th(C_9H_6NO)_3(CCl_3COO)$	see Th E, pp. 159/60
$[Th(C_9H_6NO)_3(CCl_3COO)]_2 \cdot CCl_3COOH$	see Th E, pp. 159/60
$Th(C_9H_6NO)_3(CCl_3COO) \cdot CCl_3COOH$	see Th E, pp. 159/60
$[Th(C_9H_6NO)_3(CCl_3COO) \cdot (CCl_3COOH)_2]_2 \cdot CCl_3COOH$	see Th E, pp. 159/60
$[Th(CCl_3COO)_3\{(CH_3)_3CCON(CH_3)_2\}]_2O$	see Th E, pp. 55, 58
$Th(OH)(CCl_3COO)_3 \cdot 2H_2O$	[2, 5]
$Th(OH)_2(CCl_3COO)_2 \cdot 2H_2O$	[2, 5]
$ThOCl(CCl_3COO)$	decomposition to $ThOCl_2$ at 310°C [13] or 320°C [10] in air
$Th(acac)_3(CCl_3COO) \cdot 4H_2O$	see Th E, pp. 134/5
bromoacetates	
$Th(CH_2BrCOO)_4$	decomposition to $ThO(CH_2BrCOO)_2$ at >300°C in air or 270 to 310°C in N_2, and to ThO_2 above 410°C in air or at 360°C in N_2 [18]
$Th(CH_2BrCOO)_4 \cdot 2H_2O$	loses $2H_2O$ at 110°C in air or 105°C in N_2 [18]; activation energy for decomposition to $ThO(CH_2BrCOO)_2$, $E_{act} = 122 \pm 10$ kJ/mol and for decomposition to $ThO_2 + 2ThOBr_2$, $E_{act} = 240 \pm 20$ kJ/mol [23] IR spectrum: $\nu_{as}(CO)$, 1620, 1555 cm^{-1}, $\nu_s(CO)$, 1470, 1420, 1390 cm^{-1} ($\Delta\nu(CO)$, 199), $\delta(OCO)$, 725, 705 cm^{-1}, $\varrho(COO)$, 430, 405 cm^{-1} [18]
$Th(OH)(CH_2BrCOO)_3$	decomposition to $ThO_2 \cdot 2ThO(CH_2BrCOO)_2$ at 385°C in air or 360°C in N_2, and to ThO_2 above 440°C in air or 450°C in N_2 [18]
$Th(OH)(CH_2BrCOO)_3 \cdot H_2O$	loses H_2O at 110°C in air or 105°C in N_2 to form $Th_2O(CH_2BrCOO)_6$ [18] IR spectrum: $\nu_{as}(CO)$, 1615, 1595, 1570, 1530 cm^{-1}, $\nu_s(CO)$, 1480, 1420, 1390 cm^{-1} [$\Delta\nu(CO)$, 205], $\delta(OCO)$, 730, 705, 665 cm^{-1}, $\varrho(COO)$, 460, 435 cm^{-1} [18]
$Th_2O(CH_2BrCOO)_6$	decomposition to $ThO_2 \cdot 2ThO(CH_2BrCOO)_2$ at 360°C in N_2 or 385°C in air, and to ThO_2 at 440 to 535°C in air and 450°C in N_2 [18]
$ThO(CH_2BrCOO)_2$	decomposition to $ThO_2 \cdot ThO(CH_2BrCOO)_2$ at 285 to 440°C in air or 360°C in N_2, and to ThO_2 at 520°C in air or 470°C in N_2 [18]

Table 24 (continued)

$ThO(CH_2BrCOO)_2 \cdot H_2O$	loses H_2O below 150°C in air or 250°C in N_2 [18] IR spectrum: $\nu_{as}(CO)$, 1620, 1600, 1550 cm^{-1}, $\nu_s(CO)$, 1480, 1420, 1380 cm^{-1} ($\Delta\nu(CO)$, 200), $\delta(OCO)$, 730 cm^{-1}, $\varrho(COO)$, 520, 475 cm^{-1} [18]
$Th(CHBr_2COO)_4$	m.p. >270°C; molar conductivity in dimethyl formamide 11.62 $\Omega^{-1} \cdot cm^2 \cdot mol^{-1}$ [20] IR spectrum: $\nu_{as}(CO)$, 1629, 1568, 1536 cm^{-1}, $\nu_s(CO)$, 1425 cm^{-1}, $\nu(M–O)$, 355 cm^{-1} [20]
iodoacetates	
$Th(CH_2ICOO)_4$	decomposes to $Th(CO_3)_2$ at 310 to 360°C in air, 380°C in N_2, and to ThO_2 at 350°C in vacuum, 450°C in air or 500°C in N_2 [15]; activation energy for decomposition to $Th[(CH_2COO)_2]_2$, $E_{act} = 75 \pm 6$ kJ/mol; for decomposition to $Th(CO_3)_2$, $E_{act} = 176 \pm 12$ kJ/mol and for decomposition to ThO_2, $E_{act} = 190 \pm 17$ kJ/mol [23] IR spectrum: $\nu(CH)$, 3010 cm^{-1}, $\nu_{as}(CO)$, 1630, 1580 cm^{-1}, $\nu_s(CO)$, 1440, 1390 cm^{-1} ($\Delta\nu(CO)$, 190), $\nu(CC)$, 950 cm^{-1}, $\delta(OCO)$, 700 cm^{-1}, $\nu(CI)$, 650, 540 cm^{-1}, $\varrho(OCO)$, 480 cm^{-1} [15, 18] X-ray powder diffraction data are given in [15]
$ThO(CH_2ICOO)_2 \cdot H_2O$	decomposes to $ThOCO_3$ at 250°C in N_2, 350°C in air, and to ThO_2 at 350°C in vacuum, 370°C in N_2 and 480°C in air [15]; activation energy for decomposition to $ThOCO_3$, $E_{act} = 100 \pm 8$ kJ/mol and to ThO_2, $E_{act} = 142 \pm 12$ kJ/mol [23] IR spectrum: $\nu(OH)$, 3350 cm^{-1}, $\nu(CH)$, 2990 cm^{-1}, $\nu_{as}(CO)$, 1630, 1580 cm^{-1}, $\nu_s(CO)$, 1430, 1380 cm^{-1}, $\nu(CC)$, 930 cm^{-1}, $\delta(OCO)$, 710 cm^{-1}, $\nu(CI)$, 640, 520 cm^{-1}, $\varrho(OCO)$, 470 cm^{-1} [15, 18] X-ray powder diffraction data are given in [15]

The monohydrate, $Th(CF_3COO)_4 \cdot H_2O$, has been prepared by heating anhydrous $ThCl_4$ with CF_3COOH at 50 to 60°C for 5 to 6 h and by dissolving freshly precipitated Th hydroxide in an excess of CF_3COOH, followed by evaporation of the solution to dryness. A study of its thermal decomposition has been reported [14].

The IR spectra of the halogenoacetates, $Th(RCOO)_4$ ($R = CF_3$, CH_2Cl, $CHCl_2$, or CCl_3) indicate that the COO group is probably either bidentate or bridging [6].

A number of hydrated thorium(IV) hydroxide trifluoroacetates of composition $Th(OH)_x$-$(CF_3COO)_{4-x} \cdot nH_2O$ has been reported, all of which are precipitated in the system $Th(NO_3)_4$-HNO_3-CF_3COOH-NH_4OH-H_2O at varying pH. The main product at pH 2.3 has the composition

References for 15.6.2.3.4 on pp. 63/4

$Th(OH)_{1.5}(CF_3COO)_{2.5} \cdot 2H_2O$, whereas at pH 3.7 the main product is $Th(OH)_2(CF_3COO)_2 \cdot H_2O$. At pH 4.42 the solubility increases, apparently because of the formation of $(NH_4)_2Th(CF_3COO)_6$, but at higher pH the precipitate consists of $Th(OH)_3(CF_3COO) \cdot nH_2O$ (n=1 at pH 5.00, n=4 at pH 5.28, n=2.5 at pH 6.25, and n=6 at pH 7.53) [12]. $Th(OH)_3(CF_3COO) \cdot H_2O$ probably loses water at 373 K and at 483 K a further molecule of water is lost by elimination from two OH groups, presumably with the formation of $ThO(OH)(CF_3COO)$. This product then decomposes at ca. 543 K [12].

$Th(CH_2ClCOO)_4$ has been prepared by reaction of $Th(CH_3COO)_4$ with $CH_2ClCOOH$ [6, 17] in the same way as $Th(CF_3COO)_4$; the excess $CH_2ClCOOH$ is removed by extraction with ether [17]. It is also obtained by dissolving freshly precipitated thorium hydroxide in 20% aqueous $CH_2ClCOOH$; the product precipitates from the resulting solution on standing [6]. It has also been obtained by treating the hydroxide with a 10- to 12-fold excess of the acid in aqueous solution, followed by washing the precipitate with CCl_4 [11]. Another preparative route is by heating $ThCl_4$ with $CH_2ClCOOH$ under reflux. Any free $CH_2ClCOOH$ in the product is removed by washing it with hot C_6H_6 [7]. Studies of the thermal decomposition of $Th(CH_2ClCOO)_4$ indicate that $ThO(CH_2ClCOO)_2$ is formed as the primary product [10, 11, 13].

$Th(CHCl_2COO)_4$ is prepared from $Th(CH_3COO)_4$ by reaction with $CHCl_2COOH$ in the same way as $Th(CF_3COO)_4$ (p. 55) [6, 17], the excess $CHCl_2COOH$ being removed by extraction with ether [17]. The hydrate, $Th(CHCl_2COO)_4 \cdot H_2O$, prepared by treating freshly precipitated thorium hydroxide with a 10- to 12-fold excess of $CHCl_2COOH$ as described for $Th(CH_2ClCOO)_4$ (see above) [11], loses water at ca. 130°C [10, 13] and the anhydrous salt decomposes at higher temperatures (see Table 24, p. 57) to $ThOCl(CHCl_2COO)$ [10, 11, 13] and then to $ThOCl_2$ (in air [10, 13]), or to a mixture of $ThCl_4$ and $ThOCl_2$ (in N_2 [13]) or $ThCl_4$ [11].

The basic dichloroacetate, $[Th_6(CHCl_2COO)_{12}(OH)_{12}(H_2O)_2]$, has been obtained by reaction of thorium hydroxide with an excess of $CHCl_2COOH$ followed by recrystallisation of the product from ethanol. In the structure of this complex, the six Th atoms are at the vertices of a nearly regular octahedron with an oxygen atom (presumably an OH group) above the centre of each face. The Th and O atoms are at the vertices of a rhombic dodecahedron (**Fig. 5**) and each edge of the octahedron of Th atoms is spanned by a $CHCl_2COO$ group. All Th atoms are 9-coordinated, with nine O atoms at the vertices of a monocapped square antiprism [16, 19].

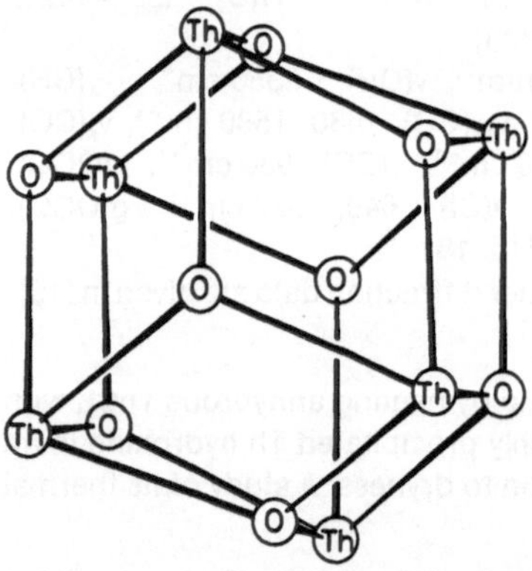

Fig. 5. The thorium and oxygen atoms which occupy the 14 vertices of the rhombic dodecahedron in the structure of the basic dichloroacetate $[Th_6(CHCl_2COO)_{12}(OH)_{12}(H_2O)_2]$ [16, 19].

$Th(CCl_3COO)_4$ has been prepared from $Th(CH_3COO)_4$ in the same way as $Th(CF_3COO)_4$ (p. 55) [6, 17]; the excess of CCl_3COOH is removed by extraction with ether [17]. It is also obtained by heating $ThCl_4$ with an excess of CCl_3COOH under reflux until the evolution of HCl ceases. This product was washed with CCl_4 to remove any free CCl_3COOH present [8].

Thermogravimetric analysis indicates decomposition to $ThCl_4$ in N_2 [10, 11, 13] or to $ThOCl(CCl_3COO)$ in air [10, 13]; this last decomposes to $ThOCl_2$ in air at higher temperatures [10, 13]. A review of published data on the decomposition of Th-halogenoacetates is given in [23]. Another report [9] suggests that the first step in the decomposition is to a thorium carbonate which then decomposes to ThO_2.

The hydrate, $Th(CCl_3COO)_4 \cdot H_2O$, is prepared in the same way as $Th(CHCl_2COO)_4 \cdot H_2O$ (p. 62) [11]. It loses H_2O at 120 to 140°C [10, 11, 13]. The dihydrate, $Th(CCl_3COO)_4 \cdot 2H_2O$, is apparently precipitated from aqueous solutions of Th^{IV} salts by 5 M CCl_3COOH at pH 0.8 to 1.0, but at pH 2.8 the product is $Th(OH)(CCl_3COO)_3 \cdot 2H_2O$ and at pH 3.8 the compound $Th(OH)_2(CCl_3COO)_2 \cdot 2H_2O$ is precipitated [2, 3, 5]. The IR spectrum of the dihydrate, $Th(CCl_3COO)_4 \cdot 2H_2O$, has been discussed in some detail [5].

The complex salts $M_2[Th(CCl_3COO)_6]$ (M^I = Na or K) are obtained when the stoichiometric quantitities of CCl_3COOM^I and $Th(CCl_3COO)_4$ are heated together in C_6H_6 under reflux for 9 to 10 h. The compounds are precipitated from the resulting solution by addition of an inert solvent such as petroleum ether. They are insoluble in CH_2Cl_2, CH_3CN, and CH_3NO_2, but are soluble in $HCON(CH_3)_2$ and in $C_6H_5NO_2$, in which the salts are reported to be monomers. Thermogravimetric analysis suggests that CCl_3COOM^I is lost in the first stage of heating. The IR spectra of the salts have been interpreted as indicating the presence of unidentate COO groups [9].

Monobromo- and monoiodoacetates have also been reported. The reaction of freshly precipitated thorium hydroxide with a 5- to 6-fold excess of $CH_2BrCOOH$, followed by heating to ca. 40°C and digestion at this temperature for 72 h, yields the basic compounds $Th(OH)(CH_2BrCOO)_3 \cdot H_2O$ and $ThO(CH_2BrCOO)_2 \cdot H_2O$ as fine white powders [18]. However, it is not clear how these two compounds are obtained separately. Evaporation to dryness of the mother liquor from this preparation yields the hydrated bromoacetate, $Th(CH_2BrCOO)_4 \cdot 2H_2O$ [18]. $Th(CHBr_2COO)_4$ forms when $ThCl_4$ is refluxed with excess of dibromoacetic acid and has been characterised by elemental analysis, molar conductance and IR spectroscopy [20].

Freshly precipitated thorium hydroxide does not dissolve in an aqueous ethanolic solution of CH_2ICOOH, but after standing for 5 to 6 days the hydroxide is converted to $Th(CH_2ICOO)_4$. When insufficient CH_2ICOOH is used in this preparation, the product is the hydrated basic salt, $ThO(CH_2ICOO)_2 \cdot H_2O$ [15].

References for 15.6.2.3.4:

[1] Hara, R.; Cady, G. H. (J. Am. Chem. Soc. **76** [1954] 4285/7).
[2] Galateanu, I. (Acad. Rep. Populare Romine Studii Cercetari Fiz. **14** [1963] 557/70).
[3] Galateanu, I.; Lapitskii, A. V.; Veiner, M.; Salimov, M. A.; Artamonova, E. P. (Radiokhimiya **6** [1964] 93/101; Soviet Radiochem. **6** [1964] 83/90).
[4] Sartori, P.; Weidenbruch, M. (Angew. Chem. **76** [1964] 376/7; Angew. Chem. Intern. Ed. Engl. **3** [1964] 376/7).
[5] Galateanu, I. (Oesterr. Chemiker-Ztg. **66** [1965] 275/85).
[6] Bagnall, K. W.; Velasquez Lopez, O. (J. Chem. Soc. Dalton Trans. **1976** 1109/13).
[7] Malhotra, K. C.; Sud, R. G. (J. Inorg. Nucl. Chem. **38** [1976] 393/6).
[8] Malhotra, K. C.; Kumar, A.; Chaudhry, S. C. (Indian J. Chem. A **18** [1979] 423/5).
[9] Malhotra, K. C.; Kumar, A.; Chaudhry, S. C. (Natl. Acad. Sci. Letters [India] **2** [1979] 383/6).
[10] Matveev, Yu. S.; Dunaeva, K. M. (Koord. Khim. **9** [1983] 361/5).

[11] Matveev, Yu. S.; Dunaeva, K. M. (Koord. Khim. **10** [1984] 511/7; Soviet J. Coord. Chem. **10** [1984] 286/92).

[12] Andryushin, V. G.; Samatov, A. V.; Chuklinov, R. N.; Shmidt, V. S. (Radiokhimiya **26** [1984] 166/70; Soviet Radiochem. **26** [1984] 152/5).
[13] Matveev, Yu. S.; Dunaeva, K. M. (Vestn. Mosk. Univ. Ser. II Khim. **25** [1984] 455/8; C.A. **102** [1985] No. 159486).
[14] Matveev, Yu. S.; Dunaeva, K. M. (Koord. Khim. **11** [1985] 207/15; Soviet J. Coord. Chem. **11** [1985] 112/20).
[15] Matveev, Yu. S.; Dunaeva, K. M. (Koord. Khim. **11** [1985] 621/7; Soviet J. Coord. Chem. **11** [1985] 352/8).
[16] Towns, E.; Smith, A. J. (Acta Cryst. A **40** [1984] C-308).
[17] Paul, R. C.; Saran, M. S.; Bains, M. S. (Indian J. Chem. **7** [1969] 384/6).
[18] Matveev, Yu. S.; Dunaeva, K. M. (Koord. Khim. **11** [1985] 766/73; Soviet J. Coord. Chem. **11** [1985] 440/6).
[19] Smith, A. J. (personal communication).
[20] Puri, J. K.; Miller, J. M. (Inorg. Chim. Acta **101** [1985] 135/40).
[21] Ahmed, I.; Bagnall, K. W. (J. Chem. Soc. Dalton Trans. **1984** 1527/9).
[22] Bagnall, K. W.; Velasquez Lopez, O.; Xing-Fu Li (J. Chem. Soc. Dalton Trans. **1983**, 1153/8).
[23] Matveev, Yu. S.; Dunaeva, K. M. (Vestn. Mosk. Univ. Ser. II Khim. **27** [1986] 389/97; C.A. **105** [1986] No. 197811).

15.6.2.3.5 **Thorium Mercaptoacetate** (Thioglycolate)

A compound with antimony thioglycolate, $Th[(SCH_2COO)_2Sb]_4$, is precipitated when an aqueous solution containing a Th^{IV} salt is added to a hot aqueous solution of $H(SCH_2COO)_2Sb$. The IR spectrum of this product exhibits features at 1560 to 1540 cm^{-1} ($\nu_{as}(CO)$) and 1380 to 1360 cm^{-1} ($\nu_s(CO)$) which suggest that the complex anion acts as a bridging chelating ligand in which both carboxylate oxygen atoms of the SCH_2COO group take part. The compound is very insoluble in the common solvents and may be polymeric [1].

Reference for 15.6.2.3.5:

[1] Sharma, H. K.; Lata, S.; Dubey, S. N.; Puri, D. M. (Indian J. Chem. A **22** [1983] 253/4).

15.6.2.3.6 Thorium Phenyl-, Phenylhydroxy-, Diphenyl-, Diphenylhydroxy-, and Naphthylacetates

The known compounds are listed in Table 25 together with physical data where available.

Table 25
Thorium Phenylacetates, Phenylhydroxyacetates(= Mandelates), Diphenylacetate, Diphenylhydroxyacetates(= Benzilates) and Naphthylacetates.

phenylacetates	
$Th(C_6H_5CH_2COO)_4$	[5]; decomposition > 100°C [3], > 190°C [21] (maximum at 247 to 270°C [21]), 216°C [6, 7], to ThO_2 at 500°C [3]; thermolysis curve illustrated in [3]
$ThCl(C_6H_5CH_2COO)_3$	decomposes > 300°C [10]

Table 25 (continued)

$ThCl_2(C_6H_5CH_2COO)_2$	decomposes >300°C [10]
$ThCl_3(C_6H_5CH_2COO)$	decomposes >300°C [10]
$Th(OH)(C_6H_5CH_2COO)_3$	[2, 12]
$Th(OH)(C_6H_5CH_2COO)_3 \cdot 2H_2O$	[1]
$Th(OH)_2(C_6H_5CH_2COO)_2$	[17]
$Th(4\text{-}NO_2C_6H_4CH_2COO)_4$	decomposes 190°C [6]
$Th(4\text{-}CNC_6H_4CH_2COO)_4$	decomposes 194°C [6]
$Th(4\text{-}CH_3COC_6H_4CH_2COO)_4$	decomposes 198°C [6]
$Th(4\text{-}ClC_6H_4CH_2COO)_4$	decomposes 209°C [6]
$Th(4\text{-}CH_3C_6H_4CH_2COO)_4$	decomposes 220°C [6]
$Th(4\text{-}CH_3OC_6H_4CH_2COO)_4$	decomposes 225°C [6]
$Th(4\text{-}(CH_3)_2NC_6H_4CH_2COO)_4$	decomposes 235°C [6]
phenylhydroxyacetates	
$Th(C_6H_5CH(OH)COO)_4 \cdot H_2O$	IR spectrum: (3300 to 450 cm^{-1}) illustrated in [11, 20]; ν(OH), 2974 cm^{-1}, ν(CO), 1590 cm^{-1}, δ_{as}(CO), 539 cm^{-1}, δ_s(CO), 473 cm^{-1} [11]
$Th(C_6H_5CH(OH)COO)_4 \cdot xH_2O$	[8]
$Th(OH)_2(C_6H_5CH(OH)COO)_2$	loses $1H_2O$ (from OH groups) at 110 to 130°C, decomposes >205°C; density, d_{obs} = 2.12 g/cm^3 at 25°C [13] dipole moment in dioxane = 2.8 D [13]
diphenylacetate	
$Th(OH)_2[(C_6H_5)_2CHCOO]_2$	[16]
diphenylhydroxyacetates	
$Th[(C_6H_5)_2C(O)COO]_2$	IR spectrum: ν(CO), 1600 cm^{-1}, ν(C(OH))(?), 1190 cm^{-1} [15]
$Th[(C_6H_5)_2C(OH)COO]_4$	decomposes to ThO_2 at ca. 850°C [19] IR spectrum: ν(OH), 2750 to 2550 cm^{-1}, ν_{as}(CO), 1670 cm^{-1}, ν_s(CO), 1415 cm^{-1} [19]
naphthylacetates	
$Th(1\text{-}C_{10}H_7CH_2COO)_4$	[18]; decomposes >280°C, to ThO_2 at 530°C; thermolysis curve illustrated in [4]
$Th(OH)_2(1\text{-}C_{10}H_7CH_2COO)_2$	[14]
$Th(OH)_3(1\text{-}C_{10}H_7CH_2COO)$	[2, 12]

References for 15.6.2.3.6 on pp. 67/8

15.6.2.3.6.1 Phenylacetates

$Th(C_6H_5CH_2COO)_4$ is precipitated when an aqueous solution of $C_6H_5CH_2COOH$ is added to a boiling dilute aqueous solution of $Th(NO_3)_4$ at a final pH of 3 to 5 [3] and when a solution of hydrated $Th(NO_3)_4$ is mixed with a concentrated aqueous solution of $C_6H_5CH_2COONa$ [7]. It has also been prepared by heating $Th(CH_3COO)_4$ (p. 47) with an excess of $C_6H_5CH_2COOH$ for 5 to 6 h, any unreacted $C_6H_5CH_2COOH$ being removed by extraction with ether [21]. The thermal decomposition of this compound [3, 6, 7] and of the analogous substituted phenylacetates, $Th(4\text{-}XC_6H_4CH_2COO)_4$ (X = NO_2, CN, CH_3CO, Cl, CH_3, CH_3O, and $(CH_3)_2N$) [6], prepared in a similar manner from aqueous solution, has been studied; the results indicate that electron-withdrawing substituents X facilitate thermal decomposition, whereas electron-donating substituents act in the opposite sense, increasing the temperature of decomposition (see Table 25, pp. 64/5) [6].

$Th(C_6H_5CH_2COO)_4$ is reported to be completely hydrolysed in aqueous media at pH 11.3 [5] and basic salts are commonly obtained under conditions which might be expected to yield $Th(C_6H_5CH_2COO)_4$. The product precipitated when boiling 4% aqueous $C_6H_5CH_2COOH$ is added to a boiling dilute aqueous solution of $Th(NO_3)_4$ containing NH_4Cl has the composition $Th(OH)(C_6H_5CH_2COO)_3 \cdot 2H_2O$ after drying to constant weight at 105°C [1] whereas precipitation from a boiling solution at pH>2.8 apparently yields the anhydrous compound $Th(OH)(C_6H_5CH_2COO)_3$, which has been recommended for the gravimetric determination of Th (as ThO_2 after ignition) [2]; see also [1]. Precipitation of Th with $C_6H_5CH_2COONH_4$ at pH 4.5 to 5.0 at 50 to 60°C is quantitative and the precipitate has the composition $Th(OH)_2(C_6H_5CH_2COO)_2$ [17]. The procedure can be used for Th analysis, after ignition to ThO_2, and for the separation of Th from Mn^{II}, Ni^{II}, Co^{II}, and Zn^{II} [17]. Precipitation of Th with $C_6H_5CH_2COOH$ can also be used for its separation from the cerite lanthanides [1]. The precipitation of Th with phenylacetic acid for analytical purposes has been discussed in some detail [9].

The reaction of $ThCl_4$ with a large excess of $C_6H_5CH_2COOH$ at 60°C for 3 h yields $ThCl_3(C_6H_5CH_2COO)$, whereas at 65°C for 6 to 7 h the product is $ThCl_2(C_6H_5CH_2COO)_2$ and at 100°C for 4 h $ThCl(C_6H_5CH_2COO)_3$ is obtained. $Th(C_6H_5CH_2COO)_4$ could not be obtained by this route [10].

15.6.2.3.6.2 Phenylhydroxyacetates

$Th(C_6H_5CH(OH)COO)_4 \cdot H_2O$ is precipitated from aqueous 2 M $C_6H_5CH(OH)COOH$(= mandelic acid) and the stoichiometric quantity of 0.5 M $Th(NO_3)_4$ at pH 1 [11, 20]. The IR spectrum of the hydrate, $Th(C_6H_5CH(OH)COO)_4 \cdot xH_2O$, indicates that the Th atom is bonded to the COO group and not to the CHOH group [8].

The basic salt, $Th(OH)_2(C_6H_5CH(OH)COO)_2$, is precipitated from aqueous $Th(NO_3)_4 \cdot 5H_2O$ by $C_6H_5CH(OH)COONa$ (molar ratio Th : Na salt = 1:1 to 1:4). It is readily soluble in acetone, dioxane, ethanol, and methanol, but is almost insoluble in benzene, chloroform, and ether. It can be purified by dissolution in dioxane and evaporation of the filtrate at 50°C [13]. The compound is a monomer in dioxane. The dielectric constant and polarisation of dioxane solutions of the compound at 25°C have also been reported [13].

15.6.2.3.6.3 Diphenylacetate

$(C_6H_5)_2CHCOOH$ precipitates thorium quantitatively at pH>2.7 as $Th(OH)_2[(C_6H_5)_2CHCOO]_2$. The precipitation can be used as a means of separating Th from Fe^{II}, but not from Al^{III}, and for the gravimetric estimation of Th as ThO_2 after ignition [16].

15.6.2.3.6.4 Diphenylhydroxyacetates

$Th\{(C_6H_5)_2C(OH)COO\}_4$ has been prepared by heating $Th(CH_3COO)_4$ (p. 47) with the stoichiometric quantity of $(C_6H_5)_2C(OH)COOH$(= benzilic acid) in ethanol under reflux for 2 h. The resulting solution was evaporated under reduced pressure to yield the crystalline product. The IR spectrum of this compound indicates that the oxygen atom of the C(OH) group is coordinated to the metal atom; ν(OH) is shifted from 3390 cm^{-1} in the spectrum of the free acid to 2750 to 2550 cm^{-1} in that of the compound [19]. A product of composition $Th\{(C_6H_5)_2C(O)COO\}_2$ is precipitated when aqueous $(C_6H_5)_2C(OH)COONa$ is added dropwise to aqueous $Th(NO_3)_4$ at pH 2.0 to 4.5. It is soluble in acetone, "butanolone", benzene, and ethyl acetate, but is insoluble in ethanol, ether, carbon tetrachloride, and petroleum ether. It is also soluble in aqueous H_2SO_4, partly soluble in aqueous HCl, but insoluble in aqueous NaOH, NH_3, or HNO_3 [15].

15.6.2.3.6.5 1-Naphthylacetates

A product of composition $Th(C_{10}H_7CH_2COO)_4$ is precipitated when 0.5% aqueous 1-$C_{10}H_7CH_2COOH$ is added to a hot (50 to 60°C) aqueous solution of a Th^{IV} salt at pH>2, followed by boiling for 5 to 7 min. This precipitation has been used as a method of separating Th, together with Ti and Zr, from Fe and Al [18].

A basic compound, $Th(OH)_3(1\text{-}C_{10}H_7CH_2COO)$, is reported to be precipitated by aqueous 1-$C_{10}H_7CH_2COOH$ from boiling aqueous solutions of Th^{IV} salts at pH>4.4. The precipitation can be used for the gravimetric estimation of Th as ThO_2 following ignition [2]. However, this preparative procedure was subsequently [4] reported to yield $Th(1\text{-}C_{10}H_7CH_2COO)_4$, and another report [12] indicates that the product is $Th(OH)(1\text{-}C_{10}H_7CH_2COO)_3$. A different basic salt, $Th(OH)_2(1\text{-}C_{10}H_7CH_2COO)_2$, is apparently precipitated when the requisite amount of hot 2% aqueous 1-$C_{10}H_7CH_2COOK$ is added to a boiling aqueous solution of a Th^{IV} salt, followed by adjustment to pH 4.65. This procedure was used for the gravimetric determination of Th as ThO_2 after ignition and it is reported to give better results than precipitation with phenyl-1,3-dioxydiacetic acid (p. 119) [14]. The diverse stoichiometries of the products obtained from aqueous solution may indicate that $Th(1\text{-}C_{10}H_7CH_2COO)_4$ is easily hydrolysed.

15.6.2.3.6.6 7-Hydroxy-4-coumarinacetate

7-Hydroxy-4-coumarinacetic acid (=$(7\text{-}HOC_9H_4O_2)CH_2COOH = C_{11}H_8O_5$) precipitates thorium completely from boiling aqueous solution at pH 2.5 to 3.5. The precipitate is probably $Th(OH)(C_{11}H_7O_5)_3 \cdot xH_2O$. The precipitation can be used to separate thorium from the cerite lanthanides and for the estimation of thorium after ignition to ThO_2 [22].

References for 15.6.2.3.6:

[1] Purushottam, A.; Raghava Rao, B. S. V. (Z. Anal. Chem. **141** [1954] 97/100).
[2] Datta, S. K.; Banerjee, G. (Anal. Chim. Acta **12** [1955] 38/46).
[3] Wendlandt, W. W. (Anal. Chem. **29** [1957] 800/2).
[4] Wendlandt, W. W. (Anal. Chim. Acta **17** [1957] 295/9).
[5] Kovalenko, K. N.; Tarasova, M. N. (Zh. Neorgan. Khim. **5** [1960] 385/92; Russ. J. Inorg. Chem. **5** [1960] 184/7).
[6] Okubo, M.; Goto, R. (Nippon Kagaku Zasshi **82** [1961] 261/3; C.A. **56** [1962] 11439).
[7] Okubo, M.; Goto, R. (Nippon Kagaku Zasshi **81** [1960] 1132/6; C.A. **56** [1962] 3342).

[8] Geleceanu, I.; Lapitskii, A. V. (Dokl. Akad. Nauk SSSR **144** [1962] 573/5; Proc. Acad. Sci. USSR Sect. **142/147** [1962] 460/2).
[9] Tserkovnitskaya, I. A.; Charykov, A. K. (Radiokhimiya **4** [1962] 184/8; Soviet Radiochem. **4** [1962] 165/8).
[10] Jaura, K. L.; Banga, H. S.; Kaushik, R. L. (J. Indian Chem. Soc. **39** [1962] 531/3).

[11] Galateanu, I. (Acad. Rep. Populare Romine Studii Cercetari Chim. **11** [1963] 343/52).
[12] Tserkovnitskaya, I. A.; Charykov, A. K. (Izv. Vysshikh Uchebn. Zavedenii Khim. Khim. Tekhnol. **7** [1964] 544/50; C.A. **62** [1965] 3385).
[13] Kovalenko, K. N.; Kazachenko, D. V. (Zh. Neorgan. Khim. **15** [1970] 1523/6; Russ. J. Inorg. Chem. **15** [1970] 781/3).
[14] Pande, C. S.; Misra, G. N.; Shukla, V. K. (Zh. Anal. Chem. **265** [1973] 31).
[15] Gupta, S. S.; Gupta, N. K. (Lab. Pract. **22** [1973] 174).
[16] Konnov, V. A. (Tr. Inst. Okeanol. Akad. Nauk SSSR **63** [1973] 199/202; C.A. **81** [1974] No. 98887).
[17] Volkov, I. I.; Konnov, V. A. (Tr. Inst. Okeanol. Akad. Nauk SSSR **63** [1973] 203/7; C.A. **81** [1974] No. 98897).
[18] Konnov, V. A. (Zh. Analit. Khim. **29** [1974] 476/9; J. Anal. Chem. [USSR] **29** [1974] 407/10).
[19] Singh, M.; Singh, A. (J. Indian Chem. Soc. **56** [1979] 1249/51).
[20] Galateanu, I. (Oesterr. Chemiker-Ztg. **66** [1965] 275/85).

[21] Paul, R. C.; Saran, M. S.; Bains, M. S. (Indian J. Chem. **7** [1969] 384/6).
[22] Eswaranarayana, N. (Recl. Trav. Chim. **72** [1953] 1003/6).

15.6.2.4 Thorium Propionates

Information on the known compounds is summarised in Table 26.

Table 26
Thorium Propionates.

$Th(C_2H_5COO)_4$	decomposes >140°C (maximum decomposition at 214 to 233°C) [3], >150°C [2], 175 to 180°C [1], >250°C [4]; differential thermal analysis and thermogravimetric analysis curves illustrated in [4] IR spectrum: ν_{as}(CO), 1560 cm^{-1}, ν_s(CO), 1385 cm^{-1}, ν(CC), 910 cm^{-1} [2]; density, d_{obs} = 2.10 g/cm^3 [4]
$ThCl(C_2H_5COO)_3$	[1]
$ThCl_2(C_2H_5COO)_2$	[1]
$ThCl_3(C_2H_5COO)$	[1]
$Th(OH)(C_2H_5COO)_3$	decomposes >210°C [4]; see also [6]
$Th(OH)(C_2H_5COO)_3 \cdot 4H_2O$	loses $4H_2O$ at 95°C [4]; differential thermal analysis and thermogravimetric analysis curves illustrated in [4]; density, d_{obs} = 2.14 g/cm^3 [4]
$Th(OH)_2(C_2H_5COO)_2$	decomposes >220°C [4]; see also [6]

Table 26 (continued)

$Th(OH)_2(C_2H_5COO)_2 \cdot H_2O$	loses H_2O at 90°C [4]; differential thermal analysis and thermogravimetric analysis curves illustrated in [4]; density, d_{obs} = 2.48 g/cm³ [4]
$Th(OH)_3(C_2H_5COO)$	decomposition >200°C [4]; loses H_2O from OH groups at 220°C and C_2H_5COO group decomposes at 350°C [6]
$Th(OH)_3(C_2H_5COO) \cdot 4H_2O$	loses $4H_2O$ at 105°C [4, 6]; differential thermal analysis and thermogravimetric analysis curves illustrated in [4]; density, d_{obs} = 2.59 g/cm³ [4]

$Th(C_2H_5COO)_4$ has been prepared by treating thorium hydroxide with C_2H_5COOH at 50°C; crystals of the compound separate from the solution on standing [2]. It is also obtained when $Th(CH_3COO)_4$ (p. 47) is heated with an excess of C_2H_5COOH for 5 to 6 h, removing the liberated CH_3COOH at low pressure. Any unreacted C_2H_5COOH is removed by washing the precipitate with ether [3]. The propionate can also be prepared by heating $ThCl_4$ with an excess of C_2H_5COOH at 160°C. Heating at 98 to 100°C for 3 h yields $ThCl(C_2H_5COO)_3$ and when $ThCl_4$ is heated with the calculated quantities of C_2H_5COOH under reflux, products of composition $ThCl_{4-x}(C_2H_5COO)_x$ (x=1 or 2) are obtained [1]. The IR spectrum of $Th(C_2H_5COO)_4$ suggests that the carboxylate groups are bidentate [2].

The hydrated basic salt, $Th(OH)(C_2H_5COO)_3 \cdot 4H_2O$, is precipitated from aqueous $Th(NO_3)_4$ containing HNO_3, NH_3 and C_2H_5COOH at pH 1.55 to 2.72, whereas $Th(OH)_2(C_2H_5COO)_2 \cdot H_2O$ is precipitated at pH 2.90 to 6.63 and $Th(OH)_3(C_2H_5COO) \cdot 4H_2O$ at pH 8.06 to 8.60 [4]. In this work it was noted that the solubility of the Th species present increased between pH 4.5 and 6.6, which may result from the formation of a soluble salt, such as $(NH_4)_2Th(C_2H_5COO)_6$ [4]. The hydrated basic salts lose water readily on heating (see Table 26) and the thermal decomposition of these compounds has been studied [4]. See also [5].

References for 15.6.2.4:

[1] Jaura, K. L.; Bajwa, P. S. (J. Sci. Ind. Res. [India] B **20** [1961] 391/4).

[2] Molodkin, A. K.; Ivanova, O. M.; Skotnikova, G. A. (Zh. Neorgan. Khim. **11** [1966] 1978/9; Russ. J. Inorg. Chem. **11** [1966] 1056/7).

[3] Paul, R. C.; Saran, M. S.; Bains, M. S. (Indian J. Chem. **7** [1969] 384/6).

[4] Andryushin, V. G.; Kozhevnikov, P. V.; Pozharskaya, M. E.; Shmidt, V. S. (Radiokhimiya **18** [1976] 185/90; Soviet Radiochem. **18** [1976] 167/71).

[5] Shmidt, V. S.; Andryushin, V. G. (Radiokhimiya **18** [1976] 506/11; Soviet Radiochem. **18** [1976] 439/43).

[6] Shmidt, V. S.; Andryushin, V. G. (Radiokhimiya **24** [1982] 601/6; Soviet Radiochem. **24** [1982] 498/503).

15.6.2.5 Thorium Compounds with Substituted Propionic Acids

15.6.2.5.1 2-Phenyl-, β-Hydroxyamino-β-phenyl-, 1- and 2-Phenoxy-, 2-*o*-Cresoxypropionates. Thiolactic Acid and 3-Methoxyphenylpyruvate-oxime

The few known compounds are listed in Table 27, p 70.

Table 27
Thorium Compounds with Substituted Propionic Acids.

2-phenylpropionic acid	
$Th(OH)_2(C_6H_5CH_2CH_2COO)_2$	[1]; decomposition >40°C, rapidly at >180°C, to ThO_2 at 475°C [3], thermal decomposition curve illustrated in [3]
1-phenoxypropionic acid	
$Th(OH)(CH_3CH(OC_6H_5)COO)_3 \cdot xH_2O$	[4, 5]
2-phenoxypropionic acid	
$Th(OH)(C_6H_5OCH_2CH_2COO)_3 \cdot xH_2O$	[4, 5]
2-o-cresoxypropionic acid	
$Th(OH)(2\text{-}CH_3C_6H_4OCH_2CH_2COO)_3 \cdot xH_2O$	[4, 5]
α- and β-aminopropionic acid (α- and β-alanine)	
$Th(ClO_4)_4 \cdot CH_3CH(NH_2)COOH \cdot 2H_2O$	see Th E, pp. 5/6
$Th(ClO_4)_4 \cdot H_2NCH_2CH_2COOH \cdot 2H_2O$	see Th E, pp. 5/6
1-mercaptopropionic acid (thiolactic acid)	
$Th[Sb(SCH(CH_3)COO)_2]_4$	[7]

The basic 2-phenylpropionate, $Th(OH)_2(C_6H_5CH_2CH_2COO)_2$, is precipitated when the aqueous acid is added to a boiling aqueous solution of a Th^{IV} salt at pH>3.6. The precipitate coagulates on boiling for ca. 10 min or on addition of a 10% aqueous solution of CH_3COONH_4. The precipitation of this compound can be used for the gravimetric estimation of Th as ThO_2 following ignition [1].

The addition of a hot 1% aqueous solution of β-hydroxyamino-β-phenylpropionic acid (or its sodium salt) to aqueous $Th(NO_3)_4$ (neutral to Congo Red), followed by the addition of 4% aqueous CH_3COONH_4, yields a flocculent precipitate which becomes granular on digestion for 2 min. The precipitate is stated to be of the form $Th(OH)_3L$, where HL is the acid [2], but no analytical data are given. Th can be separated from U by a single precipitation with this acid at pH 4.8 to 5.6 [2].

The hydrated 1-phenoxy-, 2-phenoxy-, and 2-*o*-cresoxypropionates, $Th(OH)L_3 \cdot xH_2O$, are precipitated quantitatively from hot solutions of thorium(IV) salts at pH 3.5 on addition of a hot 2% aqueous solution of the acid [4, 5]. The precipitation has been investigated for the separation of Th from the lanthanides [5].

A thiolactate derivative, $Th[Sb(SCH(CH_3)COO)_2]_4$, is precipitated when aqueous $Th(NO_3)_4$ is added to an aqueous solution of $HSb(SCH(CH_3)COO)_2$. The IR spectrum of this product is said to indicate that the COO group acts as a bridging bidentate unit, and the compound is probably polymeric [7].

A pale yellow precipitate is obtained when an ethanol solution of 4-methoxyphenylpyruvic acid oxime (=$4\text{-}CH_3OC_6H_4CH_2C(:NOH)COOH$) is added to a dilute aqueous solution of a Th^{IV} salt. This decomposes at >250°C. The frequencies of the carboxylate modes in the IR spectrum of this product are close to those of the free carboxylic acid, and replacement of the carboxylate proton is thought to be less likely than replacement of the oxime proton [6].

References for 15.6.2.5.1:

[1] Datta, S. K.; Banerjee, G. (Anal. Chim. Acta **12** [1955] 38/46).
[2] Banerjee, G. (Z. Anal. Chem. **147** [1955] 348/54).
[3] Wendlandt, W. W. (Anal. Chim. Acta **18** [1958] 316/20).
[4] Datta, S. K. (Z. Anal. Chem. **174** [1960] 104/8).
[5] Datta, S. K. (Z. Anal. Chem. **174** [1960] 109/18).
[6] Katyal, M.; Singh, R. P. (Indian J. Appl. Chem. **27** [1964] 37/9).
[7] Lata, S.; Sharma, H. K.; Dubey, S. N.; Puri, D. M. (J. Indian Chem. Soc. **61** [1984] 905/6).

15.6.2.5.2 Aminopropionic Acids

The only recorded compounds are the complexes of $Th(ClO_4)_4$ with α- or β-alanine (2- or 3-aminopropionic acid) described in Th E, pp. 5/6; see Table 27.

15.6.2.6 Thorium Butyrates, Isobutyrates, and Related Carboxylates

Information on the known compounds is summarised in Table 28.

Table 28
Thorium Butyrates.

$Th(n\text{-}C_3H_7COO)_4$	decomposes >164°C [5], 250°C [1], maximum decomposition at 242 to 268°C [5]
$Th(OH)(n\text{-}C_3H_7COO)_3$	[8]
$Th(OH)_2(n\text{-}C_3H_7COO)_2$	decomposes >198°C, density, d_{obs} = 2.27 g/cm³ [6], see also [8]; differential thermal and thermogravimetric analysis curves illustrated in [6]
$Th(OH)_3(n\text{-}C_3H_7COO)$	decomposes >230°C [6], loses H_2O from OH groups at 300°C, $n\text{-}C_3H_7COO$ group decomposition at 345°C [8]
$Th(OH)_3(n\text{-}C_3H_7COO)\cdot H_2O$	loses H_2O at 120°C, density, d_{obs} = 2.93 g/cm³ [6]; differential thermal and thermogravimetric analysis curves illustrated in [6]
$Th(OH)_3(n\text{-}C_3H_7COO)\cdot xH_2O$	loses H_2O at 120°C [8]
$Th(i\text{-}C_3H_7COO)_4$	decomposes >204°C [5], 260°C [1], maximum decomposition at 268 to 297°C [5]
$Th\{(CH_3)_2C(OH)COO\}_4\cdot H_2O$	IR spectrum illustrated in [4]
$Th\{(CH_3)_2C(OH)COO\}_4\cdot 2H_2O$	IR spectrum (3430 to 450 cm⁻¹) illustrated in [3]; ν(CO), 1630 cm⁻¹ [3]
$Th\{(CH_3)_2C(OH)COO\}_4\cdot xH_2O$	[2]

$Th(n\text{-}C_3H_7COO)_4$ is precipitated when the stoichiometric amount of $n\text{-}C_3H_7COONa$, as a concentrated aqueous solution, is mixed with aqueous $Th(NO_3)_4$; the mixture was then heated on a water bath for ca. 2 h. The isobutyrate, $Th(i\text{-}C_3H_7COO)_4$, has been obtained in the same way [1]. An alternative method of preparing both compounds is by heating $Th(CH_3COO)_4$ (p. 47) with an excess of the appropriate acid for 5 to 6 h, removing the liberated CH_3COOH under reduced pressure. Any excess of the free butyric acid was removed by washing the insoluble products with ether [5].

The basic *n*-butyrates, $Th(OH)_2(n\text{-}C_3H_7COO)_2$ and $Th(OH)_3(n\text{-}C_3H_7COO)\cdot H_2O$, are precipitated on the addition of aqueous NH_3 to a solution of $Th(NO_3)_4$ in aqueous HNO_3 containing $n\text{-}C_3H_7COOH$ at the final pH of 2.5 to 3.5 and 7.72 to 8.7, respectively [6]. The formation of $Th(OH)(n\text{-}C_3H_7COO)_3$ under similar conditions has also been reported [8]. The tendency to hydrolyse, and degree of hydrolysis, of the *n*-butyrates in aqueous solution is discussed in [7].

The hydrated α-hydroxy-iso-butyrates, $Th\{(CH_3)_2C(OH)COO\}_4\cdot xH_2O$ (x=1 [4], 2 [3] and unspecified [2]) are obtained by repeated crystallisation from 2 M $(CH_3)_2C(OH)COOH$ containing the calculated quantity of $Th(NO_3)_4$, followed by evaporation at 40°C (and pH 1) until the hydrate (x=1 [4] and 2 [3]) precipitates. The compound with x unspecified [2] appears to have been obtained in a similar manner. The three hydrates are probably the same compound. Their IR spectra indicate that the Th atom is bonded to the carboxylate group and not to the OH group (x unspecified [2]).

References for 15.6.2.6:

[1] Okubo, M.; Goto, R. (Nippon Kagaku Zasshi **81** [1960] 1132/6; C.A. **56** [1962] 3342).
[2] Geleceanu, I.; Lapitskii, A. V. (Dokl. Akad. Nauk SSSR **144** [1962] 573/5; Proc. Acad. Sci. USSR Chem. Sect. **142/147** [1962] 460/2).
[3] Galateanu, I. (Acad. Rep. Populare Romine Studii Cercetari Chim. **11** [1963] 343/52).
[4] Galateanu, I. (Oesterr. Chemiker-Ztg. **66** [1965] 275/85).
[5] Paul, R. C.; Saran, M. S.; Bains, M. S. (Indian J. Chem. **7** [1969] 384/6).
[6] Andryushin, V. G.; Slepchenko, I. G.; Chuklinov, R. N.; Shmidt, V. S. (Radiokhimiya **18** [1976] 502/5; Soviet Radiochem. **18** [1976] 435/8).
[7] Shmidt, V. S.; Andryushin, V. G. (Radiokhimiya **18** [1976] 506/11; Soviet Radiochem. **18** [1976] 439/43).
[8] Shmidt, V. S.; Andryushin, V. G. (Radiokhimiya **24** [1982] 601/6; Soviet Radiochem. **24** [1982] 498/503).

15.6.2.7 Thorium Valerates

The available information on the few known compounds is summarised in Table 29.

Table 29
Thorium Valerates.

$Th(OH)_2(n\text{-}C_4H_9COO)_2\cdot 4H_2O$	[2]
$Th(i\text{-}C_4H_9COO)_4$	decomposes >200°C, maximum decomposition at 260 to 290°C [3]
$Th\{(CH_3)_3CCOO\}_4$	decomposes >350°C [1]; activation energy for decomposition to $ThO\{(CH_3)_3CCOO\}_2$, $E_{act}=180\pm17$ kJ/mol; to $ThOCO_3$, $E_{act}=192\pm17.5$ kJ/mol and to ThO_2, $E_{act}=218\pm17$ kJ/mol [5] IR and mass spectra, thermal analysis reported in [4]

The basic valerate, $Th(OH)_2(n\text{-}C_4H_9COO)_2 \cdot 4H_2O$, is precipitated by the acid from aqueous $Th(NO_3)_4$ at pH 2 to 5.5. It can be extracted into $n\text{-}C_4H_9OH$ [2].

The isovalerate, $Th(i\text{-}C_4H_9COO)_4$, is obtained by heating $Th(CH_3COO)_4$ (p. 47) with an excess of $i\text{-}C_4H_9COOH$, removing the liberated CH_3COOH under reduced pressure. Residual $i\text{-}C_4H_9COOH$ was finally removed from the insoluble product by extraction with ether [3].

$Th\{(CH_3)_3CCOO\}_4$ (with $(CH_3)_3CCOOH$ = pivalic acid) is precipitated when aqueous $Th(NO_3)_4$ is mixed with a concentrated aqueous solution containing the stoichiometric quantity of $(CH_3)_3CCOONa$; the mixture was then heated on a waterbath for ca. 2 h [1]. It has also been prepared by treating freshly precipitated thorium hydroxide with a saturated solution of $(CH_3)_3CCOOH$ in ethanol. The IR and mass spectra, and the thermal analysis behaviour of the compound have been compared with the behaviour of $(CH_3)_3CCOOH$, $Th(CH_3COO)_4$, $U(CH_3COO)_4$ and related compounds [4].

References for 15.6.2.7:

[1] Okubo, M.; Goto, R. (Nippon Kagaku Zasshi **81** [1960] 1132/6; C.A. **56** [1962] 3342).

[2] Tserkovnitskaya, I. A.; Charykov, A. K. (Izv. Vysshikh Uchebn. Zavedenii Khim. Khim. Tekhnol. **7** [1964] 544/50; C.A. **62** [1965] 3385).

[3] Paul, R. C.; Saran, M. S.; Bains, M. S. (Indian J. Chem. **7** [1969] 384/6).

[4] Matveev, Yu. S.; Dubrovin, A. V.; Dunaeva, K. M. (Deposited Doc. VINITI-3282-84 [1984] 1/18; C.A. **103** [1985] No. 160082).

[5] Matveev, Yu. S.; Dunaeva, K. M. (Vestn. Mosk. Univ. Ser. II Khim. **27** [1986] 389/97; C.A. **105** [1986] No. 197811).

15.6.2.8 Thorium Caproates and Longer Chain Carboxylates

Information on this group of compounds is summarised in Table 30.

Table 30
Thorium Caproates and Longer Chain Carboxylates.

caproates	
$Th(n\text{-}C_5H_{11}COO)_4$	decomposes >110°C, maximum decomposition at 204 to 235°C [6]
$Th(OH)_2(n\text{-}C_5H_{11}COO)_2 \cdot 4H_2O$	[5]
heptanoate, octanoate, and nonanoate	
$Th(OH)_2(n\text{-}C_6H_{13}COO)_2$	[5]
$Th(OH)_2(n\text{-}C_7H_{15}COO)_2$	[5]
$Th(OH)_2(n\text{-}C_8H_{15}COO)_2$	[5]
laurate, myristates, palmitate, and stearates	
$Th(n\text{-}C_{11}H_{23}COO)_4$	[4]
$Th(n\text{-}C_{13}H_{27}COO)_4$	[4]
$Th(OH)_2(n\text{-}C_{13}H_{27}COO)_2$	[5]

References for 15.6.2.8 on p. 75

Table 30 (continued)

$Th(n\text{-}C_{15}H_{31}COO)_4$	[4]
$Th(n\text{-}C_{17}H_{35}COO)_4$	[4]
$Th(OH)_2(n\text{-}C_{17}H_{35}COO)_2$	decomposes >50°C, to ThO_2 at 450°C [3]; thermolysis curve illustrated in [3]
D-gluconates (D-gluconic acid = $HOCH_2(CHOH)_4COOH = C_5H_6(OH)_5COOH$)	
$ThCl(OH)(C_5H_6(O)(OH)_4COO)\cdot 5H_2O$ [a)]	[7]; possibly $ThCl(OH)_2(C_5H_6(OH)_5COO)\cdot 4H_2O$
$Th_2Cl_2(OH)_2(C_5H_6(OH)_5COO)_2(C_5H_6(O)(OH)_4COO)\cdot 3H_2O$ [a)]	[7]; possibly $Th_2Cl_2(OH)_3(C_5H_6(OH)_5COO)_3\cdot 2H_2O$
$Th(OH)_2(C_5H_6(O)(OH)_4COO)\cdot 4H_2O$ [a)]	[7]; possibly $Th(OH)_3(C_5H_6(OH)_5COO)\cdot 3H_2O$
$Th_2(OH)_4(C_5H_6(OH)_5COO)_2(C_5H_6(O)(OH)_4COO)\cdot 8H_2O$ [a)]	[7]; possibly $Th_2(OH)_5(C_5H_6(OH)_5COO)_3\cdot 7H_2O$
$KTh(OH)_2(C_5H_6(O)_2(OH)_3COO)\cdot 4H_2O$ [a)]	[7]

a) Thermogravimetric curves illustrated in [7].

The caproate, $Th(n\text{-}C_5H_{11}COO)_4$, has been prepared from $Th(CH_3COO)_4$ and $n\text{-}C_5H_{11}COOH$ in the same way as the isovalerate (p. 73) [6]. The basic caproate, $Th(OH)_2(n\text{-}C_5H_{11}COO)_2\cdot 4H_2O$, is precipitated when the acid is added to an aqueous solution of a thorium(IV) salt at pH 2 to 5.5 [5]. The anhydrous basic salts, $Th(OH)_2(RCOO)_2$, are precipitated in the same way at pH 2 to 3.5 ($R = n\text{-}C_6H_{13}$, $n\text{-}C_7H_{15}$, $n\text{-}C_8H_{17}$ [5]), pH 1.3 to 3.5 ($R = n\text{-}C_{13}H_{27}$ [5]) or pH 3 to 5 ($R = C_{17}H_{35}$ [3]). The anhydrous basic salts with $R = n\text{-}C_6H_{13}$, $n\text{-}C_7H_{15}$, and $n\text{-}C_8H_{17}$ are not extracted into *n*-butanol, whereas the hydrated caproate is readily extracted [5].

The compounds with long chain acids, $Th(RCOO)_4$ ($R = n\text{-}C_{11}H_{23}$, $n\text{-}C_{13}H_{27}$, $n\text{-}C_{15}H_{31}$, and $n\text{-}C_{17}H_{35}$), have been obtained by adding an excess of the acid, suspended in ether, to a solution of $ThCl_4$ in ether, followed by heating at 55 to 60°C for 1 h and finally heating the mixture at 15 to 25°C above the melting point of the appropriate acid until the evolution of HCl ceases. The product was cooled and then washed with ether [4].

A thorium laurate (lauric acid = $n\text{-}C_{11}H_{23}COOH$) of unspecified composition is precipitated when an aqueous solution of $n\text{-}C_{11}H_{23}COONa$ is added to aqueous $Th(NO_3)_4\cdot 4H_2O$. This product was used in a study of the formation of higher unsaturated alcohols by hydrogenation at 300 to 330°C in an autoclave [2]. It has been noted [1] that stearic acid ($= C_{17}H_{35}COOH$) precipitates thorium from aqueous solution at pH 3.5 to 6.0, but the composition of the precipitate was not reported.

Several thorium gluconates have been reported. A precipitate of composition $Th(OH)_2(C_5H_6(O)(OH)_4COO)\cdot 4H_2O$ (possibly $Th(OH)_3(C_5H_6(OH)_5COO)\cdot 3H_2O$) is obtained when aqueous 0.1M $C_5H_6(OH)_5COOH$ is added to a solution of $Th(NO_3)_4$ (molar ratio, gluconic acid: Th = 1:1) in 1N HCl, followed by addition of KOH to pH 4 to 5 [7]. $ThCl(OH)(C_5H_6(O)(OH)_4COO)\cdot 5H_2O$ (possibly $ThCl(OH)_2(C_5H_6(OH)_5COO)\cdot 4H_2O$) is precipitated when acetone is added to the 1:1 mixture of $Th(NO_3)_4$ and gluconic acid in water [7]. When the molar ratio Th : gluconic acid = 1:3, the precipitate has the approximate composition $Th_2Cl_2(OH)_2(C_5H_6(OH)_5COO)_2(C_5H_6(O)(OH)_4COO)\cdot 3H_2O$ [7] (possibly $Th_2Cl_2(OH)_3(C_5H_6(OH)_5COO)_3\cdot 2H_2O$).

When a 1N HCl solution of $Th(NO_3)_4$ and 0.1M aqueous gluconic acid (molar ratio 1:2) is adjusted to pH 3 to 4 by addition of KOH, precipitation does not occur, but dropwise addition of acetone, with stirring, yields a gelatinous precipitate of composition $Th_2(OH)_4(C_5H_6(OH)_5COO)_2(C_5H_6(O)(OH)_4COO)\cdot 8H_2O$ [7] (possibly $Th_2(OH)_5(C_5H_6(OH)_5COO)_3\cdot 7H_2O$). A hydrated precipitate in which the analytical ratio Th : $C_5H_6(OH)_5COOH$ = 3 : 4 is obtained with the molar ratio Th : $C_5H_6(OH)_5COOH$: KOH = 1 : 2 : 5 at pH 4 to 5 in the reaction medium [7].

A gelatinous precipitate of composition $KTh(OH)_2(C_5H_6(O)_2(OH)_3COO)\cdot 4H_2O$ is obtained when a solution of $Th(NO_3)_4$ in 1N HCl is added dropwise to a mixture of 0.1M gluconic acid and 0.5 M KOH (final molar ratio, Th : $C_5H_6(OH)_5COOH$: KOH = 1 : 2 : 6; pH = 9 to 10) [7]. Another product, in which the analytical ratio Th : $C_5H_6(OH)_5COOK$ = 3 : 4, was obtained by dropwise addition of $Th(NO_3)_4$ in 1N HCl to a similar mixture of gluconic acid and KOH (molar ratio, Th : gluconic acid : KOH = 1 : 2 : 7 or 1 : 3 : 8, pH > 10), followed by addition of methanol to precipitate the product [7]. The published papers on thorium gluconates (and saccharates, p. 113) are reviewed in [8].

References for 15.6.2.8:

[1] Deshmukh, G. S.; Xavier, J. (J. Indian Chem. Soc. **29** [1952] 911/4).
[2] Komori, S.; Shigeno, Y. (Technol. Rept. Osaka Univ. **3** [1953] 171/82).
[3] Wendlandt, W. W. (Anal. Chem. **29** [1957] 800/2).
[4] Prasad, S.; Kumar, S. (J. Indian Chem. Soc. **39** [1962] 444/6).
[5] Tserkovnitskaya, I. A.; Charykov, A. K. (Izv. Vysshikh Uchebn. Zavedenii Khim. Khim. Tekhnol. **7** [1964] 544/50; C.A. **62** [1965] 3385).
[6] Paul, R. C.; Saran, M. S.; Bains, M. S. (Indian J. Chem. **7** [1969] 384/6).
[7] Macarovici, C. G.; Czeglédi, L. (Rev. Roumaine Chim. **9** [1964] 411/24).
[8] Macarovici, C. G. (Mem. Sect. Stiint. Acad. Repub. Soc. Romania **4** [1981/83] 209/23).

15.6.2.9 Thorium Salts of Unsaturated Aliphatic Monocarboxylic Acids

The known compounds are listed in Table 31.

Table 31
Thorium Salts of Unsaturated Aliphatic Monocarboxylic Acids.

crotonate (= methylacrylate)	
$Th(CH_3CH{:}CHCOO)_4$	[10]
cinnamates (= phenylacrylates)	
$Th(C_6H_5CH{:}CHCOO)_4$	[10, 12]
$ThO(C_6H_5CH{:}CHCOO)_2$(?)	[11, 13]
$Th(OH)(C_6H_5CH{:}CHCOO)_3$	decomposes >180°C, to ThO_2 at 490°C; thermolysis curve illustrated in [6]
$Th(OH)(C_6H_5CH{:}CHCOO)_3\cdot 4H_2O$(?)	[1]
$Th(3\text{-}NO_2C_6H_4CH{:}CHCOO)_4$	[7]

References for 15.6.2.9 on p. 77

Table 31 (continued)

maleanilates (maleanilic acids = HOOCCH:CHCONHR)	
$Th(OH)_2(RNHC(O)CH:CHCOO)_2$	R = C_6H_5, 2- or 4-ClC_6H_4, 3- or 4-$NO_2C_6H_4$, 1- or 2-$C_{10}H_7$, 2,5-$Cl_2C_6H_3$, 2- or 4-$CH_3C_6H_4$, 2- or 4-$CH_3OC_6H_4$, 4-$C_2H_5OC_6H_4$ [8, 9], 3-$(NH_2)C_6H_4$ [9], 4,4'-$H_2NC_6H_4C_6H_4$ (derived from benzidine) [8], 4,4'-H_2N-2,2'-$(CH_3)_2C_6H_3C_6H_3$ (derived from tolidine) [9]
$[Th(OH)_2(C_{10}H_7NO_3S)H_2O]_2$ (2'-mercaptomaleanilate)	white [14], decomposes at 220 to 380°C; thermogravimetric analysis [14]

Thorium crotonate and cinnamate, $Th(CH_3CH:CHCOO)_4$ and $Th(C_6H_5CH:CHCOO)_4$, have been prepared from $ThCl_4$ and the appropriate acid in the same way as the laurate (p. 74) and other long chain carboxylates [10]. $Th(C_6H_5CH:CHCOO)_4$ has also been obtained by precipitation at low pH from aqueous solution [12]. However, it has been reported that precipitation from hot aqueous solutions of Th^{IV} salts with $C_6H_5CH:CHCOONH_4$ at ca. pH 5.2 yields a mixture of $Th(C_6H_5CH:CHCOO)_4$ and $ThO(C_6H_5CH:CHCOO)_2$ [11, 13]. This result is consistent with the reported precipitation of $Th(OH)(C_6H_5CH:CHCOO)_3 \cdot 4H_2O$ from boiling aqueous solutions of Th^{IV} salts at pH 2.0 to 2.6 [1]. This precipitation has been used for the separation of Th from the cerite group lanthanides [1, 3] and for the estimation of Th [3], presumably as ThO_2 after ignition. Anhydrous $Th(OH)(C_6H_5CH:CHCOO)_3$ is apparently obtained in a similar manner at pH 3 to 5 [6]. The precipitation of an unspecified thorium cinnamate from boiling aqueous solution at pH>1.9 has also been reported; precipitation at pH 2.0 to 2.4 under these conditions appears to be satisfactory for the separation of Th from large amounts (350-fold excess) of UO_2^{2+} [2].

Th can be extracted from aqueous solution into methylisobutyl ketone or chloroform in the presence of cinnamic acid [5], but the species present in the extract was not indentified.

The precipitation of Th from boiling aqueous solution at pH 3.8 to 4.0 by substituted cinnamic acids (2-, 3-, and 4-hydroxy- (= *o*-, *m*-, and *p*-coumaric acids), 4-methoxy-, 3,4-dimethoxy-, and 3-methoxy-4-hydroxy-(= ferulic acid)) yields products of variable composition and the method can only be used for the determination of Th after ignition to ThO_2 [4, 7]. However, 3-nitrocinnamic acid gives a precipitate of composition $Th(3\text{-}NO_2C_6H_4CH:CHCOO)_4$ under these conditions which can be weighed as such [7].

Several maleanilic acids (HOOCCH:CHC(O)NHR) precipitate Th quantitatively when a 1% solution of the acid in ethanol is added to a hot, buffered aqueous solution of a thorium salt (R = 2,5-$Cl_2C_6H_3$, 2- or 4-$CH_3C_6H_4$, 2- or 4-$CH_3OC_6H_4$, 1- and 2-$C_{10}H_7$, all at pH ca. 4.2). Precipitation is almost quantitative with the acid in which R = C_6H_5 (pH 3.2), whereas when R = 2- or 4-ClC_6H_4 (pH 3.5) the precipitate is somewhat soluble in the CH_3COONH_4 solution used as the buffer. With R = 3- or 4-$NO_2C_6H_4$ (pH ca. 3.5) and R = 4-$C_2H_5OC_6H_4$ (pH 3.5, from ethanol and acetone solution), precipitation was not quantitative. No precipitation occurred with the acids in which R = 4-$CH_3CONHC_6H_4$ or 2-$NH_2C_6H_4$. The compositions of the precipitates approximate to $Th(OH)_2(RNHC(O)CH:CHCOO)_2$ (R = C_6H_5, 2- or 4-ClC_6H_4, 3- or 4-$NO_2C_6H_4$, 4-$C_2H_5OC_6H_4$, 1- or 2-$C_{10}H_7$, 3-$NH_2C_6H_4$, and 4,4'-H_2N-2,2'-$(CH_3)_2C_6H_3C_6H_3$) but their compositions are not sufficiently defined for analytical purposes [9].

2'-mercapto-maleanilic acid (HOOCCH:CHC(O)NHC$_6$H$_4$SH = $C_{10}H_9NO_3S$) precipitates Th^{IV} as a white compound from acetate buffered solution at pH 5.5 to 7.0. The solid is dried at 130 to 135°C and weighed as $[Th(C_{10}H_7NO_3S)(OH)_2H_2O]_2$. The thermogravimetric analysis shows loss of coordinated water and water from the hydroxyl groups at 160 to 220°C. The acid can serve as a gravimetric reagent for the separation of Zr^{IV} and Th^{IV} [14].

References for 15.6.2.9:

[1] Venkateswarlu, C.; Raghava Rao, B. S. V. (J. Indian Chem. Soc. **27** [1950] 638/40).
[2] Venkateswarlu, C.; Purushottam, A.; Raghava Rao, B. S. V. (Z. Anal. Chem. **133** [1951] 151/4).
[3] Krishnamurty, K. V. S.; Venkateswarlu, C. (Recl. Trav. Chim. **71** [1952] 668/70).
[4] Verma, M. R.; Paul, S. D.; Agrawal, K. C. (Z. Anal. Chem. **152** [1956] 427/33).
[5] Hök-Bernström, B. (Acta Chem. Scand. **10** [1956] 174/85).
[6] Wendlandt, W. W. (Anal. Chem. **29** [1957] 800/2).
[7] Verma, M. R.; Agrawal, K. C.; Paul, S. D. (16th Congr. Intern. Chim. Pure Appl. Mem. Sect. Chim. Minèrale, Paris 1957 [1958], pp. 345/51; C.A. **1960** 11849).
[8] Datta, S. K. (Z. Anal. Chem. **168** [1959] 418/24).
[9] Datta, S. K. (Z. Anal. Chem. **174** [1960] 23/9).
[10] Prasad, S.; Kumar, S. (J. Indian Chem. Soc. **39** [1962] 444/6).

[11] Ostroumov, E. A.; Volkov, I. I. (Zh. Analit. Khim. **17** [1962] 461/5; J. Anal. Chem. [USSR] **17** [1962] 460/5).
[12] Tserkovnitskaya, I. A.; Charykov, A. K. (Izv. Vysshikh Uchebn. Zavedenii Khim. Khim. Tekhnol. **7** [1964] 544/50; C.A. **62** [1965] 3385).
[13] Ostroumov, E. A.; Volkov, I. I. (Zh. Analit. Khim. **23** [1968] 973/9; J. Anal. Chem. [USSR] **23** [1968] 849/54).
[14] Singh, N.; Agrawal, P. (J. Indian Chem. Soc. **61** [1984] 94/6).

15.6.2.10 Thorium Naphthenates

Naphthenic acids are carboxylic acids derived from 5- or 6-membered ring hydrocarbons. Thorium naphthenates are precipitated from aqueous media by the acids of molecular weight 160 to 270, and all have a low solubility in water (10^{-3} to 10^{-6} mol/L) [1]. The compositions of the precipitates are not known.

Reference for 15.6.2.10:

[1] Alekperov, R. A.; Akhundova, Z. A.; Efendieva, N. G. (NRV Sels'k. Khoz. Tr. 3rd. Vses. Soveshch. Izuch. Primen NRV Sels'k. Khoz., Baku 1970 [1971], pp. 88/90; C.A. **77** [1972] No. 52885).

15.6.3 Aliphatic Dicarboxylates of Thorium

15.6.3.1 Introduction

Much of the published literature on these compounds has been concerned with the use of the dicarboxylic acids for the determination of thorium by precipitation, often followed by ignition to ThO_2 for weighing. In many cases the compositions of the precipitates are uncertain and the temperatures at which the hydrated dicarboxylates lose water frequently vary from one report to another. Prolonged heating at the lower temperatures recorded in the tables in this chapter will generally suffice for this purpose, particularly since there is evidence that some decomposition also occurs at the highest dehydration temperatures that have been reported.

It has been reported [1] that dicarboxylic acids with up to ten carbon atoms generally precipitate Th at low pH and more specific details are included in each of the following sections.

Mixed oxalatocompounds are discussed on pp. 107 (tartrato-) and 122 (citrato-).

Oxalatothorium complexes in which the compounds include neutral donor ligands are described in the appropriate chapters on the complex compounds of thorium (see "Thorium" Suppl. Vol. E, 1985) and are also included in the tables listing the parent compounds.

Reference for 15.6.3.1:

[1] Tserkovnitskaya, I. A.; Charykov, A. K. (Izv. Vysshikh Uchebn. Zavedenii Khim. Khim. Tekhnol. **7** [1964] 544/50; C.A. **62** [1965] 3385).

15.6.3.2 Thorium Oxalates

The available information on these compounds is summarised in Table 32.

Table 32
Thorium Oxalates.

$Th(C_2O_4)_2$	decomposition >320°C in air or N_2 [8], >340°C in air [19], 340 to 480°C [27], 345°C in N_2 [37], 360°C in air [35], 360°C (thermogravimetry) or 370°C (differential thermal analysis) [20], 385°C [60], 400°C in N_2 [60], >400°C (in 2 stages) [25], 420°C in vacuum [60], to ThO_2 at ca. 560°C [22] or at 650°C [32] activation energy of decomposition, E_{act} = 52 kcal/mol (219 kJ/mol) (differential thermal analysis), 55 kcal/mol (231 kJ/mol) (thermogravimetry) [20], 60 kcal/mol (252 kJ/mol) [36, 56] enthalpy of decomposition, $\Delta H = -18$ kcal/mol (−76 kJ/mol) [36, 56]; enthalpy of formation (by combustion) at 303 K $\Delta H_f = -320$ kcal/mol, (−1344 kJ/mol), $\sigma = 4$ kcal/mol (16.8 kJ/mol) [55] Th–O bond energy = 77.8 kcal/mol (327 kJ/mol) [55] enthalpy of sublimation, ΔH_{subl} = 28.0 kcal/mol (118 kJ/mol) [55], $\sigma = 0.1$ kcal/mol (0.4 kJ/mol) entropy of sublimation, $\Delta S_{subl} = 34.7\ cal \cdot mol^{-1} \cdot K^{-1}$ ($146\ J \cdot mol^{-1} \cdot K^{-1}$), $\sigma = 0.9\ cal \cdot mol^{-1} \cdot K^{-1}$ ($3.8\ J \cdot mol^{-1} \cdot K^{-1}$) [55] crystallographic properties; orthorhombic, lattice parameters, a = 9.65 Å, b = 10.55 Å, c = 8.43 Å, Z = 4; d_{obs} = 3.14, $d_{calc} = 3.172\ g/cm^3$ [33] morphological properties are described in [7] IR spectrum illustrated (4000 to 400 cm^{-1}); $\nu_{as}(CO)$, ca. 1640 cm^{-1}, $\nu_s(CO)$, 1350, 1317 cm^{-1}, $\delta(OCO)$, 781 cm^{-1} [53] L emission spectrum; $L_\alpha = 74 \pm 7$, $L_{\gamma 1} = 82 \pm 11$ [64] see also "Thorium" 1955, p. 307
$Th(C_2O_4)_2 \cdot 0.75H_2O$	see "Thorium" 1955, p. 307

Table 32 (continued)

$Th(C_2O_4)_2 \cdot H_2O$	loses H_2O at 230°C (differential thermal analysis) or 260°C (thermogravimetry) [20]; loses H_2O and decomposes at 300 to 450°C [53], >340°C (in air) [19], 370 to 450°C [53], 379°C [23] or 400°C [17]; see also [24] activation energy of decomposition = 36 kcal/mol (151 kJ/mol) (thermogravimetry) or 42 kcal/mol (176 kJ/mol) (differential thermal analysis) [20] crystallographic data; orthorhombic, lattice parameters, a = 9.90 Å, b = 9.52 Å, c = 8.71 Å [33] IR spectrum illustrated (4000 to 400 cm^{-1}), $\nu_{as}(CO)$, ca. 1640 cm^{-1}, $\nu_s(CO)$, 1355, 1319 cm^{-1}, $\delta(OCO)$, 809 cm^{-1} [53] ^1H NMR spectrum reported in [52]
$Th(C_2O_4)_2 \cdot 1.5H_2O$	decomposes >610°C; thermogram illustrated in [3]; probably the monohydrate
$Th(C_2O_4)_2 \cdot 2H_2O$	loses $0.5H_2O$ at 153°C [3], loses $1H_2O$ at 120°C (differential thermal analysis) or 140°C (thermogravimetry) [20], 230 to 260°C [53], 300°C [17, 23] or 300 to 340°C in air [19]; loses $2H_2O$ at 180°C in a vacuum [60], 240°C [27, 37], 250°C [25], >260°C [27], 270°C [60], 270°C in air or N_2 [8] or at 340°C [32] decomposes >290°C [5] thermogravimetric [3, 5, 19, 20] and differential thermal analysis [20] curves illustrated activation energy of dehydration, E_{act} = 44 kcal/mol (185 kJ/mol) [36, 56]; enthalpy, ΔH = 6 kcal/mol (25 kJ/mol) [36, 56]; for loss of $1H_2O$, ΔH_{act} = 26 kcal/mol (109 kJ/mol) (thermogravimetry) or 20 kcal/mol (84 kJ/mol) (differential thermal analysis) [20]; crystallographic data; orthorhombic, lattice parameters, a = 10.51 Å, b = 9.73 Å, c = 8.45 Å [31, 33]; a = 10.504(4) Å, b = 9.735(4) Å, c = 8.506(4) Å [45]; Z = 4, d_{obs} = 3.31, d_{calc} = 3.412 g/cm^3 [33] IR spectrum illustrated (4000 to 400 cm^{-1}) [53]; $\nu_{as}(CO)$, ca. 1648 cm^{-1} [53], ca. 1640 cm^{-1} [66], $\nu_s(CO)$, 1357, 1316 cm^{-1} [53], 1358, 1319 cm^{-1} [66], $\delta(OCO)$, 802 cm^{-1} [53], 803 cm^{-1} [66] ^1H NMR spectrum [52]; solubility in liquid NH_3 = 0.51 g/100 g NH_3 [4] see also "Thorium" 1955, p. 307
$Th(C_2O_4)_2 \cdot 3H_2O$	loses $1H_2O$ at 200 to 240°C in air or dry CO_2 [19]; probably the dihydrate
$Th(C_2O_4)_2 \cdot 5H_2O$	[8]; probably the hexahydrate
$Th(C_2O_4)_2 \cdot 6H_2O$	[16, 43, 48]; loses $4H_2O$ at 70 to 130°C [53], 110°C [4, 38, 45], 120°C [37], 120 to 240°C [27], 130°C [25], 145°C [22], 150°C [17], 152°C [23], or 180°C [5]; thermogravimetric curve illustrated in [5, 24, 25, 53]; differential thermal analysis curve illustrated in [22, 24, 37 (in N_2), 56, 60]; emanation and weight loss curves illustrated in [60]; energy of activation of dehydration, E_{act} = 21 kcal/mol (88 kJ/mol); enthalpy, ΔH = 20 kcal/mol (84 kJ/mol) [36, 56]

References for 15.6.3.2 on pp. 84/6

Table 32 (continued)

	crystallographic data; triclinic, tetragonal indexing, lattice parameters, $a \approx b \approx 6.51$ Å, $c \approx 7.91$ Å; Z=1 [33], $d_{obs}=2.56$ [33], 2.596 [51], $d_{calc}=2.51$ g/cm³ [33] optical crystallographic properties: $n_g=1.575$ [25], 1.588 [51], $n_p=1.468$ [51], 1.556 [25] IR spectrum illustrated (4000 to 400 cm⁻¹) in [54], ν_{as}(CO), ca. 1649 cm⁻¹ [53], ca. 1640 cm⁻¹ [58], ν_s(CO), 1359, 1320 cm⁻¹, δ(OCO), 800 cm⁻¹ [53]; ν_7, 1730 cm⁻¹, ν_1, 1650 cm⁻¹, ν_2, 1356 cm⁻¹, ν_8, 1323 cm⁻¹, ν_9, 804 cm⁻¹, ν_{10}, 492 cm⁻¹ [54]; ν(CO)+δ(OCO), 1354, 1315 cm⁻¹, ν(CO)+δ(OCO)+ν(CC), 900 cm⁻¹, δ(OCO)+ν(ThO), 800 cm⁻¹, δ(OCO)+ν(ThO), 480 cm⁻¹ [58] electron micrography reported in [60] ^{1}H NMR spectrum [52]; binding energy, Th $5d_{5/2}=88.4$ eV [59]; solubility in water $=7.5\times10^{-5}$% [25], solubility in 0.01 to 1.0 M $HClO_4$ (constant ionic strength I=1) reported [50]; solubility product, $S=(4.2\pm0.01)\times10^{-22}$ (I=1), 1.1×10^{-25} (I=0) [50], 2.9×10^{-21} (I=0.5) [41]; in HCl (ionic strength 0.5 to 2) $1\pm0.1\times10^{-22}$ [21] see also "Thorium" 1955, pp. 305/7
$Th(C_2O_4)_2\cdot7H_2O$	thermogram illustrated in [19]; loses $4H_2O$ at 60 to 140°C in air or dry CO_2 [19]; may be the hexahydrate
$Th(C_2O_4)_2\cdot4\,n\text{-}C_4H_9NH_2\cdot2H_2O$	see Th E, pp. 5, 6
$Th(C_2O_4)_2\cdot2CO(NH_2)_2\cdot xH_2O$	x=0 or 2, see Th E, pp. 39, 42
$Th(C_2O_4)_2\cdot4CO(NH_2)_2$	see Th E, pp. 39, 42
$Th(C_2O_4)_2\cdot5CO(NH_2)_2\cdot H_2O$	see Th E, pp. 39, 42
$Th(C_2O_4)_2\cdot2dmso\cdot H_2O$	see Th E, pp. 100, 103

Anhydrous $Th(C_2O_4)_2$ has been prepared by heating $Th(C_2O_4)_2\cdot H_2O$ above 250°C or $Th(C_2O_4)_2\cdot2H_2O$ at 240°C (in N_2 [37]), 250°C [25], 260°C [27] or 270°C [22] (in air or N_2 [8]). These dehydration data are in reasonably good agreement. Other reports indicate that $Th(C_2O_4)_2\cdot H_2O$ loses water and decomposes above 300°C [53], 340°C [19], or 400°C [17] and some authors have found it difficult to obtain analytically pure $Th(C_2O_4)_2$ owing to partial decomposition [53], results which are consistent with the reported decomposition of $Th(C_2O_4)_2$ at these higher temperatures (see Table 32, p. 78). $Th(CO_3)_2$ (p. 2) appears to be formed as an intermediate in the thermal decomposition of the anhydrous oxalate [8, 37, 40, 46] prior to the final formation of ThO_2.

$Th(C_2O_4)_2$ is isomorphous with $U(C_2O_4)_2$ ("Uranium" Suppl. Vol. C 13, 1983, p. 176) [33]. The IR spectra of $Th(C_2O_4)_2$ and of the mono-, di-, and hexahydrates suggest that the C_2O_4 groups are close to ionic in character [53], but this requires confirmation.

A study of the L-emission X-ray lines due to Th in $Th(C_2O_4)_2$ indicates that the 5f states participate in the chemical bonding [64]. A study of the configuration interaction satellites in the ESCA spectrum of $Th(C_2O_4)_2$ has been reported [63]; it is not clear whether this work refers to the anhydrous oxalate or to a hydrate.

Anhydrous mixed oxalates of composition $U_xTh_{1-x}(C_2O_4)_2$ are prepared by treating the precipitate obtained when an excess of concentrated aqueous $H_2C_2O_4$ is added to a mixture of Th^{IV} and U^{IV} nitrates with an aqueous solution of $(NH_4)_2C_2O_4$, followed by evaporation to precipitate a mixed oxalatocomplex salt from which $(NH_4)_2C_2O_4$ is removed by sublimation at 150°C under reduced pressure [49].

Several hydrates of $Th(C_2O_4)_2$ have been reported, some of them on the basis of plateaux observed in the thermogravimetric and differential thermal analysis curves of higher hydrates. The best established hydrates are $Th(C_2O_4)_2 \cdot 6H_2O$ (p. 82), $Th(C_2O_4)_2 \cdot 2H_2O$ (below) and, to a lesser extent, $Th(C_2O_4)_2 \cdot H_2O$ (below). Other hydrates are almost certainly formulated incorrectly. The differences observed in the differential thermal analysis studies by different authors may be due to the influence of the gas present at the surface of the solid hydrate in the course of the dehydration, for some of the experimental work associated with the more obviously incorrect assignments can be reproduced exactly when the product is in direct contact with air, but are not reproduced in N_2 [23]. In other cases endotherms observed in the dehydration studies are wrongly ascribed; for example, an endotherm at 145°C is probably due to dehydration of $Th(C_2O_4)_2 \cdot 6H_2O$ to $Th(C_2O_4)_2 \cdot 2H_2O$ and not of $Th(C_2O_4)_2 \cdot 2H_2O$ to $Th(C_2O_4)_2 \cdot H_2O$ [22].

The existence of $Th(C_2O_4)_2 \cdot 0.75H_2O$ ("Thorium" 1955, p. 307) has not been confirmed in later work and this may be the monohydrate. Although $Th(C_2O_4)_2 \cdot H_2O$ was not observed as an intermediate phase in some thermogravimetric [8, 25, 32] and differential thermal analysis [22, 27, 37, 60] studies of the dehydration of the higher hydrates, and in a few reports some decomposition appears to occur at the dihydrate (see below) stage [5], others [17, 19, 20, 23, 24, 53] have detected the formation of $Th(C_2O_4)_2 \cdot H_2O$ in the thermal dehydration of $Th(C_2O_4)_2 \cdot 2H_2O$ (see Table 32, p. 79). X-ray powder diffraction data [33] indicate that the monohydrate is isomorphous with $U(C_2O_4)_2 \cdot H_2O$ ("Uranium" Suppl. Vol. C 13, 1983, p. 176). The IR [53] and ^{1}H NMR [52] spectra of the monohydrate have also been reported.

The formation of $Th(C_2O_4)_2 \cdot 1.5H_2O$ from the dihydrate at 153°C has been reported [3], but this seems more likely to be the monohydrate.

The dihydrate, $Th(C_2O_4)_2 \cdot 2H_2O$, is well established. It is obtained by heating $Th(C_2O_4)_2 \cdot 6H_2O$ (p. 82) at a variety of temperatures [3 to 5, 17, 22 to 25, 27, 36 to 38, 45, 52, 53, 56, 60] (see Table 32, p. 79) or $Th(C_2O_4)_2 \cdot 7H_2O$ (p. 80) at 240°C for 2 h [19]. High bulk density $Th(C_2O_4)_2 \cdot 2H_2O$ is obtained by drying (at 110°C) $Th(C_2O_4)_2 \cdot 6H_2O$ precipitated from solutions of Th salts by $H_2C_2O_4$ [38] or by soluble oxalates [39] at pH 0 to 4, both at 70 to 90°C in the presence of 0.5 to 15 moles of a "spectator" cation (Li, Na, K or NH_4) per mole of Th salt; the concentration of the cation in the precipitation medium was kept at or below 2.5 molar [38, 39]. The dihydrate is also reported to be precipitated from aqueous solutions of $Th(NO_3)_4$ by aqueous $H_2C_2O_4$ [8]; since one would expect the hexahydrate (p. 82) to be formed under these conditions, the result may be due to dehydration at the drying temperature (105°C) [8].

Ammonolysis of the dihydrate does not appear to occur in anhydrous liquid NH_3; the solubility of $Th(C_2O_4)_2 \cdot 2H_2O$ in the medium [4] has been reported in Table 32 (p. 79).

Thermogravimetry and differential thermal analysis have been used to study the kinetics of the thermal decomposition of $Th(C_2O_4)_2 \cdot 2H_2O$ [30].

X-ray powder diffraction data indicate that the dihydrate is isomorphous with $U(C_2O_4)_2 \cdot 2H_2O$ ("Uranium" Suppl. Vol. C 13, 1983, p. 176) [31, 33, 45]. The ^{1}H NMR spectrum of the dihydrate has also been reported [52].

Magnetic susceptibility data have been reported for the mixed oxalates $U_xTh_{1-x}(C_2O_4)_2 \cdot 2H_2O$, prepared by heating solid solutions of $U_xTh_{1-x}(C_2O_4)_2 \cdot 6H_2O$ (p. 82) at 105°C [47] and

References for 15.6.3.2 on pp. 84/6

by addition of aqueous solutions of Th^{IV} and U^{IV} nitrates to a continually saturated solution of $H_2C_2O_4$ [42]. See "Uranium" Suppl. Vol. C 13, 1983, p. 178.

$Th(C_2O_4)_2 \cdot 3H_2O$ may be formed as an intermediate in the thermal dehydration of $Th(C_2O_4)_2 \cdot 6H_2O$ [5] and $Th(C_2O_4)_2 \cdot 7H_2O$ [19], but this is probably the dihydrate.

A hydrate of composition $Th(C_2O_4)_2 \cdot 5H_2O$, obtained by drying (at 105°C) the precipitate formed when aqueous $(NH_4)_2C_2O_4$ is added to aqueous Th^{IV} [8] may be the hexahydrate, although there is some evidence for the possible formation of $Th(C_2O_4)_2 \cdot 5H_2O$ as an intermediate in the thermal dehydration of $Th(C_2O_4)_2 \cdot 6H_2O$ [5].

The precipitation of thorium oxalate from aqueous media has been extensively studied. In most cases the precipitate has been indentified as $Th(C_2O_4)_2 \cdot 6H_2O$, but in some papers the degree of hydration is not reported.

The hydrated oxalate which is precipitated by the addition of $H_2C_2O_4$ to a solution of $Th(NO_3)_4$ in 1M HNO_3 saturated with hexane is impossible to filter. In a study of the Th^{IV}–HNO_3–$H_2C_2O_4$–H_2O system it has been suggested that the character (filterability) of the precipitate depends on the ionic strength of the solution and, to some extent, the temperature of the precipitation medium [6]. The presence of Cl^- or NO_3^-, or the use of $(NH_4)_2C_2O_4$ as the precipitating agent, apparently tends to produce a colloidal product [12]. However, the particle size of the precipitate seems to depend less on the reagents used and more on the speed of mixing them, for rapid mixing of aqueous $Th(NO_3)_4$ and aqueous $(NH_4)_2C_2O_4$ appears to give a filterable product [13]. The influence of neutral salts on the precipitation of the hydrated oxalate has also been studied using a heterometric titration method [18].

Precipitation by $H_2C_2O_4$ is quantitative from hot HCl solutions of Th^{IV} when the HCl concentration is less than 7N, but is incomplete at higher acid concentrations [26]. Precipitation is reported to be best carried out with a 100% excess of $H_2C_2O_4$, followed by heating for 30 min [12]. Precipitation curves for the system $Th(NO_3)_4$–$Na_2C_2O_4$–H_2O have also been reported [9].

Homogeneous precipitation of Th and rare earth oxalates, using $H_2C_2O_4$ generated by the hydrolysis of methyl oxalate at 85°C, has been used as a method of separating these elements from monazite sand [1]. These elements are also precipitated quantitatively from aqueous solution using ethyl oxalate [11]. Over 90% of the Th present in monazite sand can be recovered by precipitation of the oxalate [14]. The coprecipitation of Th with rare earth, alkaline earth and Pb^{II} oxalates has been investigated as a possible separation method for Th [65].

$Th(C_2O_4)_2 \cdot 6H_2O$ is precipitated when aqueous $H_2C_2O_4$ is added to aqueous $Th(NO_3)_4$ [16, 45, 48, 52] or when aqueous $Th(NO_3)_4$ is added to aqueous $H_2C_2O_4$ [51]. In a study of the effects of the conditions of precipitation of $Th(C_2O_4)_2 \cdot 6H_2O$ on the texture of its thermal decomposition products, it was observed that the best conditions for precipitation were slow dropwise addition of saturated aqueous $H_2C_2O_4$ (slightly less than 1M) at room temperature to a boiling 1M solution of $Th(NO_3)_4$ with vigorous stirring, and continuing to boil the solution after the final addition of the acid [43]. Another report [60] recommends slow addition of ca. 0.5M $H_2C_2O_4$ to 0.4 M $Th(NO_3)_4$ at 70°C; larger crystals of the hydrate were apparently obtained by slowly cooling solutions of much lower concentration (100-fold dilution) after neutralising with aqueous NH_3 (taking care to avoid precipitation of Th^{IV} hydroxide) then adding aqueous $(NH_4)_2C_2O_4$, followed by heating to 80°C and adding dilute HCl [60]. The hexahydrate has also been prepared by a homogeneous precipitation method using methyl oxalate (see above) [22].

$Th(C_2O_4)_2 \cdot 6H_2O$ is isomorphous with $U(C_2O_4)_2 \cdot 6H_2O$ ("Uranium" Suppl. Vol. C 13, 1983, pp. 177, 179) [33]. The corresponding mixed oxalate, $U_xTh_{1-x}(C_2O_4)_2 \cdot 6H_2O$, is prepared by adding aqueous $H_2C_2O_4$ to an aqueous solution containing the two nitrates [29, 47].

Cell constants for x = 0.5 have been recorded and it has been reported that the thermogravimetric behaviour of the precipitated mixed oxalate differs somewhat from that of a simple mixture of the two oxalates [29]. In a study of the precipitation of the hydrated mixed oxalates, it was found that when increasing quantities of $H_2C_2O_4$ were added to a solution containing Th^{IV} and U^{IV} nitrates, the value of x in the product, $U_xTh_{1-x}(C_2O_4)_2 \cdot yH_2O$ (y probably = 6), varied as a function of the fraction of the theoretical quantity of $H_2C_2O_4$, whereas when the solution containing the mixed nitrates was added to $H_2C_2O_4$ the value of x was always 0.5 [44].

The triangular phase diagram for the aqueous system involving $Th(C_2O_4)_2 \cdot 6H_2O$ and the oxalato complexes $2Th(C_2O_4)_2 \cdot (NH_4)_2C_2O_4 \cdot xH_2O$ (x = 2 and 7) is illustrated in [13]. Solubility data for $Th(C_2O_4)_2 \cdot 6H_2O$ in H_2O and the pH and electrical conductivity of the solutions have been reported and the results have been compared with those obtained for $U(C_2O_4)_2 \cdot 6H_2O$ ("Uranium" Suppl. Vol. C 13, 1983, pp. 177, 179) [15]. $Th(C_2O_4)_2 \cdot 6H_2O$ does not undergo acidic dissociation in aqueous media like $U(C_2O_4)_2 \cdot 6H_2O$ [57]; the thorium compound is much less acidic in aqueous solution than its U^{IV} analogue and it does not react with pyridine [10]. The instability constant for $Th(C_2O_4)^{2+}$ has been found to be $1.5 \pm 0.5 \times 10^{-9}$ [21]. The Th compound is apparently hydrolysed at higher pH than the corresponding Zr salt [2].

The solubility of $Th(C_2O_4)_2 \cdot 6H_2O$ at 25°C in the quaternary system $Th(C_2O_4)_2$–HNO_3–$H_2C_2O_4$ (saturated)–H_2O and in the constituent ternary systems $Th(C_2O_4)_2$–$H_2C_2O_4$–H_2O and $H_2C_2O_4$–HNO_3–H_2O, as well as in the reciprocal quaternary system $Th(C_2O_4)_2$–HNO_3–H_2O (or $Th(C_2O_4)_2 + 4HNO_3 \rightleftharpoons Th(NO_3)_4 + 2H_2C_2O_4$) has been reported [34]. The solubility of $Th(C_2O_4)_2 \cdot 6H_2O$ in aqueous $H_2C_2O_4$ is very low (0.0012 to 0.0018%) and hardly changes with variations in the $H_2C_2O_4$ concentration. The solubility of the hexahydrate is slightly higher in aqueous HNO_3 at concentrations up to 15% HNO_3, and the solubility increases markedly (up to 1%) at higher HNO_3 concentrations, but addition of $H_2C_2O_4$ up to saturation in the HNO_3 solution sharply reduces the solubility (e.g. to 0.008% Th in 47% HNO_3) [34].

The solubility of $Th(C_2O_4)_2 \cdot 6H_2O$ in 0.01M to 1.0M $HClO_4$ at constant ionic strength I = 1 at 25°C has been studied and it was found that the solubility increased with $HClO_4$ concentration in mixtures of $H_2C_2O_4$ and 1.0M $HClO_4$ [50]. A study of the solubility of the oxalate at 25°C in 0.1M $HClO_4$/1.99×10^{-3} to 8.34×10^{-2}M $(NH_4)_2C_2O_4$ has been reported [41]. The solubilities in aqueous solutions of HCl, $HClO_4$, and H_2SO_4 at varying concentrations [41] and in 10^{-9} to 0.5M oxalate anion [67] have also been reported. For other solubility data, see [69]. Values are given in Table 32, p. 80.

The crystallisation field of $Th(C_2O_4)_2 \cdot 6H_2O$ has been partially defined in a study of the solubility isotherm for the aqueous system $Th(C_2O_4)_2 + 2Na_2CO_3 \rightleftharpoons Th(CO_3)_2 + 2Na_2C_2O_4$ at 25°C [28].

The exchange of oxalate anion in $Th(C_2O_4)_2 \cdot 6H_2O$ in contact with ^{14}C labeled $H_2C_2O_4$ or $K_2C_2O_4$ in aqueous solution has been investigated [48]. The ^1H NMR spectrum of $Th(C_2O_4)_2 \cdot 6H_2O$ indicates that the H_2O molecules are in two different environments, one of which involves strongly held and the other much more weakly held H_2O molecules [52].

Studies of satellite phenomena [61] and shakeup satellites [62] in the X-ray photoelectron (ESCA) spectra of $Th(C_2O_4)_2 \cdot 6H_2O$ have been reported.

Thermal dehydration studies by thermogravimetry or differential thermal analysis (see Table 32, p. 79) indicate that $Th(C_2O_4)_2 \cdot 6H_2O$ usually forms the dihydrate (p. 79) and evidence for other hydrates is less certain. Kinetic parameters for the stepwise thermal dehydration and decomposition of the hexahydrate have been evaluated by differential thermal analysis [56]. The dehydration process has also been followed by studies of the emanating power at the solid surface [17], for example using ^{85}Kr and ^{226}Rn [40].

$Th(C_2O_4)_2 \cdot 7H_2O$ has been reported [19] but this is probably the hexahydrate.

The thermal decomposition at 600, 800, and 1000°C of an unspecified Th oxalate (precipitated from aqueous $Th(NO_3)_4$) has been reported in an investigation of the powder characteristics of ThO_2 produced by a variety of methods [68].

References for 15.6.3.2:

[1] Willard, H. H.; Gordon, L. (Anal. Chem. **20** [1948] 165/9).
[2] Haïssinsky, M.; Jeng-Tsong, Yang (Anal. Chim. Acta **3** [1949] 422/7).
[3] Dupuis, T.; Duval, C. (Anal. Chim. Acta **3** [1949] 589/98).
[4] Watt, G. W.; Jenkins, W. A.; McCuiston, J. M. (J. Am. Chem. Soc. **72** [1950] 2260/2).
[5] Beckett, R.; Winfield, M. E. (Australian J. Sci. Res. A **4** [1951] 644/50).
[6] Lipkind, H.; Newton, A. S. (TID-5223 [1952] 411/4; C.A. **1957** 17554).
[7] Burakova, T. N. (Uch. Zap. Leningr. Gos. Univ. Ser. Geol. Nauk No. 4 [1954] 157/95; C.A. **1955** 8730).
[8] D'Eye, R. W. M.; Sellman, P. G. (J. Inorg. Nucl. Chem. **1** [1955] 143/8).
[9] Težak, B. (Proc. 1st Intern. Conf. Peaceful Uses At. Energy, Geneva 1955, Vol. 7, pp. 401/6).
[10] Grinberg, A. A.; Petrzhak, G. I. (Tr. Radievogo Inst. V. G. Khlopina Khim. Geokhim. **7** [1956] 50/73; C.A. **1957** 17352).

[11] Hagiwara, Z. (Kogyo Kagaku Zasshi **59** [1956] 1378/83; C.A. **1959** 1022).
[12] Kawagaki, K. (Nippon Kagaku Zasshi **77** [1956] 1459/61; C.A. **1958** 2658).
[13] Grefig, A. T. (K-1314 [1957] 1/20; C.A. **1957** 18500).
[14] Blundell, R. W. (Brit. 783628 [1957]; C.A. **1958** 3652).
[15] Grinberg, A. A.; Petrzhak, G. I.; Evteev, L. I. (Zh. Neorgan. Khim. **3** [1958] 204/11; Russ. J. Inorg. Chem. **3** No. 1 [1958] 315/26).
[16] Kurup, K. N. N.; Nair, K. V.; Moosath, S. S. (Proc. Indian Acad. Sci. A **47** [1958] 373/8).
[17] Kachi, S. (Funtai Oyobi Funmatsu Yakin **5** [1958] 37/40; C.A. **1959** 2029).
[18] Bobtelsky, M.; Ben-Bassat, A. (Bull. Soc. Chim. France **1958** 233/7).
[19] Srivastava, O. K.; Vasudeva Murthy, M. A. R. (Current Sci. [India] **29** [1960] 470/1).
[20] Padmanabham, V. M.; Saraiya, S. C.; Sundaram, A. K. (J. Inorg. Nucl. Chem. **12** [1960] 356/9).

[21] Korenman, I. M.; Korolikhin, V. V. (Tr. Khim. Khim. Tekhnol. **3** [1960] 106/9; C.A. **56** [1962] 992).
[22] Wendlandt, W. W.; George, T. D.; Horton, C. R. (J. Inorg. Nucl. Chem. **17** [1961] 273/80).
[23] Claudel, B.; Perrin, M.; Trambouze, Y. (Compt. Rend. **252** [1961] 107/9).
[24] Bussière, P.; Claudel, B.; Renouf, J.-P.; Trambouze, Y.; Prettre, M. (J. Chim. Phys. **58** [1961] 668/74).
[25] Luzhnaya, N. P.; Kovaleva, I. S. (Zh. Neorgan. Khim. **6** [1961] 1436/9; Russ. J. Inorg. Chem. **6** [1961] 736/8).
[26] Gaur, P. K.; Reddy, A. S. (J. Sci. Ind. Res. [India] B **21** [1962] 43/4).
[27] Srivastava, O. K.; Vasudeva Murthy, A. R. (J. Sci. Ind. Res. [India] B **21** [1962] 525/7).
[28] Kovaleva, I. S.; Luzhnaya, N. P. (Zh. Neorgan. Khim. **7** [1962] 1693/8; Russ. J. Inorg. Chem. **7** [1962] 873/5).
[29] Badard, A. M.; Pâris, J. M. (Compt. Rend. **257** [1963] 3421/3).
[30] Hsueh, Kwong-Hwa (Hua Hseuh Tung Pao [Huaxue Tongbao] **1963** No. 6, pp. 27/30; N.S.A. **17** [1963] No. 33773).

[31] Jenkins, I. L.; Moore, F. H.; Waterman, M. J. (Chem. Ind. [London] **1963** 35/6).
[32] Moosath, S. S.; Abraham, J.; Swaminathan, T. V. (Z. Anorg. Allgem. Chem. **324** [1963] 103/5).
[33] Bressat, R.; Claudel, B.; Trambouze, Y. (J. Chim. Phys. **60** [1963] 1265/9).
[34] Kurnakova, A. G.; Shubochkin, L. K. (Zh. Neorgan. Khim. **8** [1963] 1249/54; Russ. J. Inorg. Chem. **8** [1963] 647/50).
[35] Dollimore, D.; Griffiths, D. L.; Nicholson, D. (J. Chem. Soc. **1963** 2617/23).
[36] Ghosh-Mazumdar, A. S.; Namboodiri, M. N.; Sharma, D. H. (Proc. 3rd Intern. Conf. Peaceful Uses At. Energy, Geneva 1964, Vol. 10, pp. 286/94).
[37] Dell, R. M.; Wheeler, V. J. (React. Solids 5th Intern. Symp., Munich, FRG, 1964 [1965], pp. 395/408).
[38] Dow Chem. Co. (Brit. 952499 [1964]).
[39] Bennett, W. R.; Kline, C. W. (U.S. 3124603 [1964]; C.A. **61** [1964] 335).
[40] Ekh, C.; Zhabrova, G. M.; Roginskii, S. Z.; Shibanova, M. D. (Dokl. Akad. Nauk SSSR **164** [1965] 1343/6; Dokl. Phys. Chem. Proc. Acad. Sci. USSR **160/165** [1965] 757/60).

[41] Lu, Chao-Ta; Hsu, Shao-Ch'uan (Chung Kuo K'o Hsueh Yuan Ying Yung Hua Hsueh Yen Chiu So Chi K'an [Zhongguo Kexueyuan Yingyong Huaxue Yanjiuso Jikan] No. 14 [1965] 19/24; C.A. **64** [1966] 8981).
[42] Bressat, R. (CEA-2817 [1965] 1/67; C.A. **63** [1965] 17197).
[43] Breysse, M.; Claudel, B.; Trambouze, Y. (Bull. Soc. Chim. France **1965** 201/4).
[44] Brau, G.; Bressat, R.; Claudel, B.; Trambouze, Y.; Urbain, H. (Compt. Rend. **260** [1965] 1981/3).
[45] Jenkins, I. L.; Moore, F. H.; Waterman, M. J. (J. Inorg. Nucl. Chem. **27** [1965] 81/7).
[46] Osinovik, E. S.; Yanchuk, A. F. (Vestn. Akad. Navuk Belarussk.SSR Ser. Khim. Navuk **1966** No. 3, pp. 131/3; C.A. **66** [1967] No. 14389).
[47] Aminov, T. G.; Evdokimov, V. B.; Zelentsov, V. V. (Dokl. Akad. Nauk. SSSR **170** [1966] 615/7; Dokl. Phys. Chem. Proc. Acad. Sci. USSR **166/171** [1966] 607/8).
[48] Grzymalski, Z.; Jezowska-Trzebiatowska, B.; Ziolkowski, J. (Bull. Acad. Polon. Sci. Ser. Sci. Chim. **14** [1966] 381/7).
[49] C. E. A. (Neth. 65-15871 [1965]; C.A. **65** [1966] 12901).
[50] Moskvin, A. I.; Essen, L. N. (Zh. Neorgan. Khim. **12** [1967] 688/93; Russ. J. Inorg. Chem. **12** [1967] 359/62).

[51] Molodkin, A. K.; Balakaeva, T. A.; Kuchumova, A. N. (Zh. Neorgan. Khim. **13** [1968] 2117/23; Russ. J. Inorg. Chem. **13** [1968] 1095/8).
[52] Demarquay, J.; Tho, Pham Quang; Mentzen, B.; Claudel, B. (J. Chim. Phys. **65** [1968] 1380/5).
[53] Kharitonov, Yu. Ya.; Molodkin, A. K.; Balakaeva, T. A. (Zh. Neorgan. Khim. **14** [1969] 339/43; Russ. J. Inorg. Chem. **14** [1969] 174/7).
[54] Petrov, K. I.; Molodkin, A. K.; Saralidze, O. D.; Ivanova, O. M. (Zh. Neorgan. Khim. **14** [1969] 1227/31; Russ. J. Inorg. Chem. **14** [1969] 643/5).
[55] Athavale, V. T.; Kalyanaraman, R.; Sundaresan, M. (Indian J. Chem. **7** [1969] 386/91).
[56] Subramanian, M. S.; Singh, R. N.; Sharma, H. D. (J. Inorg. Nucl. Chem. **31** [1969] 3789/95).
[57] Petrzhak, G. I.; Stepanova, L. M.; Karago, L. V. (Radiokhimiya **12** [1970] 266/72; Soviet Radiochem. **12** [1970] 240/5).
[58] Bykhovskii, D. N.; Petrova, I. K.; Zelentsov, S. S. (Radiokhimiya **14** [1972] 171/7; Soviet Radiochem. **14** [1972] 180/5).
[59] Nefedov, V. I.; Molodkin, A. K.; Salyn', Ya. V.; Ivanova, O. M.; Porai-Koshits, M. A.; Belyakova, Z. V. (Zh. Neorgan. Khim. **19** [1974] 2628/31; Russ. J. Inorg. Chem. **19** [1974] 1435/7).
[60] Moorehead, D. R.; McCartney, E. R. (J. Australian Ceram. Soc. **12** [1976] 27/33).

[61] Allen, G. C.; Tucker, P. M. (Chem. Phys. Letters **43** [1976] 254/7).
[62] Bancroft, G. M.; Sham, T. K.; Esquivel, J. L.; Larsson, S. (Chem. Phys. Letters **51** [1977] 105/10).
[63] Bancroft, G. M.; Sham, T. K.; Larsson, S. (Chem. Phys. Letters **46** [1977] 551/7).
[64] Makarov, L. L.; Karaziya, R.; Batrakov, Yu. F.; Chibisov, N. P.; Mosevich, A. N.; Zaitsev, Yu. M.; Udris, A. I.; Shishkunova, L. V. (Radiokhimiya **20** [1978] 116/24; Soviet Radiochem. **20** [1978] 92/9).
[65] Bobrik, V. M. (Radiokhimiya **21** [1979] 465/9; Soviet Radiochem. **21** [1979] 399/403).
[66] Petrzhak, G. I.; Zelentsov, S. S. (Radiokhimiya **16** [1974] 200/3; Soviet Radiochem. **16** [1974] 201/4).
[67] Pazukhin, E. M.; Smirnova, E. A.; Krivokhatskii, A. S.; Pazukhina, Yu. L.; Kochergin, S. M. (Radiokhimiya **27** [1985] 606/11; Soviet Radiochem. **27** [1985] 567/71).
[68] Moorthy, V. K.; Kulkarni, A. K. (Trans. Indian Ceram. Soc. **22** [1963] 116/29).
[69] Monson, P. R., Jr.; Hall, R. (DP-1576 [1981] 1/15; C.A. **96** [1982] No. 189238).

15.6.3.3 Thorium Oxalato Complexes

The available information on the known complexes is summarised in Table 33.

Table 33
Thorium Oxalato Complexes.

trioxalato complexes	
$(NH_4)_2[Th(C_2O_4)_3]\cdot xH_2O$	x unspecified; see "Thorium" 1955, p. 341
$(CN_3H_6)_2[Th(C_2O_4)_3]\cdot 6H_2O$ (CN_3H_6 = guanidinium)	[27]
$(CN_3H_6)_2[Th(C_2O_4)_3]\cdot 8H_2O$	IR spectrum (4000 to 400 cm^{-1}) illustrated in [12]; ν(CO), ca. 1665 cm^{-1}, ν(CO) + ν(CC), 1443, 1427 cm^{-1}, ν(CO), 1324, 1291 cm^{-1}, ν(CO) + ν(ThO) + ν(OCO), 899 cm^{-1}, δ(OCO) + (ThO), 799, 782 cm^{-1}, ν(ThO), 487 cm^{-1} [12]
$(CN_3H_6)_2[Th(C_2O_4)_3]\cdot xH_2O$	x unspecified; binding energies, $Th5d_{5/2}$ = 88.0 eV, N1s = 400.1 eV [22]
$(C_6H_5CH_2NC_9H_7)H[Th(C_2O_4)_3]$ ($C_6H_5CH_2NC_9H_7$ = benzylquinolinium)	[20]
tetraoxalato complexes	
$Na_4[Th(C_2O_4)_4]\cdot 5.5H_2O$	loses H_2O above 60°C; thermogravimetric and differential thermal analysis curves illustrated in [16] IR spectrum (4000 to 400 cm^{-1}) illustrated in [12]; ν(CO), 1734, 1673, 1647 cm^{-1}, ν(CO) + ν(CC), 1458 cm^{-1}, ν(CO), 1361, 1322, 1296 cm^{-1}, ν(CO) + ν(ThO) + ν(OCO), ca. 899 cm^{-1}, δ(OCO) + ν(ThO), 796 cm^{-1}, ν(ThO) + δ(OCO), ca. 500, ca. 485 cm^{-1} [12]; crystal optical properties; refractive indices, n_p = 1.533, n_g = 1.573 [11] density, d_{obs} = 2.470 g/cm^3 [11]

Table 33 (continued)

$Na_4[Th(C_2O_4)_4]\cdot 6H_2O$	see "Thorium" 1955, p. 327
$K_4[Th(C_2O_4)_4]$	decomposes at 320°C [4]
$K_4[Th(C_2O_4)_4]\cdot 3.38H_2O$	diamagnetic susceptibility, $\chi_m = -235$ to -256×10^{-6} [10]
$K_4[Th(C_2O_4)_4]\cdot 4H_2O$	loses H_2O at >60°C [16], loses $4H_2O$ at 175°C [4]; thermogram [4], thermogravimetric and differential analysis [16] curves illustrated in cited references IR spectrum (4000 to 400 cm^{-1}) illustrated in [12]; ν(CO), ca. 1730, 1670, 1632 cm^{-1}, ν(CO)+ν(CC), 1437 cm^{-1}, ν(CO), 1355, 1318, 1291 cm^{-1}, ν(CO)+ν(ThO)+ν(OCO), 898 cm^{-1}, δ(OCO)+ν(ThO), 798 cm^{-1}, ν(ThO)+δ(OCO), 493, 480 cm^{-1} [12]; ν_{as}(CO), ca. 1720, 1673, 1625 cm^{-1}, ν(CO) + ν(CC), 1433 cm^{-1}, ν_s(CO), 1357, 1319, 1288 cm^{-1}, δ(OCO), 798, 788, 780 cm^{-1} [21]; ν_{as}(CO), B_1, 1640 cm^{-1}, ν(OCO), A_1, 1420 cm^{-1}, ν_s(CO), B_1, 1280 cm^{-1}, δ(OCO), B_1, 785 cm^{-1} [6] X-ray crystallographic data; monoclinic, space group $P2_1/a\text{-}C_{2h}^5$ (No. 14) [6], triclinic, space group $P\bar{1}\text{-}C_i^1$ (No. 2) [13, 23], lattice parameters, a = 9.562(18), b = 13.087(25), c = 10.387(28) Å, α = 115.75(3)°, β = 80.90(3)°, γ = 112.66(3)°; Z = 2; density, d_{calc} = 2.49 g/cm^3 [13, 23], d_{obs} = 2.48 g/cm^3 [6, 8, 13, 23], 2.524 g/cm^3 [11] crystal optical properties; refractive indices, n_g = 1.542 [4], 1.547 [11], n_m = 1.519 [4], 1.523 [11], n_p = 1.511 [4], 1.496 [11] binding energy, $Th5d_{5/2}$ = 87.9 eV [22] see also "Thorium" 1955, p. 335
$K_4[Th(C_2O_4)_4]\cdot 4.18H_2O$	diamagnetic susceptibility, $\chi_m = -224$ to -235×10^{-6} [10]
$K_4[Th(C_2O_4)_4]\cdot 4.21H_2O$	diamagnetic susceptibility, $\chi_m = -261$ to -276×10^{-6} [10]
$K_4[Th(C_2O_4)_4]\cdot 4.42H_2O$	diamagnetic susceptibility, $\chi_m = -248$ to -258×10^{-6} [10]
$(NH_4)_4[Th(C_2O_4)_4]$	see "Thorium" 1955, p. 341
$(NH_4)_4[Th(C_2O_4)_4]\cdot 3H_2O$	[11], IR spectrum (4000 to 400 cm^{-1}) illustrated in [12]; ν(CO), ca. 1645 cm^{-1}, ν(CO)+ν(CC), ca. 1457 cm^{-1}, ν(CO), 1317, 1289 cm^{-1}, ν(CO)+ν(ThO)+ν(OCO), 899 cm^{-1}, δ(OCO)+ν(ThO), 795, 785 cm^{-1}, ν(ThO)+δ(OCO), 488 cm^{-1} [12] crystal optical properties: refractive indices, n_p = 1.438, n_m = 1.545, n_g = 1.594 [11]
$(NH_4)_4[Th(C_2O_4)_4]\cdot 4H_2O$	see "Thorium" 1955, p. 342
$(NH_4)_4[Th(C_2O_4)_4]\cdot 6.5H_2O$	loses $3.5H_2O$ in air [11]

References for 15.6.3.3 on p. 96

Table 33 (continued)

$(NH_4)_4[Th(C_2O_4)_4]\cdot 7H_2O$	see "Thorium" 1955, p. 341
$(NH_4)_4[Th(C_2O_4)_4]\cdot xH_2O$	binding energies, $Th5d_{5/2}=88.0$ eV, $N1s=401.9$ eV [22]
$[NH_2(CH_3)_2]_4[Th(C_2O_4)_4]$	decomposition at 410 to 450°C [15]
$[NH_2(CH_3)_2]_4[Th(C_2O_4)_4]\cdot 2H_2O$	loses $2H_2O$ on heating [15] density, $d_{obs}=1.754$ g/cm³ [15] crystal optical properties; refractive indices, $n_g=1.528$, $n_m=1.514$, $n_p=1.501$ [15] IR spectrum; ν(CO), ca. 1650 cm⁻¹, ν(CO)+ν(CC), ca. 1470 cm⁻¹, ν(CO), 1302 cm⁻¹, ν(CO)+ν(ThO)+δ(OCO), ca. 900 cm⁻¹, δ(OCO)+ν(ThO), ca. 775 cm⁻¹, ν(ThO)+δ(rings)+δ(OCO), ca. 492 cm⁻¹ [19]
$[NH_2(n\text{-}C_3H_7)_2]_4[Th(C_2O_4)_4]\cdot 2H_2O$	effloresces in air [15]
$[NH_2(n\text{-}C_4H_9)_2]_4[Th(C_2O_4)_4]$	decomposes at 410 to 450°C [15]
$[NH_2(n\text{-}C_4H_9)_2]_4[Th(C_2O_4)_4]\cdot 4H_2O$	loses $4H_2O$ on heating [15] crystal optical properties; refractive indices, $n_g=1.492$, $n_p=1.471$ [15]
$[C_2H_4(NH_3)_2]_4[Th(C_2O_4)_4]\cdot 2.5H_2O$	probably the $[C_2H_4(NH_3)_2]_2^{2+}$ salt IR spectrum; ν(CO), ca. 1660 cm⁻¹, ν(CO)+ν(CC), ca. 1455 cm⁻¹, ν(CO), 1311, ca. 1300 cm⁻¹, ν(CO)+ν(ThO)+δ(OCO), 906 cm⁻¹, δ(OCO)+ν(ThO), ca. 800, 785 cm⁻¹, ν(ThO)+δ(rings)+δ(OCO), ca. 493, ca. 475 cm⁻¹ [19]
$(CN_3H_6)_4[Th(C_2O_4)_4]\cdot 2H_2O$	density, d_{obs} (25°C) = 2.064 g/cm³ [11] crystal optical properties; refractive indices, $n_g=1.638$, $n_m=1.629$, $n_p=1.505$ [11] IR spectrum (4000 to 400 cm⁻¹) illustrated in [12]; ν(CO), ca. 1670 cm⁻¹, ν(CO)+ν(CC), 1445, 1425 cm⁻¹, ν(CO), 1360, 1319, 1288, ca. 1279 cm⁻¹, ν(CO)+ν(ThO)+ν(OCO), 898 cm⁻¹, δ(OCO)+ν(ThO), 799, ca. 786, 782 cm⁻¹, ν(ThO)+δ(OCO), 487 cm⁻¹ [12] binding energies, $Th5d_{5/2}=87.7$ eV, $N1s=400.2$ eV [22]
$(CN_3H_6)_4[Th(C_2O_4)_4]\cdot 2.06H_2O$	diamagnetic susceptibility, $\chi_m=-289$ to -299×10^{-6} [10]
$(CN_3H_6)_4[Th(C_2O_4)_4]\cdot 2.14H_2O$	diamagnetic susceptibility, $\chi_m=-280$ to -288×10^{-6} [10]
$(CN_3H_6)_4[Th(C_2O_4)_4]\cdot 2.2H_2O$	diamagnetic susceptibility, $\chi_m=-294$ to -310×10^{-6} [10]
$(CN_3H_6)_3(NH_4)[Th(C_2O_4)_4]\cdot 3H_2O$	[11]
$Ba_2[Th(C_2O_4)_4]\cdot 11H_2O$	thermogram illustrated in [18] IR spectrum; ν_{as}(CO), ca. 1648 cm⁻¹, ν(CO)+ν(CC),

Table 33 (continued)

	1448 cm^{-1}, ν(CO) + δ(OCO), 1357, 1300 cm^{-1}, ν(CO) + δ(OCO) + ν(CC), 901 cm^{-1}, δ(OCO) + ν(ThO), 804, 787 cm^{-1}, δ(OCO)(+ν(ThO)), 480 cm^{-1} [18]
$[Co(en)_3]_4[Th(C_2O_4)_4]_3 \cdot 22H_2O$ (en = $H_2NCH_2CH_2NH_2$, ethylenediamine)	light yellow [25]
$[Co(tn)_3]_4[Th(C_2O_4)_4]_3 \cdot 3H_2O$ (tn = $H_2NCH_2CH_2CH_2NH_2$)	[25, 26]
pentaoxalato complexes	
$(NH_4)_6[Th(C_2O_4)_5] \cdot H_2O$	binding energies, Th $5d_{5/2}$ = 88.0 eV, N1s = 401.7 eV [22]
$(NH_4)_6[Th(C_2O_4)_5] \cdot 3H_2O$	density, d_{obs} (25°C) = 2.025 g/cm^3 [11] crystal optical properties; refractive indices, n_g = 1.578, n_m = 1.952 (probably 1.592), n_p = 1.449 [11] IR spectrum (4000 to 400 cm^{-1}) illustrated in [12]; ν(CO), ca. 1655 cm^{-1}, ν(CO) + ν(CC), 1430 cm^{-1}, ν(CO), 1351, 1318, 1306 cm^{-1}, ν(CO) + ν(ThO) + ν(OCO), 898 cm^{-1}, δ(OCO) + ν(ThO), 795, 786 cm^{-1}, ν(ThO) + δ(OCO), 487 cm^{-1} [12]
$(NH_4)_6[Th(C_2O_4)_5] \cdot 7.5H_2O$	loses $4.5H_2O$ in air [11]
$[Cr(NH_3)_6]_2[Th(C_2O_4)_5] \cdot 20H_2O$	yellow [25]
$[Cr(ur)_6]_2[Th(C_2O_4)_5]$ (ur = urea, $OC(NH_2)_2$)	decomposition to mixed oxides at ca. 400°C [17]
$[Cr(ur)_6]_2[Th(C_2O_4)_5] \cdot 5H_2O$	[26], greyish green; loses H_2O above 50°C, thermogravimetric analysis curve illustrated in [17] X-ray powder pattern reported in [17]
$[Co(NH_3)_6]_2[Th(C_2O_4)_5] \cdot 3H_2O$	orange [11] IR spectrum (4000 to 400 cm^{-1}) illustrated in [12]; ν(CO), ca. 1670 cm^{-1}, ν(CO) + ν(CC), 1440 cm^{-1}, ν(CO), 1297 cm^{-1}, ν(CO) + ν(ThO) + ν(OCO), ca. 896, 890 cm^{-1}, δ(OCO) + ν(ThO), ca. 795, 782, 776 cm^{-1}, ν(ThO) + δ(rings) + δ(OCO), 493 cm^{-1} [12] diamagnetic susceptibility, χ_m = −257 to −291 × 10^{-6} [10]; binding energies, Th $5d_{5/2}$ = 88.0 eV, N1s = 400.4 eV [22]
$[Co(NH_3)_6]_2[Th(C_2O_4)_5] \cdot 3.64H_2O$	diamagnetic susceptibility, χ_m = −238 to −280 × 10^{-6} [10]
pentaoxalatodithorium complexes	
$(H_3O)_2[Th_2(C_2O_4)_5] \cdot 5H_2O$	IR spectrum illustrated in [24] ν_{as}(CO) ca. 1665, 1650 cm^{-1}, ν_s(CO), ca. 1360, 1325 cm^{-1}, δ(OCO) = 810 cm^{-1} [24] binding energy, Th $5d_{5/2}$ = 88.7 eV [22]

References for 15.6.3.3 on p. 96

Table 33 (continued)

$H_2[Th_2(C_2O_4)_5]\cdot 9H_2O$	[11]; IR spectrum illustrated in [21]; $\nu_{as}(CO)$, ca. 1640 cm^{-1}, $\nu_s(CO)$, 1353, 1319 cm^{-1}, $\delta(OCO)$, 805 cm^{-1} [21] see also "Thorium" 1955, p. 308
$(NH_4)_2[Th_2(C_2O_4)_5]\cdot 4H_2O$	[9]
$(NH_4)_2[Th_2(C_2O_4)_5]\cdot 7H_2O$	loses $3H_2O$ at ca. 200°C [9] X-ray crystallographic data; orthorhombic, lattice parameters, a = 11.07, b = 9.73, c = 8.68 Å; Z = 1, density, d_{obs} = 2.04 g/cm^3 [9] see also "Thorium" 1955, p. 341
$[NH_3(n\text{-}C_4H_9)]_2[Th_2(C_2O_4)_5]\cdot 10NH_2(n\text{-}C_4H_9)$	see Th E, pp. 5/6
$[NH_2(n\text{-}C_3H_7)_2]_2[Th_2(C_2O_4)_5]\cdot NH(n\text{-}C_3H_7)_2$	see Th E, pp. 5, 7
$[NH_2(n\text{-}C_4H_9)_2]_2[Th_2(C_2O_4)_5]\cdot NH(n\text{-}C_4H_9)_2 \cdot 2H_2O$	see Th E, p. 5
other oxalato complexes	
$[NH_2(n\text{-}C_4H_9)_2]_8[Th(C_2O_4)_6]$	decomposition at 410 to 450°C [15] IR spectrum; $\nu(CO)$, 1718, ca. 1660 cm^{-1}, $\nu(CO)+\nu(CC)$, ca. 1467(?) cm^{-1}, $\nu(CO)$, ca. 1280 cm^{-1}, $\nu(CO)+\nu(ThO)+\delta(OCO)$, 889 cm^{-1}, $\delta(OCO)+\nu(ThO)$, 785, 780 cm^{-1}, $\nu(ThO)+\delta(rings)+\delta(OCO)$, 487 cm^{-1} [19] crystal optical properties; refractive indices, n_g = 1.516, n_m = 1.499, n_p = 1.483 [15]
$[NH(C_2H_5)_3]_8[Th(C_2O_4)_6]$	decomposes at 410 to 450°C [15]
$[NH(C_2H_5)_3]_8[Th(C_2O_4)_6]\cdot 3H_2O$	loses $3H_2O$ in air [15]
$[Co(en)_3]_8[Th(C_2O_4)_6]_3\cdot 42H_2O$ (en = $H_2NCH_2CH_2NH_2$)	light yellow [25, 26]
$[NH_2(C_2H_5)_2]_6[Th_2(C_2O_4)_7]$	decomposes at 410 to 450°C [15]
$[NH_2(C_2H_5)_2]_6[Th_2(C_2O_4)_7]\cdot 6H_2O$	loses $6H_2O$ on heating [15] IR spectrum; $\nu(CO)$, ca. 1680, ca. 1645 cm^{-1}, $\nu(CO)+\nu(CC)$, ca. 1460 cm^{-1}, $\nu(CO)$, 1278 cm^{-1}, $\nu(CO)+\nu(ThO)+\delta(OCO)$, 900 cm^{-1}, $\delta(OCO)+\nu(ThO)$, 789 cm^{-1}, $\nu(ThO)+\delta(rings)+\delta(OCO)$, ca. 490 cm^{-1} [19] density, d_{obs} = 1.587 g/cm^3 [15] crystal optical properties; refractive indices, n_g = 1.504, n_m = 1.484; n_p = 1.468 [15]
$[NH_2(n\text{-}C_3H_7)_2]_6[Th_2(C_2O_4)_7]$	decomposes at 410 to 450°C [15]
$[NH_2(n\text{-}C_3H_7)_2]_6[Th_2(C_2O_4)_7]\cdot 8H_2O$	loses $8H_2O$ on heating [15] IR spectrum; $\nu(CO)$, ca. 1666, ca. 1640 cm^{-1}, $\nu(CO)+\nu(CC)$, ca. 1465(?) cm^{-1}, $\nu(CO)$, ca. 1280 cm^{-1}, $\nu(CO)+\nu(ThO)+\delta(OCO)$, 895 cm^{-1}, $\delta(OCO)+$

Table 33 (continued)

	ν(ThO), 785 cm^{-1}, ν(ThO) + δ(rings) + δ(OCO), ca. 490 cm^{-1} [19] density, d_{obs} = 1.625 g/cm^3 [15] crystal optical properties; refractive indices, n_g = 1.546, n_p = 1.519 [15]
$(CN_3H_6)_6[Th_2(C_2O_4)_7] \cdot x H_2O$	x = 5, 8 or 12.5 to 13.7 [27]
$Mn_{0.67}Th(C_2O_4)_{2.67} \cdot 6 H_2O$	(≡ $Mn_2Th_3(C_2O_4)_8 \cdot 18 H_2O$); thermogram illustrated in [18] IR spectrum; ν_{as}(CO), ca. 1650 cm^{-1}, ν(CO) + ν(CC), 1464 cm^{-1}, ν(CO) + δ(OCO), 1355, 1316 cm^{-1}, ν(CO) + δ(OCO) + ν(CC), 900 cm^{-1}, δ(OCO) + ν(ThO), 810, 798 cm^{-1}, δ(OCO)(+ ν(ThO)), 478 cm^{-1} [18]
$Cd_{0.72}Th(C_2O_4)_{2.72} \cdot 6 H_2O$	(≡ $Cd_2Th_3(C_2O_4)_8 \cdot 18 H_2O$); thermogram illustrated in [18] IR spectrum; ν_{as}(CO), ca. 1638 cm^{-1}, ν(CO) + ν(CC), 1457 cm^{-1}, ν(CO) + δ(OCO), 1355, 1316 cm^{-1}, ν(CO) + δ(OCO) + ν(CC), 900 cm^{-1}, δ(OCO) + ν(ThO), 810, 798 cm^{-1}, δ(OCO)(+ ν(ThO)), 480 cm^{-1} [18]
$Ca_{0.4}Th(C_2O_4)_{2.4} \cdot 7 H_2O$	(≡ $Ca_2Th_5(C_2O_4)_{12} \cdot 35 H_2O$); thermogram illustrated in [18] IR spectrum; ν_{as}(CO), ca. 1640 cm^{-1}, ν(CO) + ν(CC), 1454 cm^{-1}, ν(CO) + δ(OCO), 1352, 1310 cm^{-1}, ν(CO) + δ(OCO) + ν(CC), 900 cm^{-1}, δ(OCO) + ν(ThO), 800 cm^{-1}, δ(OCO)(+ ν(ThO)), 482 cm^{-1} [18]
$(CN_3H_6)_8(NH_4)_{14}Th_2(C_2O_4)_{15} \cdot 9 H_2O$	density, d_{obs} (25°C) = 1.83 g/cm^3 [11] crystal optical properties; refractive indices, n_g = 1.655, n_m = 1.566, n_p = 1.414 [11]

15.6.3.3.1 Introduction

The IR spectra of a number of oxalato complex salts have been reported (see Table 33, pp. 86/91). The spectra of salts of the $[Th(C_2O_4)_3]^{2-}$ [12], $[Th(C_2O_4)_4]^{4-}$ [12, 19, 21], $[Th(C_2O_4)_5]^{6-}$ [12], $[Th(C_2O_4)_6]^{8-}$ [19], and $[Th_2(C_2O_4)_7]^{6-}$ [19] anions are consistent with the presence of bidentate coordinated C_2O_4 groups. However, in the only known structure ($K_4[Th(C_2O_4)_4] \cdot 4 H_2O$ [13, 23]) the C_2O_4 groups are bridging and are quadridentate, a result which is also consistent with the IR spectrum of the salt. The IR spectra of $H_2Th_2(C_2O_4)_5 \cdot 9 H_2O$ [21] and $(H_3O)_2Th_2(C_2O_4)_5 \cdot 5 H_2O$ [24] are very similar to those of $Th(C_2O_4)_2 \cdot 6 H_2O$ (see Tables 32, p. 80 and 33, pp. 89/90), and in these compounds the bonding of the C_2O_4 groups is considered to be predominantly ionic. It would be useful to have more structural information on these compounds in order to confirm the conclusions derived from the IR spectra.

 References for 15.6.3.3 and 15.6.3.3.1 to 15.6.3.3.6 on p. 96

15.6.3.3.2 Trioxalatothorates

The presence of the free acid $H_2[Th(C_2O_4)_3]$ has been detected in aqueous solutions of $Th(C_2O_4)_2 \cdot 6H_2O$ [14], but the acid has not been isolated. Although IR [12] and electronic binding energy data (derived from X-ray electronic spectra) [22] have been reported for hydrated guanidinium salts of this acid, preparative details for the salts were not given. However, $(CN_3H_6)_2[Th(C_2O_4)_3] \cdot 6H_2O$ is reported to be formed together with $(CN_3H_6)_6Th_2(C_2O_4)_7 \cdot 5$ (or 8) H_2O (p. 95) when freshly precipitated $(CN_3H_6)_6[Th(CO_3)_5] \cdot 4H_2O$ (p. 10) is dissolved in saturated aqueous $(NH_4)_2C_2O_4$ with the molar ratio $Th:C_2O_4^{2-}:CO_3^{2-}$ = 1:5:1 [27]. The benzylquinolinium salt, $(C_6H_5CH_2NC_9H_7)H[Th(C_2O_4)_3]$, is precipitated when an aqueous suspension of $Th(C_2O_4)_2 \cdot 6H_2O$ (p. 82), $H_2C_2O_4$ and $(C_6H_5CH_2NC_9H_7)Cl$ is heated on a water bath. The salt has an acid reaction in water [20].

15.6.3.3.3 Tetraoxalatothorates

$Na_4[Th(C_2O_4)_4] \cdot 5.5H_2O$ (the observed H_2O content was 5 to $6H_2O$) has been prepared by heating a suspension of $Th(C_2O_4)_2 \cdot 6H_2O$ (p. 82) in aqueous $Na_2C_2O_4$ until dissolution is complete, and then leaving the solution to crystallise [11]. X-ray powder diffraction data for this salt, and for the products obtained by heating it to 210°C and to 340°C, have been reported [16]. Thermogravimetric and differential thermal analysis indicate that although all of the H_2O molecules are outer sphere, the type and strength of bonding are not the same for all the H_2O molecules. The IR spectra in the C_2O_4 group regions remain unchanged in the 3 phases observed on heating, indicating that decomposition of the C_2O_4 group has not occurred. Dehydration commences at 60°C, and there are endotherms at 120 and 210°C; the X-ray powder diffraction diagram for the product at 210°C is the same as that of the initial hydrate. An exotherm at 300 to 400°C was ascribed to a structural reorganisation of the lattice [16].

The crystallisation boundary of $Na_4[Th(C_2O_4)_4] \cdot 6H_2O$ has been determined in studies of the solubility isotherm of the system $Th(C_2O_4)_2 + 2Na_2CO_3 \rightleftharpoons Th(CO_3)_2 + 2Na_2C_2O_4$ in H_2O at 25°C [5].

Anhydrous $K_4[Th(C_2O_4)_4]$ is obtained by heating the tetrahydrate at 175°C [4]; dehydration commences at 60°C [16]. $K_4[Th(C_2O_4)_4] \cdot 4H_2O$ is one of the solid phases observed in the system $K_2C_2O_4$–$Th(C_2O_4)_2$–H_2O [4]. It has been prepared by dissolving hydrated $Th(C_2O_4)_2$ in an excess of aqueous $K_2C_2O_4$, then heating the mixture to boiling and finally precipitating the salt from the cooled solution by adding absolute methanol [11] or absolute ethanol; in the latter case care must be taken to avoid precipitation of $K_2C_2O_4$ by using too much ethanol [6, 8] and it is best to add ethanol dropwise to the solution until crystallisation begins [23]. The salt also crystallises from the aqueous reaction mixture on standing without addition of an alcohol [11].

The Th atoms in the $[Th(C_2O_4)_4]^{4-}$ anion of the K salt are surrounded by ten O atoms at the vertices of a slightly irregular bicapped square antiprism in an oxalate bridged structure (**Fig. 6**) which is cross-linked by hydrogen bonding. None of the C_2O_4 groups spans any of the square edges of the antiprism, which are longer (3.11 ± 0.20 Å) than either the 8 pyramid edges (2.67 ± 0.06 Å) or the 8 equatorial edges (2.76 ± 0.17 Å). The C_2O_4 groups span positions 1 to 2, 3 to 6, 4 to 7, 5 to 9, and 8 to 10 of the polyhedron, producing a chiral structure. However, equal numbers of right- and left-handed polyhedra occur in the crystal. Adjacent polyhedra are bridged by the C_2O_4 groups at 1 to 2 and 8 to 10 (those containing the apical O atoms), producing chains parallel to [110], [13, 23]. Bond lengths and principal angles in $[Th(C_2O_4)_4]^{4-}$ (Fig. 6) are shown in Table 34.

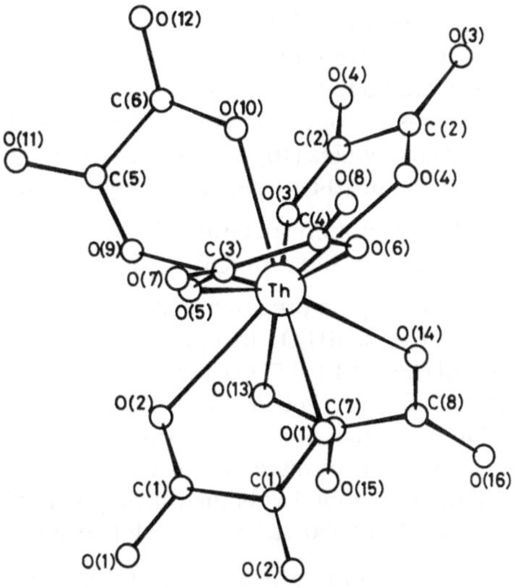

Fig. 6. The structure of $K_4[Th(C_2O_4)_4]\cdot 4H_2O$ showing five oxalate groups coordinated to one Th atom [23].

Table 34
Bond Lengths (in Å) and Principal Angles (in °) of $[Th(C_2O_4)_4]^{4-}$ in $K_4[Th(C_2O_4)_4]\cdot 4H_2O$ [23].

distances		angles at square edges		angles at pyramid edges	
Th–O(1)	2.44(2)	O(1)–Th–O(5)	75.1(7)	O(2)–Th–O(1)	62.3(7)
Th–O(2)	2.62(2)	O(5)–Th–O(9)	74.0(7)	O(2)–Th–O(5)	62.2(7)
Th–O(3)	2.49(2)	O(9)–Th–O(13)	79.8(6)	O(2)–Th–O(9)	64.0(6)
Th–O(4)	2.57(2)	O(13)–Th–O(1)	83.5(6)	O(2)–Th–O(13)	63.5(7)
Th–O(5)	2.47(2)	O(3)–Th–O(14)	78.2(6)	O(4)–Th–O(3)	63.4(6)
Th–O(6)	2.47(2)	O(14)–Th–O(6)	91.0(6)	O(4)–Th–O(14)	65.0(6)
Th–O(9)	2.41(2)	O(6)–Th–O(10)	77.9(6)	O(4)–Th–O(6)	65.7(6)
Th–O(10)	2.48(2)	O(10)–Th–O(3)	71.6(6)	O(4)–Th–O(10)	65.9(6)
Th–O(13)	2.48(2)			O(2)–Th–O(4)	174.7(6)
Th–O(14)	2.49(2)				

angles at equatorial edges		means of distances		means of O–Th–O angles	
O(1)–Th–O(6)	71.1(6)	Th–O (axial)	2.595	square adges	78.87
O(6)–Th–O(5)	64.5(7)	Th–O (other)	2.467	pyramid edges	64.00
O(5)–Th–O(10)	75.8(7)			equatorial edges	68.12
O(10)–Th–O(9)	63.7(6)			chelate edges	63.48
O(9)–Th–O(3)	71.2(6)			all polyhedron edges	70.33
O(3)–Th–O(13)	66.2(6)				
O(13)–Th–O(14)	63.6(6)				
O(14)–Th–O(1)	68.9(6)				

The IR spectrum of the salt is also consistent with the presence of C_2O_4 groups which are at least bidentate [21]. The structure of the $[Th(C_2O_4)_4]^{4-}$ anion is reported [6] to be the same as that of the $[Zr(C_2O_4)_4]^{4-}$ anion [7]. X-ray powder diffraction data for $K_4[Th(C_2O_4)_4]\cdot 4H_2O$ have also been reported [6].

References for 15.6.3.3.1 to 15.6.3.3.6 on p. 96

In a thermogravimetric and differential thermal analysis study of $K_4[Th(C_2O_4)_4]\cdot 4H_2O$ endotherms were observed at 135 and 220°C, followed by an exotherm at 300 to 400°C which was ascribed to a structural reorganisation of the lattice. The results confirm that all of the H_2O molecules are outer sphere, and evidently the type and strength of bonding is not the same for all H_2O molecules [16], differences which may be related to the hydrogen bonding which links the chains of 10-coordinate polyhedra together in the [100], [111], and [1$\bar{1}$0] directions [13].

The absorption spectrum of a Th complex oxalate, apparently $K_4[Th(C_2O_4)_4]$, is reported to be the same as those of $K_4[Zr(C_2O_4)_4]$ and $K_2[Be(C_2O_4)_2]$, with an absorbance at ca. 250 nm [1].

An attempted preparation of ammonium oxalatothorates by slow evaporation of solutions of $Th(C_2O_4)_2\cdot 6H_2O$ (p. 82) in aqueous $(NH_4)_2C_2O_4$ was unsuccessful [3]. However, $(NH_4)_4[Th(C_2O_4)_4]\cdot 6.5H_2O$ has been obtained by heating a mixture of $Th(C_2O_4)_2\cdot 6H_2O$ and $(NH_4)_2C_2O_4\cdot H_2O$ (molar ratio = 1:3) with H_2O until dissolution was complete; after cooling the solution, sufficient H_2O was added to redissolve any precipitated $(NH_4)_2C_2O_4\cdot H_2O$ and then an equal volume of methanol or ethanol was added. The initial precipitate was discarded and ethanol was added until the onset of crystallisation. The salt behaves as a 5-ion electrolyte in H_2O and the solid salt loses water to form the trihydrate in air [11].

The substituted ammonium salts, $[NH_2R_2]_4[Th(C_2O_4)_4]\cdot xH_2O$ (R = CH_3, n-C_3H_7 and x = 2; R = n-C_4H_9 and x = 4), have been prepared by heating $Th(C_2O_4)_2\cdot 6H_2O$ with a slight excess of the amine and the stoichiometric quantity of $H_2C_2O_4$ in water until dissolution was complete, after which the solution was evaporated to half volume, and the salts then crystallised. These products dehydrate in air (effloresce) or on heating. The di(n-butyl)ammonium salt is soluble in alcohols [15]. The salt reported [19] as $[C_2H_4(NH_3)_2]_4[Th(C_2O_4)_4]\cdot 2.5H_2O$ is probably the $[C_2H_4(NH_3)_2]_2^{2+}$ salt. Its IR spectrum, like that of $[NH_2(CH_3)_2]_4[Th(C_2O_4)_4]\cdot 9H_2O$, is consistent with the presence of bidentate coordinated C_2O_4 groups [19]. The corresponding guanidinium salt, $(CN_3H_6)_4[Th(C_2O_4)_4]\cdot 2H_2O$, is precipitated when the calculated quantity of $(CN_3H_6)NO_3$, either solid or as a saturated aqueous solution, is added to the solution obtained by heating $Th(C_2O_4)_2\cdot 6H_2O$ with a saturated aqueous solution of $(NH_4)_2C_2O_4\cdot H_2O$ (molar ratio = 1:2). The salt is also precipitated when the latter solution contains a 1:6 molar ratio of $Th(C_2O_4)_2\cdot 6H_2O$: $(NH_4)_2C_2O_4\cdot H_2O$, and evaporation of the mother liquor to half volume then yielded a salt of composition $(CN_3H_6)_3(NH_4)[Th(C_2O_4)_4]\cdot 3H_2O$ [11].

The barium salt, $Ba_2[Th(C_2O_4)_4]\cdot 11H_2O$, is precipitated when $Th(C_2O_4)_2$ is treated with an aqueous Ba^{2+} solution [18]. $[Co(en)_3]_4[Th(C_2O_4)_4]_3\cdot 22H_2O$ (en = $H_2NCH_2CH_2NH_2$) is precipitated from solutions of Th^{IV} in aqueous $(NH_4)_2C_2O_4$ or $K_2C_2O_4$ at pH 3 to 4 on addition of $[Co(en)_3]Cl_3$ [25]. The analogous $[Co(tn)_3]_4[Th(C_2O_4)_4]_3\cdot 3H_2O$ (tn = $H_2NCH_2CH_2CH_2NH_2$) is precipitated in a similar manner [25, 26] using $[Co(tn)_3]Cl_3$. The effect of the concentration of NaCl or NH_4Cl on the solubility of this salt in water has been investigated [26].

15.6.3.3.4 Pentaoxalatothorates

$(NH_4)_6[Th(C_2O_4)_5]\cdot 7.5H_2O$ is prepared by treating a mixture of freshly precipitated thorium hydroxide and solid $(NH_4)_2C_2O_4\cdot H_2O$ (molar ratio = 1:5) with water, then stirring until no further dissolution occurs, followed by addition of methanol or ethanol to the filtrate. The salt loses water in air to form the trihydrate [11]. This hydrate has also been obtained from the mother liquor from the preparation of $(NH_4)_4[Th(C_2O_4)_4]\cdot 6.5H_2O$ (see above), which first deposits $(NH_4)_2C_2O_4$ on standing and then, after filtration, the trihydrated complex salt separates on further standing [11].

$[Cr(ur)_6]_2[Th(C_2O_4)_5]\cdot 5H_2O$ is precipitated when aqueous $[Cr(ur)_6]Cl_3$ is added to a solution of Th^{IV} in aqueous $(NH_4)_2C_2O_4$ at pH 3 to 4. It loses H_2O above ca. 50°C [17].

$[Co(NH_3)_6]_2[Th(C_2O_4)_5] \cdot 3H_2O$ precipitates when aqueous $[Co(NH_3)_6]Cl_3$ (molar ratio Th:Co = 1:2) is added to an aqueous solution made up from solid $Th(C_2O_4)_2 \cdot 6H_2O$ (p. 82) and $(NH_4)_2C_2O_4 \cdot H_2O$ (molar ratio = 1:3) [11]. The analogous $[Cr(NH_3)_6]_2[Th(C_2O_4)_5] \cdot 20H_2O$ [25] and $[Cr(ur)_6]_2[Th(C_2O_4)_5] \cdot 5H_2O$ [26] are obtained in the same way as $[Co(en)_3]_4[Th(C_2O_4)_4]_3 \cdot 22H_2O$ (p. 94).

15.6.3.3.5 Pentaoxalatodithorium Complexes

The acid compound, $H_2[Th_2(C_2O_4)_5] \cdot 9H_2O$, separates when $H_2C_2O_4 \cdot 2H_2O$ is added to an aqueous solution of $Th(NO_3)_4 \cdot 5H_2O$ (molar ratio = 5:1) [11]. Its IR spectrum is very similar to that of $Th(C_2O_4)_2 \cdot 6H_2O$ [21], as is the IR spectrum of $(H_3O)_2[Th_2(C_2O_4)_5] \cdot 5H_2O$ [24], which is presumably the same compound.

$(NH_4)_2[Th_2(C_2O_4)_5] \cdot 7H_2O$ is mentioned in [2] as a phase in the $Th(C_2O_4)_2 \cdot 6H_2O$–$(NH_4)_2C_2O_4$–H_2O system. The salt crystallises from a diluted aqueous solution of $Th(C_2O_4)_2$ in $(NH_4)_2C_2O_4$ on standing [9]. It loses water to form an unstable tetrahydrate at ca. 200°C, and decomposes at 260°C to a product which, after rehydration, has the composition $1.69\,Th(C_2O_4)_2 \cdot 0.85(NH_4)_2C_2O_4 \cdot 0.31ThO_2 \cdot 1.22H_2O$ [9]; this may be a basic oxalato complex.

15.6.3.3.6 Other Oxalato Complexes

A few hexaoxalato complex salts have been recorded. $[NH_2(n\text{-}C_4H_9)_2]_8[Th(C_2O_4)_6]$ and $[NH(C_2H_5)_3]_8[Th(C_2O_4)_6] \cdot 3H_2O$ have been prepared in the same way as $[NH_2(CH_3)_2]_4[Th(C_2O_4)_4] \cdot 2H_2O$ (p. 94). They are soluble in water and in some alcohols. The anhydrous triethylammonium salt is obtained by heating the trihydrate or leaving it to stand in air [15]. The IR spectrum of the *n*-butylammonium salt is consistent with the presence of bidentate coordinated C_2O_4 groups [19]. $[Co(en)_3]_8[Th(C_2O_4)_6]_3 \cdot 42H_2O$ has been obtained in the same way as $[Co(en)_3]_4[Th(C_2O_4)_4]_3 \cdot 22H_2O$ (p. 94) [25]. The effect of the concentration of NaCl or NH_4Cl in the solution on the solubility of the compound has been studied [26].

A few heptaoxalatodithorate complexes are also known. $[NH_2(C_2H_5)_2]_6[Th_2(C_2O_4)_7] \cdot 6H_2O$ and $[NH_2(n\text{-}C_3H_7)_2]_6[Th_2(C_2O_4)_7] \cdot 8H_2O$ have been prepared in the same way as $[NH_2(CH_3)_2]_4[Th(C_2O_4)_4] \cdot 2H_2O$ (p. 94). Both salts dehydrate when heated [15]. Their IR spectra are consistent with the presence of bidentate coordinated C_2O_4 groups [19]. The guanidinium salt, $(CN_3H_6)_6[Th_2(C_2O_4)_7] \cdot 5$(or 8)$H_2O$, is formed in the preparation of the trioxalato complex salt, $(CN_3H_6)_2Th(C_2O_4)_3 \cdot 6H_2O$ (p. 92), by treating freshly precipitated $(CN_3H_6)_6[Th(CO_3)_5] \cdot 4H_2O$ (p. 10) with saturated aqueous $(NH_4)_2C_2O_4$ with the molar ratio Th:$C_2O_4^{2-}$:CO_3^{2-} = 1:5:1. With this molar ratio equal to 1:5:2, the product is $(CN_3H_6)_6[Th_2(C_2O_4)_7] \cdot 12.5$ to $13.7H_2O$. Mixed carbonato oxalato compounds (p. 101) are obtained at molar ratios of 1:5:3 to 1:5:5 [27].

A product of composition $(CN_3H_6)_8(NH_4)_{14}Th_2(C_2O_4)_{15} \cdot 9H_2O$ (CN_3H_6 = guanidinium) precipitates when $(CN_3H_6)NO_3$ is added to a solution of $Th(C_2O_4)_2 \cdot 6H_2O$ (p. 82) in saturated aqueous $(NH_4)_2C_2O_4$ (molar ratio = 1:14.4) at 50°C [11]. The nature of this product is uncertain.

Although $Ba_2[Th(C_2O_4)_4] \cdot 11H_2O$ (p. 94) is readily obtained, the analogous reactions with Ca^{2+}, Mn^{2+}, and Cd^{2+} in aqueous solution yield precipitates of composition $Ca_{0.4}Th(C_2O_4)_{2.4} \cdot 7H_2O$ ($\equiv Ca_2Th_5(C_2O_4)_{12} \cdot 35H_2O$), $Mn_{0.67}Th(C_2O_4)_{2.67} \cdot 6H_2O$ ($\equiv Mn_2Th_3(C_2O_4)_8 \cdot 18H_2O$), and $Cd_{0.72}Th(C_2O_4)_{2.72} \cdot 6H_2O$ ($\equiv Cd_2Th_3(C_2O_4)_8 \cdot 18H_2O$) [18]. These systems require further investigation.

References for 15.6.3.3.1 to 15.6.3.3.6 on p. 96

References for 15.6.3.3 and 15.6.3.3.1 to 15.6.3.3.6:

[1] Graddon, D. P. (Chem. Ind. [London] **1956** 80/1).
[2] Grefig, A. T. (K-1314 [1957] 1/20; C.A. **1957** 18500).
[3] Kurup, K. N. N.; Nair, K. V.; Moosath, S. S. (Proc. Indian Acad. Sci. A **47** [1958] 373/8).
[4] Luzhnaya, N. P.; Kovaleva, I. S. (Zh. Neorgan. Khim. **6** [1961] 1436/9; Russ. J. Inorg. Chem. **6** [1961] 736/8).
[5] Kovaleva, I. S.; Luzhnaya, N. P. (Zh. Neorgan. Khim. **7** [1962] 1693/8; Russ. J. Inorg. Chem. **7** [1962] 873/5).
[6] Johnson, F. A.; Larsen, E. M. (Inorg. Chem. **1** [1962] 159/65).
[7] Glen, G. L.; Silverton, J. V.; Hoard, J. L. (Inorg. Chem. **2** [1963] 250/6).
[8] Johnson, F. A.; Larsen, E. M. (Inorg. Syn. **8** [1966] 40/4).
[9] Bressat, R.; Trambouze, Y.; Claudel, B.; Lang, G.; Navarro, A. (Bull. Soc. Chim. France **1966** 2094/9).
[10] Belova, V. I.; Syrkin, Ya. K.; Molodkin, A. K.; Ivanova, O. M.; Shiporina, L. M. (Zh. Neorgan. Khim. **13** [1968] 1458/60; Russ. J. Inorg. Chem. **13** [1968] 766/7).

[11] Molodkin, A. K.; Balakaeva, T. A.; Kuchumova, A. N. (Zh. Neorgan. Khim. **13** [1968] 2117/23; Russ. J. Inorg. Chem. **13** [1968] 1095/8).
[12] Kharitonov, Yu. Ya.; Molodkin, A. K.; Balakaeva, T. A. (Zh. Neorgan. Khim. **13** [1968] 2315/8; Russ. J. Inorg. Chem. **13** [1968] 1196/8).
[13] Akhtar, M. N.; Smith, A. J. (Chem. Commun. **1969** 705/6).
[14] Petrzhak, G. I.; Stepanova, L. N.; Karago, L. V. (Radiokhimiya **12** [1970] 266/72; Soviet Radiochem. **12** [1970] 240/5).
[15] Molodkin, A. K.; Balakaeva, T. A.; Kuchumova, A. N. (Zh. Neorgan. Khim. **16** [1971] 1301/3; Russ. J. Inorg. Chem. **16** [1971] 689/90).
[16] Molodkin, A. K.; Balakaeva, T. A.; Kuchumova, A. N. (Zh. Neorgan. Khim. **16** [1971] 1892/4; Russ. J. Inorg. Chem. **16** [1971] 1004/6).
[17] Hoshi, M.; Ueno, K. (J. Inorg. Nucl. Chem. **34** [1972] 981/6).
[18] Bykhovskii, D. N.; Petrova, I. K.; Zelentsov, S. S. (Radiokhimiya **14** [1972] 171/7; Soviet Radiochem. **14** [1972] 180/5).
[19] Kharitonov, Yu. Ya.; Balakaeva, T. A.; Molodkin, A. K. (Zh. Neorgan. Khim. **18** [1973] 2869/71; Russ. J. Inorg. Chem. **18** [1973] 1526/7).
[20] Petrzhak, G. I.; V'yugina, A. F.; Karago, L. V. (Radiokhimiya **16** [1974] 120/2; Soviet Radiochem. **16** [1974] 118/9).

[21] Petrzhak, G. I., Zelentsov, S. S. (Radiokhimiya **16** [1974] 200/3; Soviet Radiochem. **16** [1974] 201/4).
[22] Nefedov, V. I.; Molodkin, A. K.; Salyn', Ya. V.; Ivanova, O. M.; Porai-Koshits, M. A.; Balakaeva, T. A.; Belyakova, Z. V. (Zh. Neorgan. Khim. **19** [1974] 2628/31; Russ. J. Inorg. Chem. **19** [1974] 1435/7).
[23] Akhtar, M. N.; Smith, A. J. (Acta Cryst. B **31** [1975] 1361/6).
[24] Kharitonov, Yu. Ya., Molodkin, A. K.; Balakaeva, T. A. (Zh. Neorgan. Khim. **21** [1976] 987/9; Russ. J. Inorg. Chem. **21** [1976] 538/40).
[25] Hoshi, M.; Ueno, K. (Radiochem. Radioanal. Letters **29** [1977] 331/9).
[26] Hoshi, M.; Ueno, K. (J. Nucl. Sci. Technol. [Tokyo] **15** [1978] 585/8).
[27] Molodkin, A. K.; Ivanova, O. M.; Skotnikova, G. A. (Zh. Neorgan. Khim. **9** [1964] 295/307; Russ. J. Inorg. Chem. **9** [1964] 162/8).

15.6.3.3.7 Thorium(IV) Nitrate Oxalate

The only recorded compound appears to be the product of composition $6Th(NO_3)_4$ $Th(C_2O_4)_2 \cdot 48H_2O$ reported in "Thorium" 1955, p. 308.

15.6.3.3.8 Thorium Halide Oxalates and Halogeno Oxalato Complexes

The fluoro oxalato complex salt, $K_4[Th(C_2O_4)_2F_4]$, has been prepared by dissolving $K_4[Th(C_2O_4)_4] \cdot 4H_2O$ (p. 92) in an aqueous solution of KF (molar ratio, Th:KF=1:4). After filtration, the mother liquor was left to crystallise in air. The salt dissolves in H_2O, but is hydrolysed, and it is also decomposed by aqueous methanol. The analogous reaction of the guanidinium salt, $(CN_3H_6)_4[Th(C_2O_4)_4] \cdot 2H_2O$ (p. 94) with aqueous KF yields products with the analytical ratios $Th:C_2O_4:F=1:1.5:5.75$ or 1:1.5:4.5 [1].

A chloride oxalate of composition $ThCl_4 \cdot 3Th(C_2O_4)_2 \cdot 2H_2O$ is described in "Thorium" 1955, p. 308.

Reference for 15.6.3.3.8:

[1] Molodkin, A. K.; Ivanova, O. M.; Skotnikova, G. A. (Zh. Neorgan. Khim. **9** [1964] 295/307; Russ. J. Inorg. Chem. **9** [1964] 162/8).

15.6.3.3.9 Thorium Sulfito Oxalato Complexes

The known compounds are listed in Table 35.

Table 35
Thorium Sulfito Oxalato Complexes.

$Na_6[Th(C_2O_4)_2(SO_3)_3] \cdot xH_2O$	x=5 to 6 [1], 6 [2]
$Na_8[Th(C_2O_4)_2(SO_3)_4] \cdot xH_2O$	x=5 [1], 6 [1, 2]
$Na_{10}[Th(C_2O_4)_2(SO_3)_5] \cdot xH_2O$	x=5 [1, 2], 5 to 6 [1], 6 [2]
$Na_{12}[Th(C_2O_4)_2(SO_3)_6] \cdot xH_2O$	x=5, 5 to 6 [1], 6, 8 [2]
$Na_{14}[Th(C_2O_4)_2(SO_3)_7] \cdot xH_2O$	x=5 [2], 5 to 6 [1]
$Na_{18}[Th(C_2O_4)_2(SO_3)_9] \cdot 6H_2O$	[2]

Products of the general composition $Na_{2n}[Th(C_2O_4)_2(SO_3)_n] \cdot xH_2O$ are reported [1, 2] to be formed by dissolving $Th(C_2O_4)_2$ in concentrated aqueous Na_2SO_3 and then pouring the solution into ethanol. The syrupy mass which separates becomes crystalline on further treatment with ethanol. When the volume of Na_2SO_3 solution was restricted to 25 mL, the product obtained with the molar ratio $Th:Na_2SO_3=1:4$ had n=3 and x=5 to 6 [1] or 6 [2], while with the molar ratios 1:5, 1:6, 1:7, 1:8, and 1:10 products with n=4, 5, 6, 7, or 9 and x=6 [1, 2]; 5 to 6 [1] or 6 [2]; 5 to 6 [1] or 8 [2]; 5 to 6 [1] or 5 [2]; and 6 [2], respectively, were obtained. With 50 mL of more dilute (half concentration) Na_2SO_3 and at molar ratios of 1:6, 1:7, and 1:8 the products had n=4, 5 or 6 with x=5 [1] or 6 [2]; 5 [1, 2] and 5 [1] or 6 [2], respectively.

Cryoscopic measurements with $Na_8[Th(C_2O_4)_2(SO_3)_4] \cdot 6H_2O$ and $Na_{12}[Th(C_2O_4)_2(SO_3)_6]$ $\cdot 6H_2O$ in water apparently indicate 9- and 13-ion electrolyte behaviour, respectively [2].

Treatment of an aqueous solution of $Na_{12}[Th(C_2O_4)_2(SO_3)_6] \cdot 6H_2O$ with aqueous $Ba(NO_3)_2$ yields a crystalline precipitate of $Ba_6[Th(C_2O_4)_2(SO_3)_6] \cdot 7H_2O$ [2].

Although these products do not appear to be mixtures [2], the systems require further investigation, for some of the compositions seem to be rather unlikely for genuine compounds.

References for 15.6.3.3.9:

[1] Gel'man, A. D.; Essen, L. N.; Zakharova, F. A.; Alekseeva, D. P.; Orlova, M. M. (Dokl. Akad. Nauk SSSR **149** [1963] 1071/3; Dokl. Chem. Proc. Acad. Sci. USSR **148/153** [1963] 317/9).
[2] Essen, L. N.; Alekseeva, D. P.; Gel'man, A. D. (Zh. Neorgan. Khim. **11** [1966] 1596/604; Russ. J. Inorg. Chem. **11** [1966] 853/7).

15.6.3.3.10 Thorium Carbonato Oxalato Complexes

Information on the known compounds is summarised in Table 36.

Table 36
Thorium Carbonato Oxalato Complexes.

carbonatooxalatothorates(IV)	
$Na_8Th(C_2O_4)(CO_3)_5$	[1]
$Na_8Th(C_2O_4)(CO_3)_5 \cdot 10$ to $11H_2O$	crystal optical properties; refractive indices, n_g = 1.556, n_p = 1.440 [4]
$Na_8Th(C_2O_4)(CO_3)_5 \cdot 11H_2O$	dehydrated at 110 to 150°C [1] crystal optical properties; refractive indices, n_g = 1.556, n_p = 1.440 [1]
$Na_8Th(C_2O_4)_2(CO_3)_4$	[1, 4]
$Na_8Th(C_2O_4)_2(CO_3)_4 \cdot 9$ to $10H_2O$	[4]
$Na_8Th(C_2O_4)_2(CO_3)_4 \cdot 9$ to $10.5H_2O$	dehydrated at 105 to 110°C [1] crystal optical properties; refractive indices, n_g = 1.500, n_m = 1.489, n_p = 1.483 [1]
$Na_8Th(C_2O_4)_2(CO_3)_4 \cdot 9.5H_2O$	[1]
$Na_8Th(C_2O_4)_2(CO_3)_4 \cdot 11H_2O$	[1]; dehydrated at 105 to 115°C [4] crystal optical properties; refractive indices, n_g = 1.500, n_m = 1.489, n_p = 1.483 [4]
$Na_{10}Th(C_2O_4)_2(CO_3)_5 \cdot 10H_2O$	[4]
$Na_{10}Th(C_2O_4)_2(CO_3)_5 \cdot 11H_2O$	crystal optical properties; refractive indices, n_g = 1.502, n_m = 1.489, n_p = 1.473 [1]
$Na_{10}Th(C_2O_4)_2(CO_3)_5 \cdot 11.5H_2O$	[4]
$Na_{10}Th(C_2O_4)_2(CO_3)_5 \cdot 16H_2O$	crystal optical properties; refractive indices, n_g = 1.502, n_m = 1.489, n_p = 1.473 [4]
$Na_{12}Th(C_2O_4)_2(CO_3)_6 \cdot 13H_2O$	[4]
$K_4Th(C_2O_4)(CO_3)_3 \cdot 4H_2O$	thermogram illustrated in [3]
$K_4Th(C_2O_4)(CO_3)_3 \cdot 6H_2O$	[1, 4]

Table 36 (continued)

$K_6Th(C_2O_4)(CO_3)_4 \cdot 6$ to $8H_2O$	[1]
$K_6Th(C_2O_4)(CO_3)_4 \cdot 8H_2O$	[4]
$K_6Th(C_2O_4)_2(CO_3)_3$	[3]
$K_6Th(C_2O_4)_2(CO_3)_3 \cdot H_2O$	dehydrated at 70 to 100°C [3]
$K_6Th(C_2O_4)_2(CO_3)_3 \cdot 4H_2O$	loses $3H_2O$ at 50°C [3] thermogram illustrated in [3]
$K_6Th_2(C_2O_4)_2(CO_3)_5 \cdot 2H_2O$	[3]
$K_6Th_2(C_2O_4)_3(CO_3)_4 \cdot 6H_2O$	[3]
$K_8Th_2(C_2O_4)_3(CO_3)_5 \cdot 13H_2O$	[3]
$K_8Th_2(C_2O_4)_3(CO_3)_5 \cdot 16H_2O$	[4]
$K_{10}Th_2(C_2O_4)_2(CO_3)_7 \cdot 8H_2O$	[3]
$K_{10}Th_2(C_2O_4)_2(CO_3)_7 \cdot 12H_2O$	[3]
$K_{10}Th_2(C_2O_4)_2(CO_3)_7 \cdot 14H_2O$	[4]
$K_{10}Th_2(C_2O_4)_4(CO_3)_5 \cdot 5H_2O$	[3]
$K_{10}Th_2(C_2O_4)_4(CO_3)_5 \cdot 7H_2O$	thermogram illustrated in [3]
$(NH_4)_4Th(C_2O_4)_2(CO_3)_2 \cdot 0.5H_2O$	decomposes in air [3]
$(NH_4)_4Th_2(C_2O_4)(CO_3)_5 \cdot 10H_2O$	[3]
$(CN_3H_6)_3(NH_4)Th(C_2O_4)_2(CO_3)_2 \cdot 3H_2O$	[2] (CN_3H_6 = guanidinium)
$(CN_3H_6)_3(NH_4)[Th(C_2O_4)(CO_3)_3] \cdot 1$ to $1.5H_2O$	[2]
$(CN_3H_6)_3(NH_4)[Th(C_2O_4)(CO_3)_3] \cdot 1.5H_2O$	[2]
$(CN_3H_6)_3(NH_4)[Th(C_2O_4)(CO_3)_3] \cdot 2$ to $3.5H_2O$	[2]
$(CN_3H_6)_3(NH_4)[Th(C_2O_4)(CO_3)_3] \cdot 3H_2O$	[2]
$(CN_3H_6)_6Th_2(C_2O_4)_2(CO_3)_5 \cdot 4H_2O$	[3]
$(CN_3H_6)_6Th_2(C_2O_4)_2(CO_3)_5 \cdot 8H_2O$	[3]
$(CN_3H_6)_6Th_2(C_2O_4)_3(CO_3)_4 \cdot 14H_2O$	[2]
$(CN_3H_6)_8Th_2(C_2O_4)(CO_3)_7$	decomposes >120°C [2]
$(CN_3H_6)_8Th_2(C_2O_4)(CO_3)_7 \cdot 5H_2O$	dehydrated above 50°C [2]
$(CN_3H_6)_8Th_2(C_2O_4)(CO_3)_7 \cdot 6H_2O$	dehydrated above 50°C [2] thermogram and thermogravimetric curve illustrated in [2]
$(CN_3H_6)_8Th_2(C_2O_4)_3(CO_3)_5 \cdot H_2O$	[2]
$(CN_3H_6)_{10}Th_2(C_2O_4)(CO_3)_8 \cdot 8H_2O$	crystal optical properties; refractive indices, n_g=1.586, n_m=1.561, n_p=1.534 [4]

References for 15.6.3.3.10 on p. 102

Table 36 (continued)

basic carbonatooxalatothorates(IV)	
$Na_4Th_2(OH)_2(C_2O_4)(CO_3)_4 \cdot 4H_2O$	thermogram illustrated in [3]
$Na_4Th_2(OH)_6(C_2O_4)(CO_3)_2 \cdot 2H_2O$	[3]
$Na_{10}Th(OH)_2(C_2O_4)_3(CO_3)_3$	[4]
$Na_{10}Th(OH)_2(C_2O_4)_3(CO_3)_3 \cdot 8$ to $9H_2O$	dehydrated at 105 to 115°C [4]
$K_2Th_2(OH)_2(C_2O_4)(CO_3)_3$	[3]
$K_2Th_2(OH)_2(C_2O_4)(CO_3)_3 \cdot 2H_2O$	loses $2H_2O$ at 100°C; thermogram illustrated in [3]
$K_5Th_2(OH)(C_2O_4)_2(CO_3)_4 \cdot 2H_2O$	thermogram illustrated in [3]

$Na_8Th(C_2O_4)_2(CO_3)_4 \cdot 11H_2O$ has been obtained by treating $Th(C_2O_4)_2 \cdot 6H_2O$ (p. 82) with Na_2CO_3 (molar ratio, Th : Na_2CO_3 = 1:5) in H_2O at room temperature, filtering and then pouring the solution into an equal or greater volume of ethanol to precipitate the compound [1]. $Na_8Th(C_2O_4)_2(CO_3)_4 \cdot 9$ to $10H_2O$ is also reported to be formed by this method [4], but the hendecahydrate was obtained when the ethanol precipitation step was omitted and the aqueous solution was left to evaporate at room temperature [4]. The ethanol precipitation procedure, but using only half the initial volume of H_2O, yields another hydrate, $Na_8Th(C_2O_4)_2(CO_3)_4 \cdot 9.5H_2O$, whereas the slow evaporation procedure without ethanol yields crystals of $Na_8Th(C_2O_4)_2(CO_3)_4 \cdot 9$ to $10.5H_2O$ [1]. The anhydrous compound is obtained by drying $Na_8Th(C_2O_4)_2(CO_3)_4 \cdot 9$ to $10H_2O$ at 105 to 110°C [1] or the hendecahydrate at 105 to 115°C [4].

Treatment of $Th(C_2O_4)_2 \cdot 6H_2O$ with Na_2CO_3 at the molar ratio Th : Na_2CO_3 = 1:6 in H_2O, followed by pouring the resulting solution into ethanol, yields a precipitate of $Na_{10}Th(C_2O_4)_2(CO_3)_5 \cdot xH_2O$ (x = 11 [1], 10, 11.5 or 16 [4]), whereas when the aqueous solution is allowed to evaporate at room temperature $Na_8Th(C_2O_4)(CO_3)_5 \cdot xH_2O$ (x = 11 [1], 10 to 11 [4]) separates after ca. 3 h (x = 10 to 11), but on prolonged standing the carbonato complex, $Na_6[Th(CO_3)_5]$ (p. 8), is formed [4]. Anhydrous $Na_8Th(C_2O_4)(CO_3)_5$ is obtained by heating the hendecahydrate at 110 to 150°C [1].

At the higher molar ratio, $Th(C_2O_4)_2 : Na_2CO_3$ = 1:8, pouring the aqueous solution into ethanol yields a precipitate of composition $Na_{12}Th(C_2O_4)_2(CO_3)_6 \cdot 13H_2O$ which can be dissolved in water and reprecipitated by pouring the solution into ethanol without any change in its composition [4].

$K_4Th(C_2O_4)(CO_3)_3 \cdot 6H_2O$ has been obtained by treating $Th(C_2O_4)_2 \cdot 6H_2O$ with K_2CO_3 in H_2O (molar ratio, Th : K_2CO_3 = 1:4), followed by pouring the resulting solution into ethanol; this procedure yielded a syrup which became crystalline on repeated washing with ethanol [1, 4]. The same procedure with a molar ratio Th : K_2CO_3 = 1:6 yielded $K_6Th(C_2O_4)(CO_3)_4 \cdot xH_2O$ (x = 6 to 8 [1], 8 [4]). The composition of the product obtained appears to depend on the initial volume of H_2O used and on the duration of the treatment with ethanol [1]. $K_6Th(C_2O_4)(CO_3)_4 \cdot 8H_2O$ behaves as a 7-ion electrolyte in H_2O [4]. With the molar ratios Th : K_2CO_3 = 1:4 and 1:5, the products $K_8Th_2(C_2O_4)_3(CO_3)_5 \cdot 16H_2O$ and $K_{10}Th_2(C_2O_4)_2(CO_3)_7 \cdot 14H_2O$ have also been obtained [4].

Precipitation from an aqueous solution at the molar ratio Th : K_2CO_3 = 1:4 by addition of methanol or acetone yields $K_{10}Th_2(C_2O_4)_2(CO_3)_7 \cdot 12H_2O$ (see the compound with $14H_2O$, above) and $K_{10}Th_2(C_2O_4)_2(CO_3)_7 \cdot 8H_2O$, respectively, whereas precipitation with a 2:1 mixture of acetone and H_2O yielded $K_4Th(C_2O_4)(CO_3)_3 \cdot 4H_2O$ (see the compound with $6H_2O$, p. 100) [3]. Addition of this acetone-H_2O mixture to a solution in which the molar ratio Th : K_2CO_3 = 1:3.5 gave $K_{10}Th_2(C_2O_4)_4(CO_3)_5 \cdot 5H_2O$ [3]. With the molar ratio Th : K_2CO_3 = 1:3, addition of acetone precipitates $K_6Th(C_2O_4)_2(CO_3)_3 \cdot 4H_2O$ (14% K_2CO_3; see the compound with $6H_2O$, above) or $K_{10}Th_2(C_2O_4)_4(CO_3)_5 \cdot 7H_2O$ (10% K_2CO_3; see the compound with $5H_2O$, p. 100), whereas addition of methanol precipitates $K_8Th_2(C_2O_4)_3(CO_3)_5 \cdot 13H_2O$ and addition of a 2:1 mixture of methanol or acetone with H_2O precipitates $K_6Th_2(C_2O_4)_3(CO_3)_4 \cdot 6H_2O$ [3]. When the basic compound $K_2Th_2(OH)_2(C_2O_4)(CO_3)_3 \cdot 2H_2O$ (p. 102) is kept in contact with the mother liquor from the preparation, the product is $K_6Th_2(C_2O_4)_2(CO_3)_5 \cdot 2H_2O$ [3]. All of these salts can be dehydrated by heating. For example, $K_6Th(C_2O_4)_2(CO_3)_3 \cdot 4H_2O$ loses $3H_2O$ to form the monohydrate at 50°C and is completely dehydrated at 70 to 100°C [3]. The analyses of many of these compounds are somewhat irreproducible, usually in respect of the H_2O content [4].

$(NH_4)_4Th_2(C_2O_4)(CO_3)_5 \cdot 10H_2O$ is precipitated when acetone is added to a filtered solution of $Th(C_2O_4)_2 \cdot 5H_2O$ (p. 82) in saturated aqueous $(NH_4)_2CO_3$ [3]. Treatment of a solution of $Th(C_2O_4)_2 \cdot 5H_2O$ in 10% aqueous $(NH_4)_2CO_3$ with an equal volume of methanol results in the separation of an oily liquid which becomes crystalline on stirring under methanol. This product analyses as $(NH_4)_4Th(C_2O_4)_2(CO_3)_3 \cdot 0.5H_2O$. It decomposes in air, losing NH_3 and CO_2, and is hydrolysed in aqueous solution [3].

The guanidinium ammonium compound, $(CN_3H_6)_3(NH_4)[Th(C_2O_4)(CO_3)_3] \cdot 1.5H_2O$, is obtained by adding aqueous $Th(NO_3)_4$ to a solution of $(CN_3H_6)_2CO_3$ in saturated aqueous $(NH_4)_2C_2O_4$ (molar ratio Th : $C_2O_4^{2-}$: CO_3^{2-} = 1:5:5). The initial precipitate redissolves on shaking and the product crystallises from the solution on standing [2]. The same compound, but with 1.0 to $1.5H_2O$, has been obtained by treating a mixture of solid $Th(C_2O_4)_2 \cdot 6H_2O$, $(NH_4)_2C_2O_4 \cdot H_2O$ and $(CN_3H_6)_2CO_3$ (molar ratio Th : $C_2O_4^{2-}$: CO_3^{2-} = 1:3:5) with water and heating to complete dissolution. The crystalline product separates on standing [2]. With the molar ratio Th : $C_2O_4^{2-}$: CO_3^{2-} = 1:5:3 the product is $(CN_3H_6)_3(NH_4)Th(C_2O_4)_2(CO_3)_2 \cdot 3H_2O$, while with the molar ratio Th : $C_2O_4^{2-}$: CO_3^{2-} = 1:5:4 $(CN_3H_6)_3(NH_4)$ $[Th(C_2O_4)(CO_3)_3] \cdot 2$ to $3.5H_2O$ is obtained, and the corresponding trihydrate results when a molar ratio of 1:5:5 is used [2]. (The preparative details for these hydrates are not very clear.)

The guanidinium compound, $(CN_3H_6)_6Th_2(C_2O_4)_2(CO_3)_5 \cdot xH_2O$ (x = 4 or 8) has been prepared by heating a mixture of $Th(C_2O_4)_2 \cdot 6H_2O$, $(NH_4)_2C_2O_4$ and $(CN_3H_6)_2CO_3$ (molar ratio Th : $C_2O_4^{2-}$: CO_3^{2-} = 1:2 to 3:2 to 5) in H_2O until dissolution is almost complete. The product separated from the filtrate (pH 8) on standing. These products dehydrate when heated [3]. $(CN_3H_6)_6Th_2(C_2O_4)_3(CO_3)_4 \cdot 14H_2O$ is obtained when freshly precipitated $(CN_3H_6)_6[Th(CO_3)_5] \cdot 4H_2O$ (p. 10) is added to saturated aqueous $(NH_4)_2C_2O_4$ until no more dissolves. The product crystallises from the filtrate on standing [2]. $(CN_3H_6)_8Th_2(C_2O_4)(CO_3)_7 \cdot 5$ (or 6) H_2O crystallises from the filtrate from the preparations of $(CN_3H_6)_2[Th(C_2O_4)_3] \cdot 6H_2O$ (p. 92) and $(CN_3H_6)_6[Th_2(C_2O_4)_7] \cdot 5$, 8 or 12.5 to $13.7H_2O$ (p. 95) [2]. It is also obtained when solid $(CN_3H_6)_6[Th(CO_3)_5] \cdot 4H_2O$ (p. 10) is added to a saturated aqueous solution of $(NH_4)_2C_2O_4 \cdot H_2O$, but $(CN_3H_6)_6Th_2(C_2O_4)_3(CO_3)_4 \cdot 14H_2O$ (see above) is apparently formed if freshly precipitated $(CN_3H_6)_6[Th(CO_3)_5] \cdot 4H_2O$ is used [2]. (The preparative details are not at all clear.)

$(CN_3H_6)_8Th_2(C_2O_4)(CO_3)_7 \cdot 5$ (or 6) H_2O dehydrates when heated in air at 50°C and the anhydrous product decomposes above 120°C. The endotherm observed at 80 to 120°C in a thermal decomposition study is due to the loss of H_2O, and the second endotherm at 120 to 150°C is related to the onset of decomposition [2].

$(CN_3H_6)_8Th_2(C_2O_4)_3(CO_3)_5 \cdot H_2O$ is apparently obtained in the same way as $(CN_3H_6)_3(NH_4)$-$Th(C_2O_4)_2(CO_3)_2 \cdot 3H_2O$ (see above) [2]. $(CN_3H_6)_{10}Th_2(C_2O_4)(CO_3)_8 \cdot 8H_2O$ has been prepared by dissolving $Th(C_2O_4)_2 \cdot 6H_2O$ in aqueous $(CN_3H_6)_2CO_3$ (molar ratio Th : CO_3^{2-} = 1:5); the product

References for 15.6.3.3.10 on p. 102

crystallises on standing for 1 h [4]. When the molar ratio CO_3^{2-}:Th exceeds 6:1, only carbonato complexes (p. 10) are obtained [4].

Several basic carbonato oxalato compounds have been reported. The mother liquor from the solution obtained by treating $Th(C_2O_4)_2 \cdot 5H_2O$ with a 10% aqueous solution of Na_2CO_3 (molar ratio Th:CO_3^{2-}=1:3) deposits a flocculent precipitate of composition $Na_4Th_2(OH)_2(C_2O_4)(CO_3)_4 \cdot 4H_2O$ and the same reaction, but with 0.5 M Na_2CO_3, yields a product of composition $Na_4Th_2(OH)_6(C_2O_4)(CO_3)_2 \cdot 2H_2O$ [3]. However, when aqueous Na_2CO_3 is added to $Th(C_2O_4)_2 \cdot 6H_2O$ (molar ratio Th:CO_3^{2-}=1:4) at room temperature, and the filtrate is poured into twice its volume of ethanol, a precipitate of composition $Na_{10}Th(OH)_2(C_2O_4)_3(CO_3)_3 \cdot 8$ to $9H_2O$ is obtained. This product was also prepared by mixing the same quantities of solid $Th(C_2O_4)_2 \cdot 6H_2O$ and Na_2CO_3, followed by treatment with H_2O and then ethanol as before. The product is completely dehydrated at 105 to 115°C [4].

$K_2Th_2(OH)_2(C_2O_4)(CO_3)_3 \cdot 2H_2O$ is obtained when solid $Th(C_2O_4)_2 \cdot 5H_2O$ (p. 82) is treated with 10% aqueous K_2CO_3 (molar ratio Th:CO_3^{2-}=1:2 or 1:3). The same product is obtained when the aqueous mixture with the molar ratio 1:2 is heated, whereas the mixture with the molar ratio 1:3 yields $K_5Th_2(OH)(C_2O_4)_2(CO_3)_4 \cdot 2H_2O$ on heating [3]. These basic products require further investigation.

References for 15.6.3.3.10:

[1] Essen, L. N.; Alekseeva, D. P. (Dokl. Akad. Nauk SSSR **146** [1962] 380/2; Proc. Acad. Sci. USSR Chem. Sect. **142/147** [1962] 825/7).

[2] Molodkin, A. K.; Ivanova, O. M.; Skotnikova, G. A. (Zh. Neorgan. Khim. **9** [1964] 295/307; Russ. J. Inorg. Chem. **9** [1964] 162/8).

[3] Molodkin, A. K.; Skotnikova, G. A. (Zh. Neorgan. Khim. **9** [1964] 555/61; Russ. J. Inorg. Chem. **9** [1964] 308/11).

[4] Essen, L. N.; Alekseeva, D. P.; Gel'man, A. D. (Zh. Neorgan. Khim. **11** [1966] 1596/1604; Russ. J. Inorg. Chem. **11** [1966] 853/7).

15.6.3.3.11 Thorium Phosphato Oxalato Complex

Physical data for the only known complex are summarised in Table 37.

Table 37
Thorium Phosphato Oxalato Complex.

$K_4[Th(HPO_4)_2(C_2O_4)_2(H_2O)_2]$	decomposes above 170°C [1]
$K_4[Th(HPO_4)_2(C_2O_4)_2(H_2O)_2] \cdot 4H_2O$	loses $4H_2O$ at 40 to 170°C [1], thermogram and thermogravimetric curve illustrated in [1]
	crystal optical properties; refractive indices, n_g=1.537, n_p=1.525 [1]
	IR spectrum: ν(CO), 1730 cm^{-1}, ν(CO)+δ(H_2O), 1640 to 1680 cm^{-1}, ν(CO)+ν(CC), 1437 cm^{-1}, ν(CO), 1295 cm^{-1}, ν(CO)+ν(ThO)+ν(OCO), 898 cm^{-1}, δ(OCO)+ν(ThO), 782 cm^{-1}, ν(ThO)+δ(rings)+ δ(OCO)+ δ(PO_4), 510 to 520 cm^{-1} [1]

$K_4[Th(HPO_4)_2(C_2O_4)_2(H_2O)_2]\cdot 4H_2O$ is precipitated when solid $K_3PO_4\cdot 7H_2O$ is added to the solution obtained by dissolving $Th(C_2O_4)_2\cdot 6H_2O$ in warm (ca. 70 to 80°C) aqueous $K_2C_2O_4\cdot H_2O$ and $H_2C_2O_4\cdot 2H_2O$ (molar ratio Th : $K_2C_2O_4$: $H_2C_2O_4$: K_3PO_4 = 1.5 : 6 : 2 to 3 : ca. 2 to 3). The IR spectrum indicates that the C_2O_4 groups are bidentate. The compound loses $4H_2O$ to form $K_4[Th(HPO_4)_2(C_2O_4)_2(H_2O)_2]$ at 40 to 170°C and decomposes at 170°C [1]. The presence of $2H_2O$ within the coordination sphere requires confirmation.

Reference for 15.6.3.3.11:

[1] Molodkin, A. K.; Balakaeva, T. A.; Kuchumova, A. N. (Zh. Neorgan. Khim. **15** [1970] 1152/3; Russ. J. Inorg. Chem. **15** [1970] 589/90).

15.6.3.4 Thorium Malonates and Related Compounds

The known compounds are listed in Table 38.

Table 38
Thorium Malonates.

$Th(CH_2(COO)_2)_2\cdot 2H_2O$	[2]
$K_2[Th(CH_2(COO)_2)_3]$	[2]

A thorium malonate of unknown composition is precipitated when aqueous $CH_2(COOK)_2$ is added to a solution of $Th(NO_3)_4$ in 90% aqueous ethanol [1]. $Th(CH_2(COO)_2)_2\cdot 2H_2O$ is obtained by treating aqueous $Th(NO_3)_4$ with $CH_2(COOK)_2$ (molar ratio 1:1 or 2), and the malonato complex salt, $K_2[Th(CH_2(COO)_2)_3]$, has been prepared in the same way with the molar ratio Th : $CH_2(COOK)_2$ = 1 : 3 [2]. The preparative details are not very clear. Precipitation of thorium malonate on mixing a 3×10^{-3} M boiling aqueous solution of the acid with aqueous $Th(NO_3)_4$ begins at pH 2.1 (molar ratio, Th : acid = 1 : 5) and with substituted malonic acids precipitation under these conditions begins at pH 2.2 (methyl-), pH 0.7 (dimethyl-), pH 1 (diethyl-), pH 1.2 (di-n-propyl-), pH 1 (di-n-butyl-), pH 1.7 (di-i-propyl-), pH 1.6 (phenyl-), pH 1 (benzyl-), pH 2.4 (dichloro-), and pH 2.1 (p-methoxybenzyl-malonic acid) [3]. Dihydroxymalonic (mesoxalic) acid is also mentioned in [3]. The compositions of the substituted malonates are not reported in [3], but the products are evidently hydrated, for thermogravimetric analysis indicates that H_2O is lost at 80 to 120°C and the products decomposed at 200 to 400°C [3].

The attempted precipitation of complex malonatothorates(IV) from aqueous media by $[Co(NH_3)_6]Cl_3$, $[Co(en)_3]Cl_3\cdot 3H_2O$, $[Co(H_2N(CH_2)_3NH_2)_3]Cl_3$ or $[Cr(ur)_6]Cl_3\cdot H_2O$ was unsuccessful [4].

References for 15.6.3.4:

[1] Bobtelsky, M.; Bar-Gadda, I. (Bull. Soc. Chim. France **1953** 382/6).
[2] Aggarwal, R. C.; Srivastava, T. N.; Agrawal, S. P. (J. Prakt. Chem. **10** [1960] 305/10).
[3] Jauker, C.; Pietsch, R. (Anal. Chim. Acta **90** [1977] 349/54).
[4] Hoshi, M.; Ueno, K. (J. Nucl. Sci. Technol. [Tokyo] **17** [1980] 370/6).

15.6.3.5 Thorium Succinates

The known compounds are listed in Table 39.

Table 39
Thorium Succinates.

succinic acid = $C_2H_4(COOH)_2$	
$Th(C_2H_4(COO)_2)_2$	[5, 6]
$(ThOH)_2(C_2H_4(COO)_2)_3 \cdot 8H_2O$	[7]
$Th(OH)_2(HOOCC_2H_4COO)_2$	decomposes above 55°C [4]

Hot aqueous succinic acid precipitates Th from aqueous solution, but the precipitation is incomplete [1]. Aqueous succinic acid does not precipitate Th from neutral or slightly acid solution, and precipitation at 70 to 80°C is best (97 to 98%) at a final pH of 3.9 to 4.5 [2]. Other evidence for the formation of an insoluble succinate, probably $Th(C_2H_4(COO)_2)_2$, was obtained in a heterometric titration study [3], and this compound has been identified as the product precipitated when aqueous $Th(NO_3)_4$ is treated with $C_2H_4(COOK)_2$ (molar ratio = 1:1) [5], but the preparative details are not clear. The compound is best prepared by heating an ethereal solution of $ThCl_4$ with an excess of $C_2H_4(COOH)_2$, suspended in ether, at 55 to 60°C for 1 h, followed by heating at ca 15 to 25°C above the melting point of the free acid until the evolution of HCl ceases [6].

A basic succinate of composition $(ThOH)_2(C_2H_4(COO)_2)_3 \cdot 8H_2O$ is precipitated when the calculated quantity of the acid is added to an aqueous solution of $Th(NO_3)_4$ at pH 2.2 to 2.3. Solubility data are available for this compound in aqueous $(H, Na)ClO_4$ solutions at varying concentrations (ionic strength 0.1) at 25°C. The solubilities of the basic succinate, in 10^{-4} mol/L (concentrations of $HClO_4$ in 10^{-3} mol/L in parentheses) were found to be: 49.1 (104.0), 36.4 (52.0), 28.8 (41.6), 27.4 (31.2), 26.6 (20.8), 19.9 (10.4), 10.15 (6.24), 5.92 (4.16), 4.23 (3.12), and 3.045 (2.08) [7]. The thermal decomposition of another basic compound, formulated as $Th(OH)_2(OOC(CH_2)_2COOH)_2$, has been reported [4]; this was prepared by the procedure described in [2] (see above).

It has been noted that precipitated Th succinate dissolves in an excess of the acid, presumably because of the formation of a succinato complex [9], but the attempted preparation of Th succinato complex salts by precipitation with $[Co(NH_3)_6]Cl_3$, $[Co(en)_3]Cl_3 \cdot 3H_2O$, $[Co(H_2N(CH_2)_3NH_2)_3]Cl_3$ or $[Cr(ur)_6]Cl_3 \cdot 3H_2O$ (ur = urea = $OC(NH_2)_2$) was unsuccessful [8].

References for 15.6.3.5:

[1] Gordon, L.; Vanselow, C. H. (Anal. Chem. **21** [1949] 1323/5).
[2] Suryanarayana, T. V. S.; Raghava Rao, B. S. V. (J. Indian Chem. Soc. **28** [1951] 511/4).
[3] Bobtelsky, M.; Bar-Gadda, I. (Bull. Soc. Chim. France **1953** 382/6).
[4] Wendlandt, W. W. (Anal. Chim. Acta **17** [1957] 295/9).
[5] Aggarwal, R. C.; Srivastava, T. N.; Agrawal, S. P. (J. Prakt. Chem. **10** [1960] 305/10).
[6] Prasad, S.; Kumar, S. (J. Indian Chem. Soc. **39** [1962] 444/6).
[7] Merkusheva, S. A.; Skorik, N. A.; Serebrennikov, V. V. (Radiokhimiya **10** [1968] 731/3; Soviet Radiochem. **10** [1968] 718/9).
[8] Hoshi, M.; Ueno, K. (J. Nucl. Sci. Technol. [Tokyo] **17** [1980] 370/6).
[9] Tserkovnitskaya, I. A.; Charykov, A. K. (Izv. Vysshikh Uchebn. Zavedenii Khim. Khim. Tekhnol. **7** [1964] 544/50; C.A. **62** [1965] 3385).

15.6.3.6 Thorium Malates and Malato Complexes

The known compounds are listed in Table 40.

Table 40
Thorium Malates and Malato Complexes.

malic acid = hydroxysuccinic acid = HOOCCH(OH)CH_2COOH=$C_4H_6O_5$	
Th_2Cl_2[OOCCH(OH)CH_2COO$]_3$	[6]
Th(OOCCH(OH)CH_2COO$)_2$·1 to 2H_2O	[2]
Th(OOCCH(OH)CH_2COO$)_2$·2H_2O	IR spectrum (3300 to 500 cm^{-1}) illustrated in [3, 4]; ν(OH) of acid group, 3208 cm^{-1} [3]
(ThOH$)_2$(OOCCH(OH)CH_2COO$)_3$·4H_2O	[1]
NaTh(OH)(OOCCH(OH)CH_2COO$)_2$·6H_2O	[1]
Na_2[Th(OH$)_2$(OOCCH(OH)CH_2COO$)_2$]·4H_2O	[1]
Na_2[ThO(OOCCH(OH)CH_2COO$)_2$]·6H_2O	see "Thorium" 1955, p. 328
K_2[ThO(OOCCH(OH)CH_2COO$)_2$]·4H_2O	see "Thorium" 1955, p. 336
$(NH_4)_2$[ThO(OOCCH(OH)CH_2COO$)_2$]·4H_2O	see "Thorium" 1955, p. 342
ThO[MoO_3(OOCCH(OH)CH_2COO$)_2$]·3H_2O	[5]

Th_2Cl_2(OOCCH(OH)CH_2COO$)_3$ has been prepared by reaction of $ThCl_4$ with the acid in the same way as the succinate, Th(C_2H_4(COO$)_2)_2$ (p. 104) [6].

Th(OOCCH(OH)CH_2COO$)_2$·2H_2O has been obtained by adding 0.5 M aqueous Th($NO_3)_4$ to 2.5 M aqueous malic acid at pH ca. 0.5, followed by addition of methanol (half the volume of the aqueous phase) to precipitate the compound [3]. The hydrate with 1 to 2 molecules of H_2O has also been reported as being synthesised from acid media [3]. Its IR spectrum is said to indicate that the Th atom is bonded to the carboxylate group and not to the OH group [2].

(ThOH$)_2$(OOCCH(OH)CH_2COO$)_3$·4H_2O has been obtained by adding methanol to an aqueous mixture of Th($NO_3)_4$ and malic acid, and also by mixing aqueous solutions of HOCHCH_2-(COONa$)_2$ and Th($NO_3)_4$ (molar ratio = 3:2). The concentration of a saturated aqueous solution of this product at room temperature is 7.8×10^{-4} M and the conductivity of the solution corresponds to that expected for a 2-ion electrolyte [1].

The addition of methanol to a solution containing Th($NO_3)_4$ and HOCHCH_2(COONa$)_2$ (molar ratio = 1:3, pH ca. 4) yields a precipitate which, after drying at 100°C and exposure to air at room temperature, has the composition Na[Th(OH)(OOCCH(OH)CH_2COO$)_2$]·6H_2O. This product is relatively soluble in water and the conductivity of the solution corresponds to that expected of a 2-ion electrolyte [1]. However, when aqueous Th($NO_3)_4$ is added to a solution of malic acid and 6 equivalents of NaOH "in an amount equal to half the quantity of malic acid", addition of methanol precipitates Na_2[Th(OH$)_2$(OOCCH(OH)CH_2COO$)_2$]·4H_2O, which behaves in water as a 3-ion electrolyte [1]. This compound is probably the same as the previously reported Na_2[ThO(OOCCH(OH)CH_2COO$)_2$]·4H_2O ("Thorium" 1955, p. 328).

A malatomolybdate of composition ThO[(MoO_3(OOCCH(OH)CH_2COO)]·3H_2O is precipitated when an aqueous solution of the sodium malatomolybdate is added to aqueous Th($NO_3)_4$ [5]. The attempted precipitation of malato complex salts with [Co($NH_3)_6$]Cl_3, [Co(en$)_3$]Cl_3·3H_2O, [Co(H_2N($CH_2)_3$$NH_2)_3$]$Cl_3$ or [Cr(ur$)_6$]Cl_3·3H_2O was unsuccessful [7].

References for 15.6.3.6 on p. 106

References for 15.6.3.6:

[1] Zvyagintsev, O. E.; Khromenko, L. G. (Zh. Neorgan. Khim. **6** [1961] 593/600; Russ. J. Inorg. Chem. **6** [1961] 301/6).
[2] Geleceanu, I.; Lapitskii, A. V. (Dokl. Akad. Nauk SSSR **144** [1962] 573/5; Proc. Acad. Sci. USSR Chem. Sect. **142/147** [1962] 460/2).
[3] Galateanu, I. (Acad. Rep. Populare Romine Studii Cercetari Chim. **11** [1963] 343/52).
[4] Galateanu, I. (Oesterr. Chemiker-Ztg. **66** [1965] 275/85).
[5] Prasad, S.; Pandey, L. P. (J. Indian Chem. Soc. **42** [1965] 783/8).
[6] Prasad, S.; Kumar, S. (J. Indian Chem. Soc. **39** [1962] 444/6).
[7] Hoshi, M.; Ueno, K. (J. Nucl. Sci. Technol. [Tokyo] **17** [1980] 370/6).

15.6.3.7 Thorium Tartrates and Tartrato Complexes

The known compounds are listed in Table 41.

Table 41
Thorium Tartrates and Tartrato Complexes.

tartaric acid = dihydroxysuccinic acid = $(CHOH)_2(COOH)_2$	
$Th\{(CHOH)_2(COO)_2\}_2$	dirty brown [3]
$Th\{(CHO)_2(COO)_2\}\cdot 9H_2O$	[5]
$Th(OH)_2\{(CHOH)_2(COO)_2\}\cdot 3H_2O$	[5]
$Th_3(OH)_4\{(CHOH)_2(COO)_2\}_4\cdot xH_2O$	x = 0, 2, 5, 18 or 22; see "Thorium" 1955, p. 309
$KTh(OH)\{(CHOH)_2(COO)_2\}_2\cdot 7H_2O$	[5]
$K_2Th(OH)_2\{(CHOH)_2(COO)_2\}_2\cdot 8H_2O$	[5]
$KTh(OH)\{(CHO)_2(COO)_2\}\cdot 10H_2O$	[5]
$K_2Th(OH)_2\{(CHO)_2(COO)_2\}\cdot 4H_2O$	[5]
$Na_2ThO\{(CHOH)_2(COO)_2\}_2\cdot 8H_2O$	see "Thorium" 1955, p. 328
$K_2ThO\{(CHOH)_2(COO)_2\}_2\cdot 8H_2O$	see "Thorium" 1955, p. 336
$(NH_4)_2ThO\{(CHOH)_2(COO)_2\}_2\cdot 3H_2O$	see "Thorium" 1955, p. 342
$K_2ThO\{(CHO)_2(COO)_2\}\cdot 4H_2O$	see "Thorium" 1955, p. 336
$(NH_4)_2ThO\{(CHO)_2(COO)_2\}\cdot 4H_2O$	see "Thorium" 1955, p. 342
$K_4[Th\{(CHOH)_2(COO)_2\}_2(C_2O_4)_2]\cdot 3H_2O$	decomposes at 243°C [1]

$Th\{(CHOH)_2(COO)_2\}_2$ has been prepared from $ThCl_4$ and the acid in the same way as the succinate, $Th((CH_2)_2(COO)_2)_2$ (p. 104), except that the product was finally extracted into ether and recovered from the extract by evaporation [3]. The colour of this product (dirty brown [3]) suggests that some decomposition had occurred. A product of composition $Th\{(CHO)_2(COO)_2\}\cdot 9H_2O$ is precipitated when a solution containing equimolar amounts of $Th(NO_3)_4$ and the acid in water is treated with KOH (molar ratio Th : KOH = 1 : 4) [5]. This may be a basic salt, $ThO\{(CHOH)_2(COO)_2\}\cdot 8H_2O$.

A basic tartrate, $Th(OH)_2\{(CHOH)_2(COO)_2\}\cdot 3H_2O$, is precipitated when aqueous $Th(NO_3)_4$ is treated with a solution of the acid or its mono- or dialkali metal salts [5]. The previously reported basic tartrates ("Thorium" 1955, p. 309), $Th_3(OH)_4\{(CHOH)_2(COO)_2\}_4\cdot xH_2O$ (x = 0, 2, 5, 18 or 22) can also be formulated as $Th_3O_2\{(CHOH)_2(COO)_2\}_4\cdot(x+2)H_2O$.

The precipitation of Th from aqueous solution by tartaric acid is mentioned in [2]. It has also been noted that Th tartrate (composition not specified) is not precipitated from aqueous solution by NH_3, NaOH or KOH when the tartaric acid concentration is ≧1% [4].

The complex salt $KTh(OH)\{(CHO)_2(COO)_2\}\cdot 10H_2O$ is prepared by treating an aqueous solution containing equimolar amounts of $Th(NO_3)_4$ and the acid with KOH (molar ratio, Th:KOH = 1:5). The initial precipitate redissolves slowly, and the product is precipitated from the filtrate with methanol. The conductivity of the salt in water is consistent with 2-ion electrolyte behaviour [5]. The same preparative procedure, but with the molar ratio Th:KOH = 1:6, yields a product of composition $K_2Th(OH)_2\{(CHO)_2(COO)_2\}\cdot 4H_2O$ which behaves as a 3-ion electrolyte in water [5]. This compound is probably the same as the previously reported ("Thorium" 1955, p. 336) $K_2ThO\{(CHO)_2(COO)_2\}\cdot 4H_2O$.

$KTh(OH)\{(CHOH)_2(COO)_2\}_2\cdot 7H_2O$ has been prepared by treating an aqueous solution of $(CHOH)_2(COOK)_2$ (3 volumes) with aqueous $Th(NO_3)_4$ (1 volume) of the same concentration. The initial precipitate redissolves, and the product is then precipitated from the clear solution by adding methanol. It behaves as a 2-ion electrolyte in water [5]. $K_2Th(OH)_2\{(CHOH)_2(COO)_2\}_2\cdot 8H_2O$ was obtained in the same way, except that aqueous KOH (molar ratio Th:KOH = 1:2) was added to the clear solution before precipitating the product with methanol. It behaves as a 3-ion electrolyte in water [5]. This compound is probably the same as the previously reported ("Thorium" 1955, p. 336) $K_2ThO\{(CHOH)_2(COO)_2\}_2\cdot 8H_2O$.

The tartrato oxalato complex salt, $K_4[Th\{(CHOH)_2(COO)_2\}_2(C_2O_4)_2]\cdot 3H_2O$, has been prepared by dissolving $Th(C_2O_4)_2\cdot 6H_2O$ (p. 82) in aqueous $(CHOH)_2(COOK)_2$ (molar ratio = 1:4) at 60°C, followed by addition of ethanol to the cooled solution to precipitate the salt [1].

References for 15.6.3.7:

[1] Grinberg, A. A.; Petrzhak, G. I.; Lozhkina, G. S. (Radiokhimiya **13** [1971] 836/40; Soviet Radiochem. **13** [1971] 862/5).
[2] Jauker, C.; Pietsch, R. (Anal. Chim. Acta **90** [1977] 349/54).
[3] Prasad, S.; Kumar, S. (J. Indian Chem. Soc. **39** [1962] 444/6).
[4] Haïssinsky, M.; Yang Jeng-Tsong (Anal. Chim. Acta **3** [1949] 422/7).
[5] Zvyagintsev, O. E.; Khromenkov, L. G. (Zh. Neorgan. Khim. **6** [1961] 874/82; Russ. J. Inorg. Chem. **6** [1961] 445/50).

15.6.3.8 Thorium Compounds with Other Substituted Succinic Acids

There has been very little definitive work published concerning Th compounds with other substituted succinic acids.

Hot aqueous dibromosuccinic acid precipitates Th from aqueous solution, but the precipitation is incomplete [1] and the composition of the precipitate has not been recorded.

Aminosuccinic acid, $HOOCCH_2CH(NH_2)COOH$ (= aspartic acid) yields the compounds formulated as $[Th(OOCCH_2CH(NH_2)COO)(H_2O)_3](NO_3)_2$ and $[Th(OOCCH_2CH(NH_2)COO)(HOOCCH_2CH(NH_2)COO)(H_2O)]NO_3\cdot 3H_2O$ with aqueous $Th(NO_3)_4$ at pH 1 with the molar ratio Th:acid = 1:1 and 1:2, respectively [2].

N-(β-naphthyl)succinamic acid has been used for the gravimetric estimation of Th by precipitation from aqueous solution at pH 4, followed by ignition to ThO_2 because the composition of the precipitate, which appears to be a basic compound in which the molar ratio Th : acid group = ca. 1:1, is not well defined. The best conditions for precipitation appear to be addition of a 1% aqueous solution of the acid to a hot (60 to 70°C) solution of hydrated $Th(NO_3)_4$ in HNO_3 at pH 2 to which 1% CH_3COONH_4 solution has been added (pH 4 best, pH 3.5 to 5.0 satisfactory) [3]. The thermogravimetric curve for the precipitate is illustrated in [3]. Precipitation with this acid can also be used for the separation of Th from the lanthanides [3].

References: for 15.6.3.8

[1] Gordon, L.; Vanselow, C. H. (Anal. Chem. **21** [1949] 1323/5).
[2] Sergeev, G. M.; Korshunov, I. A. (Radiokhimiya **16** [1974] 787/90; Soviet Radiochem. **16** [1974] 771/4).
[3] De Laiseca, N. P.; Casal, A. R. (Afinidad **34** [1977] 359/62).

15.6.3.9 Thorium Compounds with Unsaturated C_4 Dicarboxylic Acids and Their Derivatives

Information on the known compounds is summarised in Table 42.

Table 42
Thorium Compounds with Unsaturated C_4 Dicarboxylic Acids and Their Derivatives.

fumarates	
$Th(OH)_2$(*trans*-$C_4H_2O_4$)	decomposes at 146°C [6]
$Th(OH)_2$(*trans*-$C_4H_2O_4$)·$4H_2O$	[5]
$Th(OH)_2$(*trans*-$C_4H_3O_4$)$_2$·$5H_2O$	loses $5H_2O$ at 31 to 127°C; differential thermal analysis and thermogravimetric analysis curves reported in [6] (possibly Th(*trans*-$C_4H_2O_4$)$_2$·$7H_2O$)
maleates	
Th(*cis*-$C_4H_2O_4$)$_2$·$2H_2O$	decomposes above 170°C [7]
$Th(OH)_4$[(*cis*-$C_4H_2O_4$)$Th(OH)_2$]$_4$	decomposes above 287°C [6]
$Th(OH)_4$[(*cis*-$C_4H_2O_4$)$Th(OH)_2$]$_4$·$6H_2O$	loses $6H_2O$ at 35 to 213°C; differential thermal analysis and thermogravimetric analysis curves reported in [6]
$Na_2Th(OH)_2$(*cis*-$C_4H_2O_4$)$_2$·$4H_2O$	decomposes at 315 K [9]
$Na_2Th(OH)_4$(*cis*-$C_4H_2O_4$)·$4H_2O$	decomposes at 315 K [9]
K_2ThO(*cis*-$C_4H_2O_4$)$_2$·$2H_2O$	decomposes at 319 K [9]; IR spectrum, ν(Th–O), 921 cm^{-1} [9]
dihydroxymaleate	
Th(*cis*-OOCC(OH) : C(OH)COO)$_2$	bright yellow; solubility 0.0035 g/100 cm^3 H_2O, 0.0038 g/100 cm^3 C_2H_5OH; solubility product, S = 2.3 (or 2.99) × 10^{-13} in H_2O [4]

Table 42 (continued)

2′-carboxymaleanilate	
$[Th(OH)_2(cis\text{-}C_{11}H_7NO_5)(H_2O)]_2$	[8], (2′-carboxymaleanilic acid = *cis*-HOOCCH=CHCONH-(C_6H_4-2′-COOH))

Precipitation curves for the aqueous systems $Th(NO_3)_4$-dipotassium maleate and dipotassium fumarate have been reported [3] (maleic acid = *cis*-HOOCCH : CHCOOH = *cis*-$C_4H_4O_4$ and fumaric acid = *trans*-HOOCCH : CHCOOH = *trans*-$C_4H_4O_4$). The compositions of the products are not given in [3]. The pyrolysis curve for a Th fumarate (composition not given) is illustrated in [1]. It decomposes to ThO_2 at ca. 405°C [1].

The basic fumarate, $Th(OH)_2$(*trans*-$C_4H_2O_4$)·4H_2O, is precipitated by the acid from aqueous solutions of Th^{IV} salts at pH 2 to 5 [5]. The precipitate obtained in the system $Th(NO_3)_4$–Na_2(*trans*-$C_4H_2O_4$)–H_2O is reported to be $Th(OH)_2$(*trans*-$C_4H_3O_4$)$_2$·5H_2O [6], but this could also be written as Th(*trans*-$C_4H_2O_4$)$_2$·7H_2O. In the analogous system $Th(NO_3)_4$–K_2(*cis*-$C_4H_2O_4$)–H_2O the precipitate which formed is described as "$(OH)_2Th(OH)_2$[(*cis*-$C_4H_2O_4$)$Th(OH)_2$]$_4$·6H_2O" [6] (= Th_5(*cis*-$C_4H_2O_4$)$_4$$(OH)_{12}$·6$H_2O$). However, the normal maleate, Th(*cis*-$C_4H_2O_4$)$_2$·2H_2O, is obtained when freshly precipitated Th^{IV} hydroxide is dissolved in the minimum quantity of aqueous maleic acid in the cold and the solution is then left to evaporate slowly at room temperature in a vacuum desiccator over $CaCl_2$ [7].

The maximum optical density in the system $Th(NO_3)_4$–Na_2(*cis*-$C_4H_2O_4$)–H_2O is observed at the molar ratio Th : maleate = 1 : 2 and precipitation begins at the molar ratio 1 : 3; there is some evidence for the formation of the maleato complex salt, Na_2Th(*cis*-$C_4H_2O_4$)$_3$ [2], but this salt was not isolated.

$Na_2Th(OH)_2$(*cis*-$C_4H_2O_4$)$_2$·4H_2O, K_2ThO(*cis*-$C_4H_2O_4$)$_2$·2H_2O, and $Na_2Th(OH)_4$(*cis*-$C_4H_2O_4$)·4H_2O form from solutions containing Th^{IV}, maleic acid and KOH or NaOH, respectively. The precipitates are dried over $CaCl_2$. All three compounds are white, insoluble in water and common organic solvents. IR spectra from 1000 to 400 cm^{-1} are illustrated [9].

The dihydroxymaleate, Th(*cis*-OOCC(OH) : C(OH)COO)$_2$, is precipitated when a freshly prepared aqueous 0.1N solution of *cis*-$C_4H_2(OH)_2O_4$ in 50% excess is added to a 10% solution of $Th(NO_3)_4$ in H_2O at 60 to 70°C. It can be dried to constant weight at 80 to 90°C and the precipitation can be used for the quantitative determination of Th [4].

The 2′-carboxymaleanilate (2′-carboxymaleanilic acid = *cis*-HOOCCH : CHCONH(C_6H_4-2′-COOH) = *cis*-$C_{11}H_9NO_5$), $[Th(OH)_2(cis\text{-}C_{11}H_7NO_5)(H_2O)]_2$, is precipitated when aqueous Na_2(*cis*-$C_{11}H_7NO_5$) is added to an aqueous solution of a Th^{IV} salt at 40 to 50°C and pH 5.5 to 6.8. The compound can be dried at 130 to 135°C and the precipitation can be used for the gravimetric determination of Th [8]. Although the compound is written as a dimer, no evidence for this is provided in [8].

References for 15.6.3.9:

[1] Dupuis, T.; Duval, C. (Anal. Chim. Acta **3** [1949] 589/98).
[2] Bobtelsky, M.; Bar-Gadda, I. (Bull. Soc. Chim. France **1953** 382/6).
[3] Težak, B. (Proc. 1st. Intern. Conf. Peaceful Uses At. Energy, Geneva 1955, Vol. 7, pp. 401/6).
[4] Sadzhaya, N. D. (Zh. Analit. Khim. **18** [1963] 1028/9; J. Anal. Chem. [USSR] **18** [1963] 887/8).
[5] Tserkovnitskaya, I. A.; Charykov, A. K. (Izv. Vysshikh Uchebn. Zavedenii Khim. Khim. Tekhnol. **7** [1964] 544/50; C.A. **62** [1965] 3385).

[6] Bilinski, H.; Despotovic, Z. (Croat. Chem. Acta **39** [1967] 165/74).
[7] Bhat, T. R.; Mathur, B. S.; Shankar, J. (Indian J. Chem. **8** [1970] 275/81).
[8] Singh, N.; Gupta, C. S. (J. Indian Chem. Soc. **58** [1981] 357/9).
[9] Bilinski, H.; Sjöberg, S.; Kežić, S.; Brničević, N. (Acta Chem. Scand. A **39** [1985] 317/25).

15.6.3.10 Thorium Glutarates

The only recorded glutarates (glutaric acid = $HOOC(CH_2)_3COOH = C_3H_6(COOH)_2 = C_5H_8O_4$) are the basic compounds $(ThOH)_2(C_5H_6O_4)_3 \cdot 8H_2O$ and $Th(OH)_2(C_5H_6O_4) \cdot 4H_2O$. The first of these is precipitated when the calculated quantity of $C_5H_8O_4$ is added to an aqueous solution of $Th(NO_3)_4$ at pH 2.2 to 2.3. Solubility data have been reported for this compound in aqueous $(H,Na)ClO_4$ solutions of varying concentration (ionic strength = 0.1) at 25°C. The solubilities of the basic glutarate in 10^{-4} mol/L (concentrations of $HClO_4$ in 10^{-3} mol/L in parentheses) were found to be: 61.72 (104.0), 36.63 (62.4), 30.41 (41.6), 29.97 (29.1), 29.90 (24.9), 27.53 (20.8), 23.75 (16.6), 17.32 (12.4), 11.10 (8.32), and 5.50 (4.16) [1]. The second compound is apparently obtained in the same way at ca. pH 2 [2].

References for 15.6.3.10:

[1] Merkusheva, S. A.; Skorik, N. A.; Serebrennikov, V. V. (Radiokhimiya **10** [1968] 731/3; Soviet Radiochem. **10** [1968] 718/9).
[2] Tserkovnitskaya, I. A.; Charykov, A. K. (Izv. Vysshikh Uchebn. Zavedenii Khim. Khim. Tekhnol. **7** [1964] 544/50; C.A. **62** [1965] 3385).

15.6.3.11 Thorium Compounds with Substituted Glutaric Acids

Information on the known compounds is summarised in Table 43.

Table 43
Thorium Compounds with Substituted Glutaric Acids.

trihydroxyglutarates (trihydroxyglutaric acid = $HOOC(CHOH)_3COOH = C_5H_8O_7$)	
$Th(C_5H_6O_7)_2 \cdot H_2O$	[2, 4]; decomposes >80°C [3] IR spectrum (3285 to 475 cm^{-1}) illustrated in [3], ν(CO), 1578 cm^{-1}, ν(CO) + δ(OH), 1306 cm^{-1}, δ_{as}(CCO), 560 cm^{-1}, δ(OCO), 499 cm^{-1}, δ_s(CCO), 470 cm^{-1} [3]
$(ThOH)_2(C_5H_6O_7)_3 \cdot H_2O$	solubility in water, 3.9×10^{-4} M [1], solubility product $pK_s = 56.0 \pm 0.4$, stability constant, log $\beta = 8.40$ [5]
$Th(OH)(C_5H_5O_7) \cdot 2H_2O$	[1]
$NaTh(OH)(C_5H_6O_7)_2 \cdot H_2O$	[1]
$NaTh(OH)_2(C_5H_5O_7) \cdot H_2O$	[1]
$Na_2Th(OH)_2(C_5H_6O_7)_2$	[1]
glutamates (aminoglutarates)	
$[Th(C_5H_7NO_4)(H_2O)_3](NO_3)_2$	[6]
$[Th(C_5H_7NO_4)(C_5H_8NO_4)(H_2O)]NO_3 \cdot 3H_2O$	[6]

The trihydroxyglutarate, $Th(C_5H_6O_7)_2 \cdot H_2O$, is precipitated when a 3 M aqueous solution of $C_5H_8O_7$ is added to an acidified (ca. pH 0.5) solution of $Th(NO_3)_4$ in methanol [3]. See also [2, 4]. Its IR spectrum indicates that the Th atom is bonded to the COO groups and not to the OH groups of the acid [2].

The basic compound, $(ThOH)_2(C_5H_6O_7)_3 \cdot H_2O$, is precipitated when aqueous $Th(NO_3)_4$ is mixed with a 1 M aqueous solution of $C_5H_8O_7$ or its Na salt [1] and by mixing aqueous $Th(NO_3)_4$ with aqueous $C_5H_8O_7$ at pH 2.3 [5]. Its molar conductivity in H_2O is consistent with 5-ion electrolyte behaviour ($\lambda = 533\ \Omega^{-1} \cdot cm^2 \cdot mol^{-1}$) [1]. The solubility of the compound in 0.1M aqueous solutions of (H, Na)ClO_4 has been found to be 10.72×10^{-4} mol/L for $[H^+] = 9.610 \times 10^{-2}$ mol/L decreasing to 1.04×10^{-4} mol/L for $[H^+] = 0.961 \times 10^{-2}$ mol/L from which the solubility product and a stability constant was determined, see Table 43, p. 110 [5].

$Th(OH)(C_5H_5O_7) \cdot 2H_2O$ (possibly $Th(OH)_2(C_5H_6O_7) \cdot H_2O$) has been obtained by adding aqueous $Th(NO_3)_4$ dropwise to a mixture of $C_5H_8O_7$ and NaOH in H_2O (molar ratio Th : $C_5H_8O_7$: NaOH = 1 : 1 : 4). After leaving the mixture to stand until the initial precipitate had redissolved, filtering if necessary, methanol was added to precipitate the compound. It is readily soluble in H_2O, unlike $(ThOH)_2(C_5H_6O_7)_3$ (see above), and behaves as a non-electrolyte in aqueous solution ($\lambda = 41\ \Omega^{-1} \cdot cm^2 \cdot mol^{-1}$) [1].

$NaTh(OH)(C_5H_6O_7)_2 \cdot H_2O$ is precipitated when methanol is added to the clear solution obtained when aqueous $Th(NO_3)_4$ and $Na_2C_5H_6O_7$ (molar ratio = 1 : 3) are mixed. It is readily soluble in H_2O, in which it behaves as a 2-ion electrolyte ($\lambda = 103\ \Omega^{-1} \cdot cm^2 \cdot mol^{-1}$) [1]. A product formulated as $NaTh(OH)_2(C_5H_5O_7) \cdot H_2O$ (possibly $NaTh(OH)_3(C_5H_6O_7)$) is obtained in exactly the same way as $Th(OH)(C_5H_5O_7) \cdot 2H_2O$ (see above) but with the molar ratio Th : $C_5H_8O_7$: NaOH = 1 : 1 : 5. Its conductivity in H_2O corresponds to that expected for a 2-ion electrolyte ($\lambda = 103\ \Omega^{-1} \cdot cm^2 \cdot mol^{-1}$) [1].

$Na_2Th(OH)_2(C_5H_6O_7)_2$ precipitates when methanol is added to an aqueous solution containing $Th(NO_3)_4$, $Na_2C_5H_6O_7$, and NaOH (molar ratio = 1 : 3 : 2). This compound behaves as a 3-ion electrolyte in H_2O ($\lambda = 223\ \Omega^{-1} \cdot cm^2 \cdot mol^{-1}$) [1].

The glutamates (= aminoglutarates) formulated as $[Th(C_5H_7NO_4)(H_2O)_3](NO_3)_2$ and $[Th(C_5H_7NO_4)(C_5H_8NO_4)(H_2O)](NO_3) \cdot 3H_2O$ are obtained by treating aqueous $Th(NO_3)_4$ with glutamic acid at pH 1 with the molar ratios Th : acid = 1 : 1 and 1 : 2, respectively [6].

References for 15.6.3.11:

[1] Zvyagintsev, O. E.; Khromenkov, L. G. (Zh. Neorgan. Khim. **6** [1961] 1074/83; Russ. J. Inorg. Chem. **6** [1961] 548/54).
[2] Geleceanu, I.; Lapitskii, A. V. (Dokl. Akad. Nauk SSSR **144** [1962] 573/5; Proc. Acad. Sci. USSR Chem. Sect. **142/147** [1962] 460/2).
[3] Galateanu, I. (Acad. Rep. Populare Romine Studii Cercetari Fiz. **14** [1963] 557/70).
[4] Galateanu, I. (Oesterr. Chemiker-Ztg. **66** [1965] 275/85).
[5] Poskrebysheva, L. M.; Skorik, N. A.; Serebrennikov, V. V. (Radiokhimiya **11** [1969] 113/5; Soviet Radiochem. **11** [1969] 108/10).
[6] Sergeev, G. M.; Korshunov, I. A. (Radiokhimiya **16** [1974] 787/90; Soviet Radiochem. **16** [1974] 771/4).

15.6.3.12 Thorium Compounds with C_6 and Higher Aliphatic Dicarboxylic Acids

The available information on these compounds is summarised in Table 44, p. 112.

Table 44
Thorium Compounds with C_6 and Higher Aliphatic Dicarboxylic Acids.

adipates	
$Th\{OOC(CH_2)_4COO\}_2$	[8, 11]
$ThO\{OOC(CH_2)_4COO\} \cdot 3H_2O$	[4]
$Th(OH)_3\{OOC(CH_2)_4COOH\}$	decomposes to ThO_2 at 450°C, thermal decomposition curve illustrated in [7] possibly $Th(OH)_2\{OOC(CH_2)_4COO\} \cdot H_2O$
$(ThOH)_2\{OOC(CH_2)_4COO\}_3 \cdot 4.5H_2O$	[13]
mucates	
$(ThOH)_2\{OOC(CHOH)_4COO\}_3 \cdot 6H_2O$	[10]
$NaTh(OH)\{OOC(CHOH)_2(CHO)_2COO\} \cdot 2H_2O$	stable to 140°C [10] probably $NaTh(OH)_3\{OOC(CHOH)_4COO\}$
$NaTh(OH)\{OOC(CHOH)_4COO\}_2 \cdot 6H_2O$	[10]
$Na_2Th(OH)_2\{OOC(CHOH)_4COO\}_2 \cdot 10H_2O$	[10]
D-saccharates	
$Th(OH)Cl\{OOC(CHOH)_4COO\} \cdot 3H_2O$	thermogravimetric curve illustrated in [12]
$K_2Th(OH)_2\{OOC(CHO)_2(CHOH)_2COO\} \cdot 7H_2O$	thermogravimetric curve illustrated in [12]
azelaate	
$(ThOH)_2\{OOC(CH_2)_7COO\}_3 \cdot 4H_2O$	[13]
sebacate	
$Th\{OOC(CH_2)_8COO\}_2$	[1, 11, 14, 18]; decomposes >125°C, to ThO_2 at 650°C; pyrolysis curve illustrated in [2]
camphorates	
$Th(C_8H_{14}(COO)_2)_2$	camphoric acid (= $C_8H_{14}(COOH)_2$ = 1,2,2-trimethyl-1,3-cyclopentanedicarboxylic acid) decomposes to ThO_2 at 475°C; thermal decomposition curve illustrated in [7]
$ThO(C_8H_{14}(COO)_2) \cdot 3H_2O(?)$	[15]

The adipate $Th\{OOC(CH_2)_4COO\}_2$ (adipic acid = $HOOC(CH_2)_4COOH$) is reported to be formed when equimolar amounts of $Th(NO_3)_4$ and $(CH_2)_4(COOK)_2$ are mixed in aqueous solution [8]. It has also been prepared from $ThCl_4$ and the acid in the same way as the succinate, $Th\{OOCCH_2CH_2COO\}_2$ (p. 104) [11].

Addition of aqueous adipic acid and solid CH_3COONH_4 to an aqueous solution of a Th^{IV} salt which has been neutralised to Congo Red, followed by boiling for 15 min, yields a gelatinous precipitate of slightly variable composition which appears to be $ThO\{OOC(CH_2)_4COO\} \cdot 3H_2O$. The precipitation is best carried out in the pH range 4.2 to 4.4 [4]. A product described as $Th(OH)_3\{OOC(CH_2)_4COOH\}$ [7] was prepared in the same way and is probably the same compound. Both are likely to be compounds of the type $Th(OH)_2\{OOC(CH_2)_4COO\} \cdot xH_2O$ with x=1 and 2, respectively.

The basic compound $(ThOH)_2\{OOC(CH_2)_4COO\}_3\cdot4.5H_2O$ is precipitated when the calculated quantity of adipic acid is added to an aqueous solution of $Th(NO_3)_4$ at pH 2.2 to 2.3. Solubility data are available for its solution in aqueous $(H,Na)ClO_4$ at varying concentrations (ionic strength = 0.1) at 25°C. The solubilities of the basic adipate, in 10^{-4} mol/L (concentrations of $HClO_4$ in 10^{-3} mol/L in parentheses), were determined to be: 39.96 (104.0), 33.74 (62.4), 27.53 (41.6), 25.09 (29.1), 24.42 (24.9), 22.87 (20.8), 21.09 (16.6), 16.65 (12.4), 15.54 (8.32), and 5.772 (4.16) [13].

An unidentified gelatinous precipitate is reported to be formed when adipic acid is added to an aqueous solution of a Th^{IV} salt [3]. This is presumably a basic adipate.

The corresponding mucate (mucic acid = $HOOC(CHOH)_4COOH$), $(ThOH)_2\{OOC(CHOH)_4COO\}_3\cdot6H_2O$, is precipitated when 0.05 M aqueous solutions of $Th(NO_3)_4$ and the acid or its Na salt are mixed. It has a low solubility in water ($\equiv1.13\times10^{-4}$ M solution) and its molar conductivity in water indicates dissociation to 5 ions [10].

The complex salt formulated as $NaTh(OH)\{OOC(CHOH)_2(CHO)_2COO\}\cdot2H_2O$ (possibly $NaTh(OH)_3\{OOC(CHOH)_4COO\}$) is obtained by adding an aqueous solution of the acid and NaOH dropwise, with stirring, to aqueous $Th(NO_3)_4$ (molar ratio, Th : acid : NaOH = 1 : 1 : 5). After standing for several days to complete dissolution, methanol was added to the clear solution to precipitate the compound. It is sparingly soluble in H_2O and its molar conductivity corresponds to that expected for a 2-ion electrolyte. The compound hydrolyses when heated in H_2O [10].

$NaTh(OH)\{OOC(CHOH)_4COO\}_2\cdot6H_2O$ is precipitated when methanol is added to an aqueous solution of $Th(NO_3)_4$ and Na mucate (molar ratio = 1 : 3) at pH $\leqq$ 5. It is sparingly soluble in H_2O and behaves as a 2-ion electrolyte in solution [10]. $Na_2Th(OH)_2\{OOC(CHOH)_4COO\}_2\cdot10H_2O$ is obtained in the same way, but with additional NaOH present in the solution (molar ratio, Th : Na mucate : NaOH = 1 : 3 : 2). This product is soluble in H_2O, in which it behaves as a 3-ion electrolyte [10].

The D-saccharate of composition $Th(OH)Cl\{OOC(CHOH)_4COO\}\cdot3H_2O$ (D-saccharic acid = $HOOC(CHOH)_4COOH$) is precipitated when acetone is added to an equimolar mixture of $Th(NO_3)_4$ in 1N HCl and 0.1 M D-saccharic acid [12]. The same procedure with the molar ratios of Th : D-saccharic acid equal to 1 : 2 and 1 : 3 yields precipitates in which the analytical ratios of Th : D-saccharic acid are 7 : 8 and 5 : 6, respectively. Thermogravimetric curves for these products are illustrated in [12]. A product of composition $K_2Th(OH)_2\{OOC(CHO)_2(CHOH)_2COO\}\cdot7H_2O$ is precipitated when methanol is added to the clear solution obtained by dropwise addition of a solution of $Th(NO_3)_4$ in 1N HCl to a mixture of 0.05 M potassium hydrogen D-saccharate and 0.5 M KOH (molar ratio, Th : D-saccharic acid : KOH = 1 : 2 : 7, pH > 9) [12]. All of these D-saccharates begin to lose water at 40 to 50°C [12].

The basic azelaate, $(ThOH)_2\{OOC(CH_2)_7COO\}_3\cdot4H_2O$ (azelaic acid = $HOOC(CH_2)_7COOH$), is precipitated under the same conditions as the basic adipate, $(ThOH)_2\{OOC(CH_2)_4COO\}_3\cdot4.5H_2O$ (see above) [13]. Data on its solubility in aqueous $(H,Na)ClO_4$ at varying concentrations (ionic strength = 0.1) at 25°C are given in [13]. The solubilities of the basic azelaate, in 10^{-4} mol/L (concentrations of $HClO_4$ in 10^{-3} mol/L in parentheses), were determined to be: 10.15 (104.0), 8.46 (52.0), 7.614 (41.6), 6.768 (31.2), 5.076 (20.8), 4.230 (10.4), 3.384 (6.24), 3.045 (4.16), 2.707 (3.12), and 2.199 (2.08) [13].

The sebacate, $Th\{OOC(CH_2)_8COO\}_2$, has been prepared from $ThCl_4$ and the acid in the same way as the succinate, $Th\{C_2H_4(COO)_2\}_2$ (p. 104) [11]. It is precipitated when a mixture of an aqueous Th^{IV} salt solution and a 3% methanol solution of methyl sebacate is heated [14] (product written as $Th(C_4H_2COO)_2$ in [14]) and from aqueous solutions of a Th^{IV} salt and the acid [1, 2]. It can be dried at 70 to 125°C [1, 2]. The weighing of the precipitated sebacate has

References for 15.6.3.12 on p. 114

been recommended for the gravimetric estimation of Th [1, 2, 14] after drying at 70 to 125°C [1, 2]; alternatively, the precipitate is ignited to ThO_2 and weighed [1, 2].

The precipitation of Th from aqueous solution by sebacic acid at pH 1.88 can be used as a method of separating Th from UO_2^{2+} [5].

Precipitation of Th also occurs when a solution of sebacic (or azelaic) acid in 96% ethanol is mixed with aqueous $Th(NO_3)_4$ [6], but the compositions of the precipitates were not recorded. Solubility data have been reported for the precipitate obtained when aqueous $Th(NO_3)_4$ is mixed with the "calculated" quantity of sebacic acid [9], but no analytical data were published. The sebacate is formed as a white precipitate [18].

A boiling 1% aqueous solution of camphoric acid (=1,2,2-trimethyl-1,3-cyclopentanedicarboxylic acid = $C_8H_{14}(COOH)_2$) precipitates Th quantitatively from a boiling aqueous solution of $Th(NO_3)_4$ containing CH_3COONH_4 as buffer (pH ≧ 4.2). The compound precipitated at an initial pH of ca. 5.4 was first described as the basic salt, $ThO(C_8H_{14}(COO)_2) \cdot 3H_2O$. It can be dried to constant weight at 105°C and is soluble in a concentrated aqueous solution of CH_3COONH_4 [15]. However, a later report [7] indicates that it is actually the normal salt, $Th(C_8H_{14}(COO)_2)_2$. A double precipitation with camphoric acid (initial pH ca. 5.4, final pH 4.4) can be used to separate Th from the cerite lanthanides and the precipitation can also be used for the determination of Th as ThO_2 after ignition [15]. A Th camphorate of unspecified composition is reported to be less soluble in H_2O than the La salt [16], a result consistent with the data reported in [15].

A resin based on 1,8-naphthalenedioxydiacetic acid (= $C_{10}H_6$-1,8-$(OCH_2COOH)_2$) has been used for the separation of Th^{IV} and Zr^{IV}, but no solid compounds with the acid have been recorded [17].

References for 15.6.3.12:

[1] Dupuis, T.; Duval, C. (Compt. Rend. **228** [1949] 401/2).
[2] Dupuis, T.; Duval, C. (Anal. Chim. Acta **3** [1949] 589/98).
[3] Gordon, L.; Vanselow, C. H.; Willard, H. H. (Anal. Chem. **21** [1949] 1323/5).
[4] Suryanarayana, T. V. S.; Raghava Rao, B. S. V. (J. Indian Chem. Soc. **28** [1951] 511/4).
[5] Nageswara Rao, M.; Raghava Rao, B. S. V. (Z. Anal. Chem. **142** [1954] 27/30).
[6] Kuan Pan; Hwa-Sheng Chen (J. Chinese Chem. Soc. [Taipei] [2] **3** [1956] 1/12).
[7] Wendlandt, W. W. (Anal. Chim. Acta **17** [1957] 295/9).
[8] Aggarwal, R. C.; Srivastava, T. N.; Agrawal, S. P. (J. Prakt. Chem. **10** [1960] 305/10).
[9] Wenger, P. E.; Kapetanidis, I. (Recl. Trav. Chim. **79** [1960] 567/73).
[10] Zvyagintsev, O. E.; Khromenkov, L. G. (Zh. Neorgan. Khim. **6** [1961] 2663/71; Russ. J. Inorg. Chem. **6** [1961] 1345/9).

[11] Prasad, S.; Kumar, S. (J. Indian Chem. Soc. **39** [1962] 444/6).
[12] Macarovici, C. Gh.; Czeglédi, L. (Rev. Roumaine Chim. **9** [1964] 411/24).
[13] Merkusheva, S. A.; Skorik, N. A.; Serebrennikov, V. V. (Radiokhimiya **10** [1968] 731/3; Soviet Radiochem. **10** [1968] 718/9).
[14] Singh, H. (J. Indian Chem. Soc. **51** [1974] 663).
[15] Murty, D. S. N.; Raghava Rao, B. S. V. (J. Indian Chem. Soc. **28** [1951] 218/20).
[16] Dodonov, Ya. Ya.; Pirkes, S. B. (Zh. Obshch. Khim. **26** [1956] 379/81; J. Gen. Chem. [USSR] **26** [1956] 401/2).
[17] Blasius, E.; Kynast, G. (J. Radioanal. Chem. **2** [1969] 55/71).
[18] Gonzales Portal, A.; Bermejo-Martinez, F.; Baluja-Santos, G.; Sixto-Pajaro, M. (Acta Quim. Compostelana **6** No. 2 [1982] 77/92; C.A. **100** [1984] No. 95641).

15.6.3.13 Thorium Oxodiacetates

Information on the known compounds is summarised in Table 45.

Table 45
Thorium Oxodiacetates.

oxodiacetic acid = $O(CH_2COOH)_2 = C_4H_6O_5$	
$[Th(C_4H_4O_5)_2]_n$	IR spectrum (2000 to 400 cm^{-1}) illustrated in [1]; ν_{as}(CO), 1585 cm^{-1}, ν_s(CO), 1426 cm^{-1}, ν_s(COC), 1118 cm^{-1}, ν_{as}(COC), 1040 cm^{-1}, δ(COO) + δ(ring), 723 cm^{-1}, π(COO) + δ(ring), 680 cm^{-1}, δ(ring) + ν(ThO), 474, 357, 325 cm^{-1}, ν(ThO) + δ(ring), 246, 213, 177 cm^{-1} [1]
$[Th(C_4H_4O_5)_2(H_2O)_4] \cdot 6H_2O$	crystallographic data: tetragonal, space group $P4_12_12-D_4^4$ (No. 92), lattice parameters, a = 10.335(2), c = 20.709(5) Å; Z = 4; density, d_{calc} = 2.03 g/cm^3 [3]
$\{[Th(C_4H_4O_5)(SO_4)(H_2O)_2] \cdot H_2O\}_n$	IR spectrum (two vibrational species, A and B): ν_{as}(CO), 1590 cm^{-1}, δ(CH_2) + ν_s(CO), 1470 cm^{-1}, ν_s(CO) + δ(CH_2), 1435 cm^{-1} (1422 cm^{-1} for B), ω(CH_2) + ν_{as}(CO), 1365 cm^{-1} (A), ν_{as}(CO), ω(CH_2), 1045 cm^{-1} (B), ν(CC) + δ(CO), 972 cm^{-1} (B), δ(CO) + ν_s(CO), 747 cm^{-1} (A), δ(CO) + δ(CCO) + ν_s(CO), 728 cm^{-1} (B), δ(CO) + ν_{as}(CO) + ν(CC), 620 cm^{-1} (B), δ(CO) + ν(ThO) + ν_s(CO), 555 cm^{-1} (A), ν_s(ThO) + ν_s(CO), 470 cm^{-1} (A), ν(ThO) + δ(COC), 363 cm^{-1} (A), ν_{as}(ThO) + δ(COO), 361 cm^{-1} (B), ν_{as}(ThO) + δ(CCO), 325 cm^{-1} (B), ν(ThO), 236 cm^{-1} (A), ν(ThO) + δ(ring), 200, 120 (B), 177, 108 cm^{-1} (A) [2] crystallographic data: monoclinic, space group $C2-C_2^3$ (No. 5), lattice parameters, a = 10.67(1), b = 8.35(1), c = 6.73(1) Å, β = 110.96(2)°; Z = 2; density, d_{calc} = 3.05 g/cm^3 [2]
$Th(C_4H_4O_5)_2 \cdot 3C_5H_5NO \cdot 3H_2O$	see Th E, pp. 69, 74
$Na_2[Th(C_4H_4O_5)_3] \cdot nH_2O$	n = 2 or 3; IR spectrum illustrated (2000 to 400 cm^{-1}) in [1]; ν_{as}(CO), 1645 cm^{-1}, ν_s(CO), 1406 cm^{-1}, ν_s(COC), 1107 cm^{-1}, ν_{as}(COC), 1042 cm^{-1}, δ(COO) + δ(ring), 722 cm^{-1}, π(COO) + δ(ring), 598 cm^{-1}, δ(ring) + ν(ThO), 471, 357, 316 cm^{-1} [1]
$Na_2[Th(C_4H_4O_5)_3] \cdot 2NaNO_3$	IR spectrum illustrated (2000 to 400 cm^{-1}) in [1]; ν_{as}(CO), 1653 cm^{-1}, ν_s(CO), 1409 cm^{-1}, ν_s(COC), 1113 cm^{-1}, ν_{as}(COC), 1042 cm^{-1}, δ(COO) + δ(ring), 722 cm^{-1}, π(COO) + δ(ring), 598 cm^{-1}, δ(ring) + ν(ThO), 471, 357, 314 cm^{-1} [1] crystallographic data: monoclinic, space group $C2/c-C_{2h}^6$ (No. 15), lattice parameters, a = 17.096(5), b = 9.451(2), c = 16.245(4) Å, β = 107.8(1)°; Z = 4; density, d_{calc} = 2.24 g/cm^3 [3]

Polymeric $[Th(C_4H_4O_5)_2]_n$ is reported to be formed when the aqueous acid (oxodiacetic acid = $HOOCCH_2OCH_2COOH = C_4H_6O_5$) is added to aqueous $Th(NO_3)_4$ (molar ratio = 2:1) and

References for 15.6.3.13 on p. 119

the mixture is then allowed to stand for several days. Precipitation is immediate and quantitative if the reaction is carried out in methanol [1]. The hydrated compound $[Th(C_4H_4O_5)_2(H_2O)_4]\cdot 6H_2O$ is, however, obtained in the same way from aqueous solution [3]. The Th atom in this hydrated compound is 10-coordinate, with the two $C_4H_4O_5$ groups tridentate to Th, with 4 H_2O molecules completing the coordination sphere; the coordination geometry is a bicapped square antiprism (**Fig. 7**) [3]. Bond distances and principal angles in $[Th(C_4H_4O_5)_2(H_2O)_4]\cdot 6H_2O$ (Fig. 7) are shown in Table 46.

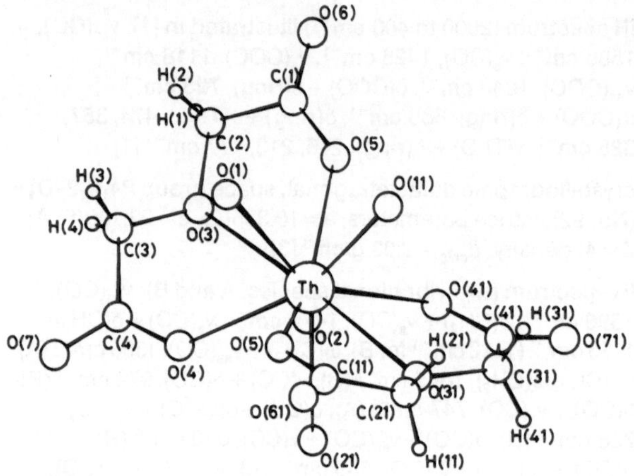

Fig. 7. The structure of thorium oxodiacetate $[Th(C_4H_4O_5)_2(H_2O)_4]\cdot 6H_2O$ viewed down a [3].

Table 46
Bond Distances (in Å) and Principal Angles (in °) of $[Th(C_4H_4O_5)_2(H_2O)_4]\cdot 6H_2O$ (Fig. 7) [3].

distances (O(1), O(2) from water, O(3) from ether)			
Th–O(3)	2.697(9)	Th–O(1)	2.479(10)
Th–O(4)	2.411(8)	Th–O(2)	2.486(8)
Th–O(5)	2.414(9)		
angles (atoms with the second figure 1 are those of symmetry y, x, −z)			
O(5)–Th–O(3)	60.0(3)	O(4)–Th–O(2)	79.8(4)
O(3)–Th–O(4)	59.6(3)	O(1)–Th–O(3)	65.7(3)
O(4)–Th–O(41)	140.3(3)	O(1)–Th–O(2)	129.4(4)
O(4)–Th–O(51)	75.6(3)	O(1)–Th–O(11)	68.9(3)
O(4)–Th–O(31)	119.4(3)	O(1)–Th–O(31)	116.6(3)
O(4)–Th–O(11)	142.0(3)	O(3)–Th–O(31)	177.5(3)
O(2)–Th–O(3)	63.7(3)		

The sulfato oxodiacetate, $\{[Th(C_4H_4O_5)(SO_4)(H_2O)_2]\cdot H_2O\}_n$, is prepared by adding an excess of $Na_2C_4H_4O_5$, as a dilute solution in aqueous H_2SO_4, to an aqueous solution of $Th(NO_3)_4$. The Th atom in this compound is 9-coordinate, with monocapped square antiprismatic coordination geometry (**Fig. 8**); the $C_4H_4O_5$ and SO_4 groups are shared between different Th atoms to form a polymeric network [2].

Interatomic distances and interbond angles in $\{[Th(C_4H_4O_5)(SO_4)(H_2O)_2]\cdot H_2O\}_n$ (Fig. 8) are shown in Table 47.

Fig. 8. The structure of the polymeric sulfato oxodiacetate $\{[Th(C_4H_4O_5)(SO_4)(H_2O)_2]\cdot H_2O\}_n$ [2].

Table 47
Interatomic Distances (in Å) and Interbond Angles (in °) of $\{[Th(C_4H_4O_5)(SO_4)(H_2O)_2]\cdot H_2O\}_n$ (Fig. 8) [2].

co-ordination					
Th–O(1^{I})	2.41(1)	O(3)–Th–O(2)	60.2(2)	O(6)–Th–O(1^{I})	68.7(5)
Th–O(2)	2.44(1)	O(3)–Th–O(4)	68.5(5)	O(6)–Th–O(1^{III})	69.4(4)
Th–O(3)	2.63(1)	O(2)–Th–O(1^{III})	73.7(4)	O(6)–Th–O(6^{II})	96.6(5)
Th–O(6)	2.44(1)	O(2)–Th–O(4)	77.1(7)	O(6)–Th–O(4^{II})	72.8(6)
Th–O(4)	2.38(1)	O(2)–Th–O(6^{II})	84.1(4)	O(1^{I})–Th–O(4)	72.9(7)
		O(2)–Th–O(4^{II})	81.9(6)	O(1^{I})–Th–O(1^{III})	115.0(4)
oxodiacetato ligand					
Th–O(1^{I})–C(1^{I})	156(1)	Th–O(2)–C(1)	129(1)	Th–O(3)–C(2)	123(1)
sulfate					
Th–O(4)–S	163(1)				

roman numeral superscripts refer to the following coordinate transformations:
I $-\frac{1}{2}+x$, $\frac{1}{2}+y$, z; II $\bar{x}$, y, $1-z$; III $\frac{1}{2}-x$, $\frac{1}{2}+y$, $1-z$

References for 15.6.3.13 on p. 119

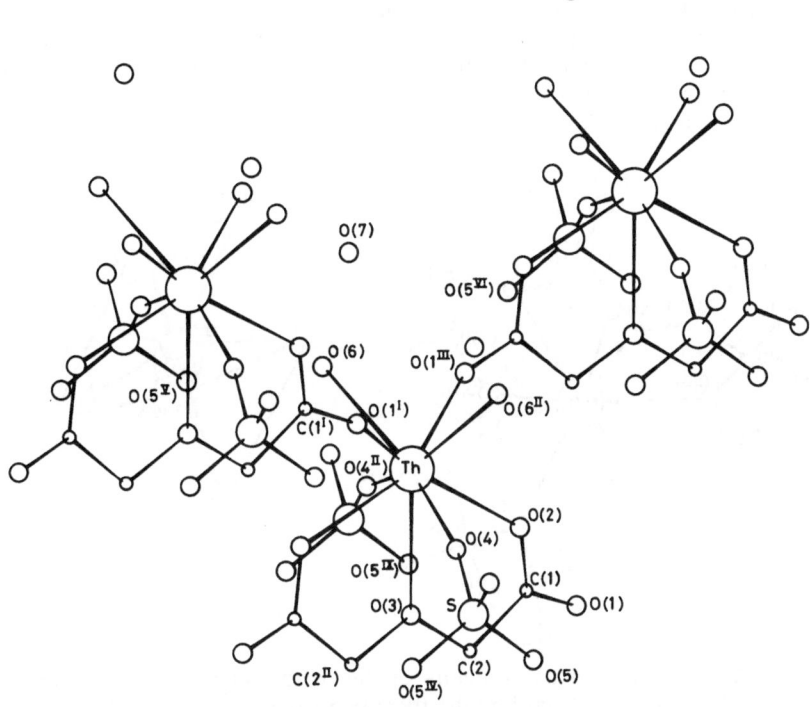

The oxodiacetato complex salt, $Na_2[Th(C_4H_4O_5)_3]\cdot nH_2O$ (n = 2 or 3), is precipitated when ether is added to the solution obtained by stirring the stoichiometric quantity of $Na_2C_4H_4O_5$ with a suspension of $[Th(C_4H_4O_5)_2]_n$ (p. 115) in aqueous methanol [1]. The salt $Na_2[Th(C_4H_4O_5)_3]\cdot 2NaNO_3$ is precipitated when a methanol solution of $Th(NO_3)_4$ is added to a solution of $Na_2C_4H_4O_5$ (molar ratio = 1:3) in the same solvent. $[Th(C_4H_4O_5)_2]_n$ is obtained if the molar ratios are 1:1 or 1:2 [1]. The Th atom in $Na_2[Th(C_4H_4O_5)_3]\cdot 2NaNO_3$ is 9-coordinate, with tricapped trigonal prismatic geometry (**Fig. 9**). Bond distances and principal angles in $Na_2[Th(C_4H_4O_5)_3]\cdot 2NaNO_3$ (Fig. 9) are shown in Table 48.

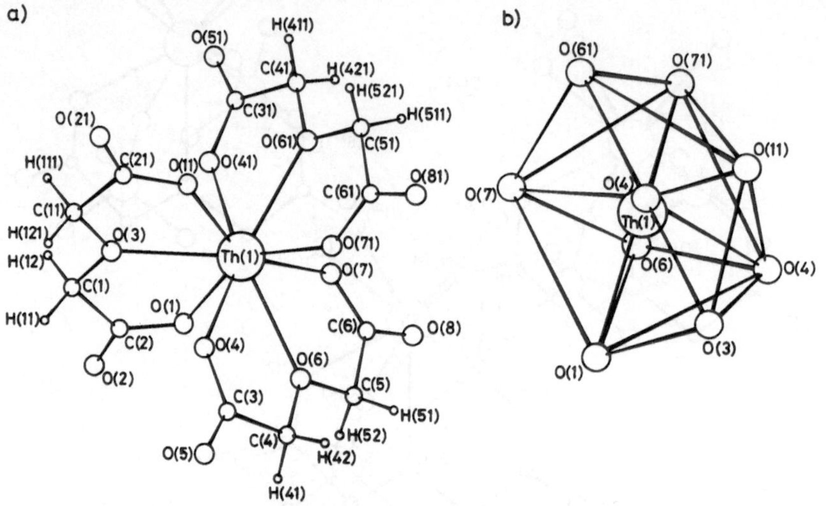

Fig. 9. View down a of the structure of the oxodiacetato complex anion in $Na_2[Th(C_4H_4O_5)_3]\cdot 2NaNO_3$ (a) and the tricapped trigonal prismatic geometry of the anion (b) [3].

Table 48

Bond Distances (in Å) and Principal Angles (in °) of $Na_2[Th(C_4H_4O_5)_3]\cdot 2NaNO_3$ (Fig. 9) [3].

thorium co-ordination polyhedron					
Th–O(1)	2.384(3)	Th–O(4)	2.402(4)	Th–O(7)	2.402(4)
Th–O(3)	2.559(5)	Th–O(6)	2.562(4)		
thorium environment (atoms having 1 as the second figure are those at 1−x, y, ½−z)					
O(1)–Th–O(3)	62.0(1)	O(1)–Th–O(4)	84.8(1)	O(4)–Th–O(7)	123.5(1)
O(4)–Th–O(6)	61.6(1)	O(1)–Th–O(6)	75.2(1)	O(4)–Th–O(41)	146.3(1)
O(6)–Th–O(7)	61.9(1)	O(1)–Th–O(7)	80.6(1)	O(4)–Th–O(61)	139.4(1)
O(1)–Th–O(11)	124.0(1)	O(3)–Th–O(4)	73.2(1)	O(4)–Th–O(71)	83.2(1)
O(1)–Th–O(41)	79.5(1)	O(3)–Th–O(6)	119.4(1)	O(6)–Th–O(61)	121.3(1)
O(1)–Th–O(61)	135.8(1)	O(3)–Th–O(7)	138.4(1)	O(6)–Th–O(71)	74.8(1)
O(1)–Th–O(71)	149.9(1)	O(4)–Th–O(6)	61.6(1)	O(7)–Th–O(71)	83.2(1)
oxodiacetato ligands					
Th–O(3)–C(1)	122.3(3)	Th–O(6)–C(4)	122.8(3)	Th–O(6)–C(5)	122.6(4)
C(2)–O(1)–Th	129.8(4)	C(3)–C(4)–Th	130.3(4)	C(6)–C(7)–Th	129.6(4)

The $C_4H_4O_5$ groups are tridentate and the $[Th(C_4H_4O_5)_3]^{2-}$ anions are connected to each other in a 3-dimensional array through carboxylic O atoms coordinated to Na atoms which are also interconnected through the NO_3 groups [3].

References for 15.6.3.13:

[1] Sbrignadello, G.; Tomat, G.; Battiston, G.; de Paoli, G.; Magon, L. (Inorg. Chim. Acta **18** [1976] 195/9).

[2] Graziani, R.; Battiston, G. A.; Casellato, U.; Sbrignadello, G. (J. Chem. Soc. Dalton Trans. **1983** 1/7).

[3] Benetollo, F.; Bombieri, G.; Tomat, G.; Bisi Castellani, C.; Cassol, A.; Di Bernardo, P. (Inorg. Chim. Acta **95** [1984] 251/61).

15.6.3.14 Thorium Phenyl-1,3-dioxodiacetate

A basic compound of composition close to $Th(OH)_2(C_{10}H_8O_6)$ is quantitatively precipitated when a hot 2% aqueous solution of phenyl-1,3-dioxodiacetic acid (= *m*-phenylenedioxodiacetic acid = $C_{10}H_{10}O_6$) is added to a boiling aqueous solution of a Th^{IV} salt and the pH is adjusted to ca. 4.5 by addition of NaOH. The composition of the precipitate is variable, but the precipitation of Th with this acid can be used for the gravimetric determination of the element following ignition to ThO_2 [1].

Reference for 15.6.3.14:

[1] Pande, C. S.; Srivastava, T. S. (Z. Anal. Chem. **167** [1959] 332/5).

15.6.3.15 Thorium Iminodiacetates and Related Compounds

The available information on these compounds is summarised in Table 49.

Table 49
Thorium Iminodiacetates and Related Compounds.

iminodiacetates (iminodiacetic acid = $HN(CH_2COOH)_2 = C_4H_7NO_4$)	
$[ThO(C_4H_5NO_4)]\cdot 2H_2O$	IR spectrum: $\nu_{as}(CO)$, 1614 cm^{-1} [4]
$[Th(C_4H_5NO_4)_2]\cdot 4H_2O$	[4]
4,4'-diaminodiphenylmethane-N,N'-diacetate (acid = $HOOCCH_2NH(C_6H_4CH_2C_6H_4)NHCH_2COOH = C_{17}H_{18}N_2O_4$)	
$Th_2(C_{17}H_{16}N_2O_4)_4\cdot 8H_2O$	decomposes >100°C, to ThO_2 at 200 to 670°C [3] IR spectrum: $\nu_{as}(CO)$, 1615 cm^{-1}, $\nu_s(CO)$, 1430 cm^{-1} [3]
N,N'-bis(2-hydroxyphenyl)ethylenediamine-N,N'-diacetate (acid = $2\text{-}HOC_6H_4(CH_2COOH)NCH_2CH_2N(CH_2COOH)C_6H_4\text{-}2\text{-}OH = C_{18}H_{20}N_2O_6$) (reported as $C_{18}H_{22}N_2O_6$ in [2])	
$[Th(C_{18}H_{16}N_2O_6)]\cdot 6H_2O$	[2] (reported as $[Th(C_{18}H_{18}N_2O_6)]\cdot 6H_2O$)

Table 49 (continued)

uramil diacetate (acid = 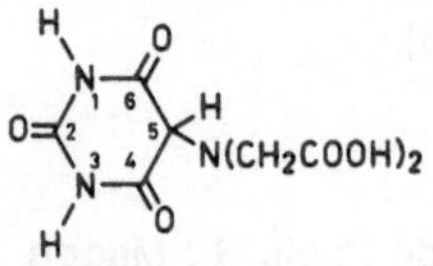

$= C_8H_9N_3O_7$)

$(NH_4)_2[Th(C_8H_6N_3O_7)_2]\cdot 4H_2O$ (?) [1]

3-aminomethylalizarin-N,N-diacetate
[3,4-dihydroxy-9,10-dioxo-9,10-dihydro-[2]-anthrylmethylimino]-diacetic acid
$= COC_6H_4COC_6H(OH)_2CH_2N(CH_2COOH)_2 = C_{19}H_{15}NO_8$)

$Th(C_{19}H_{11}NO_8)$ (?) (possibly $Th(OH)_2(C_{19}H_{13}NO_8)$)	IR spectrum: ν(OH, phenolic), 3420 cm^{-1}, ν(CO, carboxyl), 1722 cm^{-1}, ν(CO, quinone), 1620 cm^{-1}, ν(CN), 1365 cm^{-1}, ν(ThN), 510 cm^{-1}, ν(ThO), 348 cm^{-1} [5]

$[Th(C_4H_5NO_4)_2]\cdot 4H_2O$ is precipitated when methanol is added to a solution of freshly precipitated Th hydroxide in aqueous iminodiacetic acid ($= HOOCCH_2NHCH_2COOH = C_4H_7NO_4$) [4]. The basic compound, $[ThO(C_4H_5NO_4)]\cdot 2H_2O$, is precipitated immediately when aqueous $Th(NO_3)_4$ is added to an aqueous solution of $Na_2C_4H_5NO_4$ (molar ratio = 1:1 or 1:2); this product may be dimeric via OH or O bridges [4].

The 4,4'-diaminodiphenylmethane-N,N'-diacetate (acid = $HOOCCH_2NH(C_6H_4CH_2C_6H_4)$-$NHCH_2COOH = C_{17}H_{18}N_2O_4$), formulated as $Th_2(C_{17}H_{16}N_2O_4)_4\cdot 8H_2O$, separates slowly (1 to 3 h), when an ethanol, or aqueous ethanol, solution of $ThCl_4$ is added to the stoichiometric quantity of $Na_2C_{17}H_{16}N_2O_4$ in ethanol [3]. The evidence for the dimeric formulation is not clear.

Aqueous N,N'-bis(2-hydroxyphenyl)ethylenediamine-N,N'-diacetic acid (= 2-HOC_6H_4-$(CH_2COOH)NCH_2CH_2N(CH_2COOH)$-2-$C_6H_4OH = C_{18}H_{20}N_2O_6$ (reported as $C_{18}H_{22}N_2O_6$ in [2])) reacts with aqueous $Th(NO_3)_4$ (molar ratio = 1:1, pH 0.9) to form $[Th(C_{18}H_{16}N_2O_6)]\cdot 6H_2O$ (reported as $[Th(C_{18}H_{18}N_2O_6)]\cdot 6H_2O$ in [2]), which separates from the solution on the addition of a 4:1 mixture of methanol and ether [2]. The phenolic and carboxylate groups are presumably all bonded to the Th atom.

The uramyl (or uramil) diacetate, $(NH_4)_2[Th(C_8H_6N_3O_7)_2]\cdot 4H_2O$ (formula of the acid = $C_8H_9N_3O_7$, see Table 49, above) is obtained from aqueous solutions of Th^{IV} salts and the acid (molar ratio = 1:2) at pH 8.2. It presumably involves enolisation of the 2,4,6-trioxo-5-pyrimidinyl group and replacement of one enolic and two carboxylic acid protons by the Th atom. The salt behaves as a 3-ion electrolyte in H_2O [1].

Alizarin complexone (= 3-aminomethyl-alizarin-N,N-diacetic acid = $COC_6H_4COC_6H(OH)_2$-$CH_2N(CH_2COOH)_2 = C_{19}H_{15}NO_8$, see Table 49, above) yields a solid 1:1 complex, presumably either $Th(OH)_2(C_{19}H_{13}NO_8)$ or $Th(C_{19}H_{11}NO_8)$, when a 0.1M aqueous solution of $Th(NO_3)_4$ is mixed with a 0.1M aqueous solution of the acid. The IR spectrum of the product seems to indicate displacement of the phenolic protons, supporting the latter formulation, but no analytical data are reported [5].

References for 15.6.3.15:

[1] Ryabchikov, D. I.; Belyaeva, V. K. (Zh. Analit. Khim. **12** [1957] 166/74; J. Anal. Chem. [USSR] **12** [1957] 163/71).

[2] Ryabchikov, D. I.; Volynets, M. P. (Zh. Neorgan. Khim. **10** [1965] 619/27; Russ. J. Inorg. Chem. **10** [1965] 334/9).
[3] Macarovici, C. G.; Chis, E. (Rev. Roumaine Chim. **22** [1977] 657/64).
[4] Battiston, G. A.; Sbrignadello, G.; Bandoli, G.; Clemente, D. A.; Tomat, G. (J. Chem. Soc. Dalton Trans. **1979** 1965/71).
[5] Issa, R. M.; Etaiw, S. H.; El-Assy, N. B. (Ann. Chim. [Rome] **70** [1980] 33/42).

15.6.3.16 Thorium Compounds with Aliphatic Polycarboxylic Acids

15.6.3.16.1 Introduction

The acids derived from N-substituted amines (e.g., nitrilotriacetic acid, $N(CH_2COOH)_3$) or polyamines (e.g. 1,2-diaminoethane-N,N,N'N'-tetraacetic acid = ethylenediamine tetraacetic acid, H_4(edta), $(HOOCCH_2)_2NCH_2CH_2N(CH_2COOH)_2$) act as multidentate ligands and are important in analytical chemistry for the determination of Th. In many cases the published work deals mainly with the solution chemistry of the thorium/acid systems which is described in "Thorium" Suppl. Vol. D 1, 1988, pp. 89/103 and, apart from preparative details and some infrared data, little definitive information is available on the structures of the compounds formed with these acids.

15.6.3.16.2 Thorium Compounds with Tricarboxylic Acids

The known compounds and their properties are listed in Table 50.

Table 50
Thorium Compounds with Tricarboxylic Acids.

citrates (citric acid = $(HO)C(CH_2COOH)_2COOH = C_6H_8O_7$)	
$Th_3(C_6H_5O_7)_4$	brown [4]
$Th_3(C_6H_5O_7)_4 \cdot 7.5\,H_2O$	loses H_2O at 79 to 185°C, decomposes at 460°C [8]; heating curve illustrated in [8]
$(ThOH)[(ThOH)_2(C_6H_5O_7)_3]$	[2, 8]
$[Co(NH_3)_6]_2[Th(C_6H_5O_7)_2]_3 \cdot 10\,H_2O$	orange [11]
$K_4[Th(C_6H_5O_7)_2(C_2O_4)] \cdot 3\,H_2O$	decomposes at 279°C [9] IR spectrum illustrated in [9]
$ThO\{MoO_3(C_6H_6O_7)\} \cdot H_2O$	[6]
nitrilotriacetate (acid = $N(CH_2COOH)_3 = C_6H_9NO_6$)	
$(NH_4)_2[Th(C_6H_6NO_6)_2] \cdot 4\,H_2O$	[3]
(N'-2-hydroxyethyl)ethylenediamine-N,N,N'-triacetate (acid = $(HOC_2H_4)(HOOCCH_2)NCH_2CH_2N(CH_2COOH)_2 = C_{10}H_{18}N_2O_7$)	
$[Th(C_{10}H_{14}N_2O_7)] \cdot 2\,H_2O$	[5]

References for 15.6.3.16.2 on p. 123

Thorium is precipitated by citric acid (= $(HO)C(CH_2COOH)_2COOH$=$C_6H_8O_7$) from aqueous solution at pH 11.4 (pH adjusted with NaOH) but is not precipitated by aqueous NH_3 from its solution in an excess of the acid; hydrolysis does not occur at high pH, in contrast to the behaviour of Zr [1]. The composition of the precipitate is not given in [1].

The normal citrate, $Th_3(C_6H_5O_7)_4$, has been prepared from $ThCl_4$ and the acid in the same way as the succinate, $Th\{C_2H_4(COO)_2\}_2$ (p. 104), except that the product was extracted into ether and the extract evaporated to dryness [4]. The brown colour of this product suggests that some decomposition has occurred.

The hydrated citrate, $Th_3(C_6H_5O_7)_4 \cdot 7.5H_2O$, is obtained by mixing equimolar aqueous solutions of $Th(NO_3)_4$ and citric acid; it loses water at 79 to 185°C [8]. The solubility product, $pK_s = 56.37 \pm 0.30$, was derived from solubility data in 0.1M (H, Na)ClO_4 at 25°C in which the solubility of the compound was found to decrease from 1.68×10^{-3}M when $[H^+] = 4.336 \times 10^{-2}$M to 1.006×10^{-4}M when $[H^+] = 1.951 \times 10^{-3}$M [8].

A product of empirical composition $[Th(C_6H_5O_7)]_n$ is precipitated when aqueous $Th(NO_3)_4 \cdot 4H_2O$ is titrated against $Na_3C_6H_5O_7 \cdot 2H_2O$. It redissolves when the molar ratio Th : $(C_6H_5O_7)^{3-}$ = 2 : 3, possibly with the formation of the complex anion $[(ThOH)_2(C_6H_5O_7)_3]^{3-}$ [2]. The insoluble product may be $[Th(OH)][(ThOH)_2(C_6H_5O_7)_3]$ [2] (or, simply, $Th(OH)(C_6H_5O_7)$) and the formation of a compound of this composition has been confirmed [8]. A study of the solubility of this basic citrate in 0.1M solutions of (H, Na)ClO_4 at 25°C suggests that the complex cation $[Th(C_6H_5O_7)]^+$ is formed [8].

The citrato complex salt, $[Co(NH_3)_6]_2[Th(C_6H_5O_7)_2]_3 \cdot 10H_2O$, is precipitated when aqueous $[Co(NH_3)_6]Cl_3$ is added to an aqueous solution containing a Th^{IV} salt and citric acid. The complex salt appears to be best precipitated at pH 6.0 from approximately 0.02 M $C_6H_8O_7$ solution [11].

The oxalato citrato complex salt, $K_4[Th(C_6H_5O_7)_2(C_2O_4)] \cdot 3H_2O$, is precipitated when ethanol (3 vols.) is added to the solution obtained by dissolving solid $Th(C_2O_4)_2 \cdot 6H_2O$ (p. 82) in 0.1M $K_3C_6H_5O_7$ at 60°C [9].

A molybdocitrate of composition $ThO\{MoO_3(C_6H_6O_7)\} \cdot H_2O$ is precipitated when aqueous $Th(NO_3)_4$ is added to an aqueous solution of sodium molybdocitrate [6].

The nitrilotriacetato complex salt, $(NH_4)_2[Th(C_6H_6NO_6)_2] \cdot 4H_2O$, is precipitated from aqueous solutions of Th^{IV} salts and the acid (= $N(CH_2COOH)_3$ = $C_6H_9NO_6$) in the molar ratio Th : acid = 1 : 2 at pH 8.2 on the addition of methanol and ether. It behaves as a 3-ion electrolyte in 10^{-3}M aqueous solution [3]. There is also solution chemistry evidence for the formation of the $[Th(C_6H_6NO_6)]^+$ cation [7] (see "Thorium" Suppl. Vol. D 1, 1988, p. 98). The attempted precipitation of a Th nitrilotriacetato complex salt with $[Co(NH_3)_6]Cl_3$, $[Co(en)_3]Cl_3 \cdot 3H_2O$, $[Co(H_2N(CH_2)_3NH_2)_3]Cl_3$ or $[Cr(ur)_6]Cl_3 \cdot 3H_2O$ was unsuccessful [12].

The (N'-2-hydroxyethyl)ethylenediamine-N, N, N'-triacetate (acid = $(HOC_2H_4)(HOOCCH_2)$-$NCH_2CH_2N(CH_2COOH)_2$ = $C_{10}H_{18}N_2O_7$), $[Th(C_{10}H_{14}N_2O_7] \cdot 2H_2O$, is obtained by treating an equimolar mixture of aqueous $Th(NO_3)_4$ and the acid (at pH 0.9) with a 4 : 1 mixture of methanol and ether [5]. When N, N''-bis(2-hydroxybenzyl)-N, N', N''-diethylenetriamine triacetic acid (= $(2\text{-}HOC_6H_4CH_2)N(CH_2COOH)CH_2CH_2N(CH_2COOH)CH_2CH_2N(CH_2COOH)(CH_2C_6H_4\text{-}2\text{-}OH)$) is added to a suspension of freshly precipitated Th^{IV} hydroxide, the latter dissolves on stirring or warming. A thorium compound separates when this solution is evaporated to small volume and left to stand [10], but the composition of this product is not stated. Other ligands of the type $(2\text{-}HOC_6H_4CH_2)\text{-}[N(CH_2COOH)CH_2CH_2]_x\text{-}N(CH_2COOH)(CH_2C_6H_4\text{-}2\text{-}OH)$ (x = 0 or 2) are mentioned in [10] without preparative or analytical details.

References for 15.6.3.16.2:

[1] Haïssinsky, M.; Jeng-Tseng, Yang (Anal. Chim. Acta **3** [1949] 422/7).
[2] Bobtelsky, M.; Graus, B. (J. Am. Chem. Soc. **76** [1954] 1536/9).
[3] Ryabchikov, D. I.; Belyaeva, V. K. (Zh. Analit. Khim. **12** [1957] 166/74; J. Anal. Chem. [USSR] **12** [1957] 163/71).
[4] Prasad, S.; Kumar, S. (J. Indian Chem. Soc. **39** [1962] 444/6).
[5] Ryabchikov, D. I.; Volynets, M. P. (Zh. Neorgan. Khim. **10** [1965] 619/27; Russ. J. Inorg. Chem. **10** [1965] 334/9).
[6] Prasad, S.; Pandey, L. P. (J. Proc. Inst. Chemists [India] **37** [1965] 207/11; C.A. **64** [1966] 6072).
[7] Skorik, N. A.; Kumok, V. N.; Serebrennikov, V. V. (Zh. Neorgan. Khim. **12** [1967] 3381/4; Russ. J. Inorg. Chem. **12** [1967] 1788/90).
[8] Skorik, N. A.; Kumok, V. N.; Serebrennikov, V. V. (Radiokhimiya **9** [1967] 515/7; Soviet Radiochem. **9** [1967] 499/500).
[9] Grinberg, A. A.; Petrzhak, G. I.; Lozhkina, G. S. (Radiokhimiya **13** [1971] 836/40; Soviet Radiochem. **13** [1971] 862/5).
[10] Martell, A. E. (U.S. 3632637 [1972]; C.A. **76** [1972] No. 126614).

[11] Hoshi, M.; Ueno, K. (Radiochem. Radioanal. Letters **30** [1977] 145/53).
[12] Hoshi, M.; Ueno, K. (J. Nucl. Sci. Technol. [Tokyo] **17** [1980] 370/6).

15.6.3.16.3 Thorium Compounds with Aliphatic Amine-N-Tetracarboxylic Acids

Information on the known compounds is summarised in Table 51.

Table 51
Thorium Compounds with Aliphatic Amine-N-Tetracarboxylic Acids.

ethylenediamine-N, N, N′, N′-tetraacetates (acid = $(HOOCCH_2)_2NCH_2CH_2N(CH_2COOH)_2 = C_{10}H_{16}N_2O_8$)	
$Th(C_{10}H_{12}N_2O_8)$	[1]; decomposes >260°C [7]
$Th(C_{10}H_{12}N_2O_8) \cdot 0.5H_2O$	loses $0.5H_2O$ at 214°C [1]
$Th(C_{10}H_{12}N_2O_8) \cdot 1.45H_2O$(?)	loses all H_2O at 160°C/10 d [7]
$Th(C_{10}H_{12}N_2O_8) \cdot 2H_2O$	[8]; loses $0.55H_2O$ at 110°C/6 h [7], loses $1.5H_2O$ at 140°C [1] IR spectrum: ν(CO), 1610, 1575 cm^{-1} [7]
$Th(C_{10}H_{12}N_2O_8) \cdot 3H_2O$	[9]
$Th(C_{10}H_{12}N_2O_8) \cdot 6H_2O$	[18]
$Th(C_{10}H_{12}N_2O_8) \cdot 9H_2O$	IR spectrum illustrated in [3] (1800 to 800 cm^{-1}) and (4) (2920 to 740 cm^{-1}), ν_{as}(CO), 1630, 1535 cm^{-1}, ν_s(CO), 1410, 1395 cm^{-1}, ν(CN), 1095 cm^{-1} [3, 4]; pK(stability constant) = 23.2 [3]
$Na[Th(OH)(C_{10}H_{12}N_2O_8)] \cdot 4H_2O$	[2]; IR spectrum, ν(CO), 1610, 1575 cm^{-1} [7]
$NH_4[Th(OH)(C_{10}H_{12}N_2O_8)] \cdot 4H_2O$	[2]

References for 15.6.3.16.3 on pp. 126/7

Table 51 (continued)

$(CN_3H_6)_4[Th(C_{10}H_{12}N_2O_8)(CO_3)_2]$ (CN_3H_6 = guanidinium)	[14]
$(CN_3H_6)_3[Th(C_{10}H_{12}N_2O_8)F_3]$	[14]; crystallographic data; monoclinic, space group $P2_1$-C_2^2 (No. 4), lattice parameters, a = 11.010(3), b = 11.617(3), c = 11.451(3) Å, β = 112.39(2)°; Z = 2 [16]
1,2-diaminopropane-N,N,N',N'-tetraacetate (acid = $(HOOCCH_2)_2NCH(CH_3)CH_2N(CH_2COOH)_2 = C_{11}H_{18}N_2O_8$)	
$Th(C_{11}H_{14}N_2O_8)\cdot xH_2O$(?)	[11]
1,6-diaminohexane-N,N,N',N'-tetraacetates (acid = $(HOOCCH_2)_2N(CH_2)_6N(CH_2COOH)_2 = C_{14}H_{24}N_2O_8$)	
$Th(C_{14}H_{20}N_2O_8)$	decomposes >280°C [10], decomposes to ThO_2 at 225 to 425°C [6]
$Th(C_{14}H_{20}N_2O_8)\cdot 3H_2O$(?)	loses $3H_2O$ at 50 to 170°C [6] thermogravimetric curve illustrated in [6]
$Th(C_{14}H_{20}N_2O_8)\cdot 4.5H_2O$	loses $4.5H_2O$ at 50 to 150°C [10]; solubility product, $S = 9.79 \pm 0.56 \times 10^{-32}$ at 20°C in (H, $NaClO_4)_{aq}$ (μ = 0.1) [10]
1,2-diaminocyclohexane-N,N,N',N'-tetraacetates (acid = 1,2-$(HOOCCH_2)_2NC_6H_{10}N(CH_2COOH)_2 = C_{14}H_{22}N_2O_8$)	
$Th(C_{14}H_{18}N_2O_8)$	decomposes to ThO_2 at 270°C [5]
$Th(C_{14}H_{18}N_2O_8)\cdot H_2O$	loses H_2O at 105°C [5] thermogram illustrated in [5]
$NH_4[Th(OH)(C_{14}H_{18}N_2O_8)]\cdot 5H_2O$	[2]

The anhydrous compound with ethylenediamine tetraacetic acid (= $(HOOCCH_2)_2NCH_2CH_2$-$N(CH_2COOH)_2 = C_{10}H_{16}N_2O_8$), $Th(C_{10}H_{12}N_2O_8)$, is obtained by heating the hydrate, $Th(C_{10}H_{12}N_2O_8)\cdot xH_2O$ at 160°C/10 d (x = 2 [7]) or at 214°C (x = 0.5 [1]). The hydrate with x = 0.5 is obtained when the dihydrate is dried at 140°C [1], and an intermediate with x = 1.45 appears to be formed when the dihydrate is heated at 110°C/6 h [7]. The last (x = 1.45) and the anhydrous salt are hygroscopic [7].

$Th(C_{10}H_{12}N_2O_8)\cdot 2H_2O$ is prepared by adding $Th(NO_3)_4$ to a boiling aqueous solution of $C_{10}H_{16}N_2O_8$ and then concentrating the filtrate [1]. It is also precipitated when the stoichiometric quantity of solid $ThCl_4$ is added to a near-boiling solution of $Na_2H_2(C_{10}H_{12}N_2O_8)\cdot 2H_2O$ (final pH ca. 1.5) and crystallisation can be enhanced by the addition of ethanol or acetone [7, 8] or by concentrating the solution [8]. The compound is thought to be polymeric, $[Th(C_{10}H_{12}N_2O_8)]_n\cdot 2nH_2O$, with bridging $C_{10}H_{12}N_2O_8$ groups [8].

A trihydrate, $Th(C_{10}H_{12}N_2O_8)\cdot 3H_2O$, is apparently precipitated when a 4:1 mixture of methanol and ether is added to an aqueous solution containing equimolar amounts of $Th(NO_3)_4$ and the complexone at pH 0.9 [9]. The hexahydrate, $Th(C_{10}H_{12}N_2O_8)\cdot 6H_2O$, is obtained when the guanidinium salt, $(CN_3H_6)_4[Th(C_{10}H_{12}N_2O_8)(CO_3)_2]$ (p. 125), is treated with 4 equivalents of HBr [18]. Another hydrate, $Th(C_{10}H_{12}N_2O_8)\cdot 9H_2O$, is reported to be formed when the stoichiometric quantity of $Na_2H_2(C_{10}H_{12}N_2O_8)\cdot 2H_2O$ is added to an aqueous solution of $Th(NO_3)_4$

$\cdot 4H_2O$, followed by neutralisation with the stoichiometric quantity of $NaHCO_3$ and adjustment to pH 9. After evaporation to small volume, methanol was added to induce crystallisation and the solution was cooled in an ice bath. This hydrate can be recrystallised from aqueous ethanol [4]. The IR spectrum of this hydrate indicates that the bonding to the $C_{10}H_{12}N_2O_8$ group is primarily covalent [3]. The IR and ^1H NMR spectra of the 1:1 Th complex in D_2O have also been reported [17].

The complex salts $M^I[Th(OH)(C_{10}H_{12}N_2O_8)]\cdot 4H_2O$ (M^I = Na [2, 7], NH_4 [2]) have been obtained from an aqueous solution containing equimolar quantities of Th^{IV} and $C_{10}H_{12}N_2O_8^{4-}$ at pH 7.0 [2]. The Na salt is also precipitated from a solution of $Th(C_{10}H_{12}N_2O_8)\cdot 2H_2O$ in aqueous NaOH at pH 7 on the addition of ethanol or acetone [7]. The attempted precipitation of salts of ethylenediamine tetraacetato Th^{IV} complexes with $[Co(NH_3)_6]Cl_3$, $[Co(en)_3]Cl_3\cdot 3H_2O$, $[Co(H_2N(CH_2)_3NH_2)_3]Cl_3$ or $[Cr(ur)_6]Cl_3\cdot 3H_2O$ was unsuccessful [13].

Mixed carbonato and fluoro complex salts are also known. The guanidinium compound $(CN_3H_6)_4[Th(C_{10}H_{12}N_2O_8)(CO_3)_2]\cdot 6H_2O$ and the anhydrous compound have been obtained by treating $Th(NO_3)_4\cdot 5H_2O$ with the stoichiometric quantities of $(CN_3H_6)_2CO_3$ and $C_{10}H_{16}N_2O_8$. The analogous fluoride complex salt, $(CN_3H_6)_3[Th(C_{10}H_{12}N_2O_8)F_3]$ has also been reported [18]. Both compounds are isomorphous with their U^{IV} analogues [14]. The Th atom in the trifluoro complex anion is 9-coordinate, bonded to three F, two N, and four O atoms in a monocapped tetragonal antiprismatic arrangement (**Fig. 10**) in which the skeleton of the hexadentate $C_{10}H_{12}N_2O_8$ group has the gauche conformation [14, 16]. Bond distances and angles in $(CN_3H_6)_3[Th(C_{10}H_{12}N_2O_8)F_3]$ (Fig. 10) are given in Table 52, p. 126.

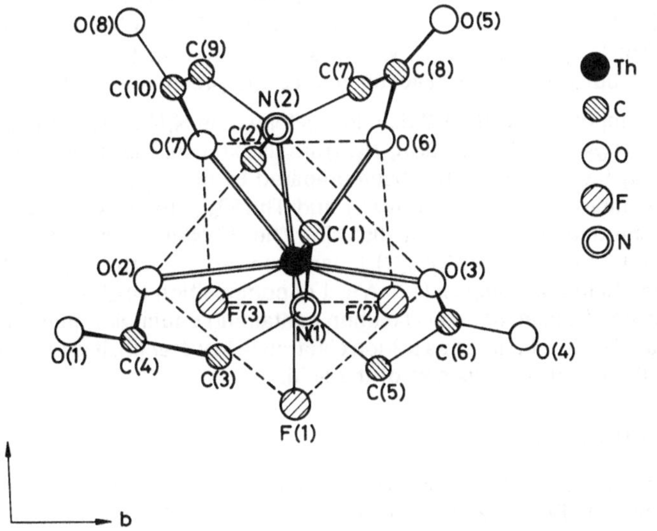

Fig. 10. The structure of the complex anion in the guanidinium salt $(CN_3H_6)_3[Th(C_{10}H_{12}N_2O_8)F_3]$ [16].

Potentiometric studies indicate that mixed $C_{10}H_{12}N_2O_8^{4-}$/amino acid complex anions of the type $[Th(C_{10}H_{12}N_2O_8)A]^{2-}$, where A represents the aspartate or glutamate anion, are formed in aqueous solution [15], but salts of these anions were not isolated.

An aqueous solution of 1,2-diaminopropane-N,N,N',N'-tetraacetic acid (= $(HOOCCH_2)_2$-$NCH(CH_3)CH_2N(CH_2COOH)_2$ = $C_{11}H_{18}N_2O_8$) dissolves freshly precipitated Th^{IV} hydroxide with

References for 15.6.3.16.3 on pp. 126/7

difficulty, and the addition of aqueous NH_3 to the solution does not reprecipitate the hydroxide [11]; the product may be $Th(C_{11}H_{14}N_2O_8) \cdot xH_2O$.

Table 52
Bond Distances (in Å) and Angles (in °) of
$(CN_3H_6)_3[Th(C_{10}H_{12}N_2O_8)F_3]$ (Fig. 10, p. 125) [16].

Th–O(2)	2.50(2)	O(3)–Th–O(6)	77.5(9)
Th–O(3)	2.48(3)	O(7)–Th–O(2)	66.6(8)
Th–O(6)	2.46(3)	O(7)–Th–O(6)	73.3(8)
Th–O(7)	2.50(3)	F(1)–Th–O(2)	80.1(8)
Th–F(1)	2.27(1)	F(1)–Th–O(3)	77.6(8)
Th–F(2)	2.29(2)	F(2)–Th–O(7)	67.3(9)
Th–F(3)	2.30(2)	F(3)–Th–O(3)	76.4(8)
Th–N(1)	2.78(1)	F(3)–Th–O(6)	74.3(8)
Th–N(2)	2.77(2)	F(2)–Th–F(3)	75.9(8)
		F(2)–Th–F(1)	79.2(8)

Addition of an aqueous solution of the disodium salt of 1,6-diaminohexane-N,N,N',N'-tetraacetic acid (= $(HOOCCH_2)_2N(CH_2)_6N(CH_2COOH)_2 = C_{14}H_{24}N_2O_8$) to aqueous $Th(NO_3)_4$ precipitates $Th(C_{14}H_{20}N_2O_8) \cdot 3H_2O$, which is dehydrated at 50 to 170°C [6]. Another hydrate, of composition $Th(C_{14}H_{20}N_2O_8) \cdot 4.5H_2O$, which is dehydrated at 50 to 150°C, is obtained in a similar manner, but using the tetraacetic acid as the precipitant [10].

Thorium compounds with 1,2-diaminocyclohexane-N,N,N',N'-tetraacetic acid (=1,2-$(HOOCCH_2)_2NC_6H_{10}N(CH_2COOH)_2 = C_{14}H_{22}N_2O_8$) are also known. The monohydrate, $Th(C_{14}H_{18}N_2O_8) \cdot H_2O$, apparently the *cis* isomer, has been prepared by mixing aqueous solutions containing equimolar quantities of $Na_2H_2(C_{14}H_{18}N_2O_8)$ and $Th(NO_3)_4$, then evaporating the solution to crystallise the product. It loses H_2O at 105°C [5]. The ^{13}C NMR spectrum of a solution species, described as $[Th(\textit{trans}\text{-}C_{14}H_{18}N_2O_8)(H_2O)_n]$, has been reported; the $C_{14}H_{18}N_2O_8$ group appears to be hexadentate in this complex [12]. A salt of composition $NH_4[Th(OH)(C_{14}H_{18}N_2O_8)] \cdot 5H_2O$ has been obtained from an aqueous solution containing equimolar quantities of Th^{IV} and the complexone at pH 9.0. It behaves as a 2-ion electrolyte in water [2]. It is not clear whether this is derived from the *cis* or *trans* isomer of the acid.

References for 15.6.3.16.3:

[1] Brintzinger, H.; Thiele, H.; Müller, U. (Z. Anorg. Allgem. Chem. **251** [1943] 285/94).
[2] Ryabchikov, D. I.; Belyaeva, V. K. (Zh. Analit. Khim. **12** [1957] 166/74; J. Anal. Chem. [USSR] **12** [1957] 163/71).
[3] Sawyer, D. T. (Ann. N.Y. Acad. Sci. **88** [1960] 307/21).
[4] Sawyer, D. T.; McKinnie, J. M. (J. Am. Chem. Soc. **82** [1960] 4191/6).
[5] Saraiya, S. C.; Sundaram, A. K. (J. Sci. Ind. Res. [India] B **21** [1962] 264/5).
[6] Ryabchikov, D. I.; Volynets, M. P.; Zarinskii, V. A. (Zh. Analit. Khim. **18** [1963] 542/4; J. Anal. Chem. [USSR] **18** [1963] 466/8).
[7] Langer, H. G. (J. Inorg. Nucl. Chem. **26** [1964] 59/72).
[8] Langer, H. G. (U.S. 3198817 [1965]; C.A. **63** [1965] 9515).
[9] Ryabchikov, D. I.; Volynets, M. P. (Zh. Neorgan. Khim. **10** [1965] 619/27; Russ. J. Inorg. Chem. **10** [1965] 334/9).

[10] Podkorytova, N. V.; Skorik, N. A. (Radiokhimiya **11** [1969] 115/6; Soviet Radiochem. **11** [1969] 111/2).

[11] Vicente-Perez, S.; Hernandez, L.; Hernandez Mendez, J. (Inform. Quim. Anal. **26** [1972] 30/43).
[12] Howarth, O. W.; Moore, P.; Winterton, N. (J. Chem. Soc. Dalton Trans. **1975** 360/8).
[13] Hoshi, M.; Ueno, K. (J. Nucl. Sci. Technol. [Tokyo] **17** [1980] 370/6).
[14] Shchelokov, R. N.; Mikhailov, Yu. N.; Orlova, I. M.; Sergeev, A. V. (Dokl. Akad. Nauk SSSR **272** [1983] 380/2; Dokl. Chem. Proc. Acad. Sci. USSR **268/273** [1983] 297/9).
[15] Saxena, A. K.; Singh, M. K.; Tiwari, R. C.; Srivastava, M. N. (J. Chem. Soc. Pakistan **6** [1984] 231/4).
[16] Mikhailov, Yu. N.; Lobanova, G. M.; Kanishcheva, A. S.; Sergeev, A. V.; Bolotova, G. T.; Shchelokov, R. N. (Koord. Khim. **11** [1985] 545/50; Soviet J. Coord. Chem. **11** [1985] 312/7).
[17] Martell, A. E.; Fried, A. R., Jr. (J. Am. Chem. Soc. **93** [1971] 4695/700).
[18] Shchelokov, R. N.; Bolotova, G. T.; Sergeev, A. V.; Mikhailov, Yu. N.; Zhilov, V. I. (Koord. Khim. **11** [1985] 498/502).

15.6.3.16.4 Thorium Compounds with Aliphatic Amine-N-Pentacarboxylic Acids

The known compounds are listed in Table 53.

Table 53
Thorium Compounds with Aliphatic Amine-N-Pentacarboxylic Acids.

diethylenetriaminepentaacetates	
(acid = $(HOOCCH_2)_2NCH_2CH_2N(CH_2COOH)CH_2CH_2N(CH_2COOH)_2 = C_{14}H_{23}N_3O_{10}$)	
$H[Th(C_{14}H_{18}N_3O_{10})] \cdot H_2O$	[1]; IR spectrum: ν(CO), 1600 cm^{-1}, ν(CN), 1085 cm^{-1} [2]
$Na[Th(C_{14}H_{18}N_3O_{10})] \cdot 4H_2O$	[3]

Ultracentrifugation experiments have provided evidence for the formation of the diethylene triamine pentaacetate, $HTh(C_{14}H_{18}N_3O_{10})$ (acid = $(HOOCCH_2)_2NCH_2CH_2N(CH_2COOH)CH_2CH_2$-$N(CH_2COOH)_2 = C_{14}H_{23}N_3O_{10}$) [1], and the monohydrated compound has been obtained by adding aqueous $Th(NO_3)_4 \cdot 4H_2O$ to an aqueous solution of $C_{14}H_{23}N_3O_{10}$, followed by evaporation of the solution to 1/9 of its original volume and addition of 6 volumes of acetone to precipitate the compound [2]. The ^1H NMR and IR spectra of the 1:1 complex in D_2O have also been reported [4].

The sodium salt, $Na[Th(C_{14}H_{18}N_3O_{10})] \cdot 4H_2O$, is salted out when an aqueous solution containing equimolar quantities of $Th(NO_3)_4$ and the sodium salt of the complexone at pH 0.9 is treated with a 4:1 mixture of methanol and ether [3].

References for 15.6.3.16.4:

[1] Gustafson, R. L.; Martell, A. E. (J. Am. Chem. Soc. **82** [1960] 5610/6).
[2] Sievers, R. E.; Bailar, J. C., Jr. (Inorg. Chem. **1** [1962] 174/82).
[3] Ryabchikov, D. I.; Volynets, M. P. (Zh. Neorgan. Khim. **10** [1965] 619/27; Russ. J. Inorg. Chem. **10** [1965] 334/9).
[4] Martell, A. E.; Fried, A. R., Jr. (J. Am. Chem. Soc. **93** [1971] 4695/700).

15.6.3.16.5 Thorium Compound with an Aliphatic Amine-N-Hexacarboxylic Acid

The IR and ^1H NMR spectra of a 1:1 complex of Th^{IV} with triethylene tetramine hexaacetic acid ($=(HOOCCH_2)_2N(CH_2)_2N(CH_2COOH)(CH_2)_2N(CH_2COOH)(CH_2)_2N(CH_2COOH)_2 = C_{18}H_{30}N_4O_{12}$) in D_2O have been reported [1], but the complex, which is probably $H_2Th(C_{18}H_{24}N_4O_{12})$, was not isolated.

Reference for 15.6.3.16.5:

[1] Fried, A. R., Jr.; Martell, A. E. (J. Coord. Chem. 1 [1971] 47/56).

15.6.4 Aromatic Carboxylates of Thorium

15.6.4.1 Introduction

The precipitation of Th from aqueous solution by aromatic monocarboxylic acids has been extensively investigated for possible application to the analytical determination of Th and for its separation from the lanthanides. In many cases the precipitates were not analysed, but merely ignited to ThO_2 for weighing, so that there is some scope for further work in this area.

The effect of the ring positions of the substituents in substituted benzoic acids on the constitutions of the Th carboxylates formed with them has been discussed in some detail. Substitution by NO_2, Cl, Br or SO_3H in any position, or by OH, OR, NH_2 or NHR in the *m*- or *p*-positions of C_6H_5COOH does not influence the composition of the complex or its extractibility into the higher alcohols, whereas substitution by OH, OR, NH_2 or NHR in the *o*-position results in the formation of compounds of the type $Th(OH)_2L_2$ (or ThL_2), where HL is the acid involved. These are potentially extractible by the higher alcohols [1].

Reference for 15.6.4.1:

[1] Tserkovnitskaya, I. A.; Charykov, A. K. (Izv. Vysshikh Uchebn. Zavedenii Khim. Khim. Tekhnol. 7 [1964] 544/50).

15.6.4.2 Thorium Benzoates

Information on the known compounds is summarised in Table 54.

Table 54
Thorium Benzoates.

$Th(C_6H_5COO)_4$	[2, 4, 11]; decomposes >155°C [8], 300°C [5], 300 to 400°C [10], to ThO_2 at 510°C [12], maximum decomposition, 388 to 420°C [8] thermogravimetric analysis reported in [8], thermolysis curve illustrated in [12] diamagnetic susceptibility, $\chi_{mol} = -214 \times 10^{-6}$ [6]
$Th(OH)_2(C_6H_5COO)_2$	[4]

The benzoate, $Th(C_6H_5COO)_4$, is precipitated when a concentrated aqueous solution of C_6H_5COONa is mixed with an aqueous solution of $Th(NO_3)_4$ [5, 9]; the mixture was then heated on a water bath for ca. 2 h [5]. The optimum pH for the precipitation is reported to be 2.2 to 2.6 [11]. It has also been prepared by heating $Th(CH_3COO)_4$ (p. 47) with an excess of C_6H_5COOH, removing the liberated CH_3COOH under reduced pressure. Unreacted C_6H_5COOH was removed from the product by washing it with ether [8]. The compound is also obtained by heating the basic benzoate, $Th(OH)_2(C_6H_5COO)_2$ (see below), with C_6H_5COOH in H_2O on a steam bath for 5 h [4]. It is also reported to be formed in the thermal decomposition of the 3-nitro-N-*m*-tolylbenzohydroxamate (see Th E, p. 170) [10]. The factors affecting the crystal form and anisometry of Th benzoate, which crystallises as prisms, have also been studied [7].

The precipitation of Th by benzoic acid is incomplete below pH 1.8, and in an alternative preparation of $Th(C_6H_5COO)_4$ 2% aqueous C_6H_5COOH was added to a boiling aqueous solution of a Th^{IV} salt at pH 2.2 to 2.6. The precipitate became crystalline on further boiling for 10 min. The product had a tendency to hydrolyse when washed with H_2O [2] and ignition of the precipitate to ThO_2 is recommended for analytical purposes [2, 3]. The precipitation of the benzoate in this way has been used for the separation of Th from the lanthanides [3]. Th has also been separated from the cerite lanthanides by precipitation, probably as $Th(C_6H_5COO)_4$, by adding hot 1% aqueous C_6H_5COOH to a boiling aqueous solution containing Th^{IV}, followed by addition of 5% aqueous CH_3COONH_4 to neutrality, with subsequent redissolution in hot, dilute HNO_3 and reprecipitation [1]. This separation was also achieved by adding cold 3% aqueous $C_6H_5COONH_4$ to a solution containing Th^{IV} which had been acidified with glacial CH_3COOH. After standing on a water bath, the precipitate was redissolved in dilute HCl, neutralised with aqueous NH_3, and the precipitation repeated [1].

$Th(OH)_2(C_6H_5COO)_2$ is precipitated when 2 M C_6H_5COONa is added to aqueous $Th(NO_3)_4$ at pH 6.0 [4].

The thermal decomposition at 600, 800, and 1000°C of an unspecified Th benzoate (precipitated from aqueous $Th(NO_3)_4$) has been reported in an investigation of the powder characteristics of the ThO_2 produced by thermal decomposition of a variety of Th compounds [13].

References for 15.6.4.2:

[1] Venkataramaniah, M.; Satyanarayanamurthy, T. K.; Raghava Rao, B. S. V. (J. Indian Chem. Soc. **27** [1950] 81/6).
[2] Venkataramaniah, M.; Rao, C. L.; Raghava Rao, B. S. V. (Analyst **77** [1952] 103/5).
[3] Wengert, G. B.; Walker, R. C.; Loucks, M. F.; Stenger, V. A. (Anal. Chem. **24** [1952] 1636/8).
[4] Baldwin, W. H. (ORNL 2864 [1960] 1/6; C.A. **1960** 11731).
[5] Okubo, M.; Goto, R. (Nippon Kagaku Zasshi **81** [1960] 1132/6).
[6] Belova, V. L.; Syrkin, Ya. K.; Molodkin, A. K.; Ivanova, O. M.; Shiporina, I. M. (Zh. Neorgan. Khim. **13** [1968] 1458/60; Russ. J. Inorg. Chem. **13** [1968] 766/7).
[7] Packter, A.; Chauhan, P.; Saunders, D. F. (Krist. Tech. **4** [1969] 45/55).
[8] Paul, R. C.; Saran, M. S.; Bains, M. S. (Indian J. Chem. **7** [1969] 384/6).
[9] Kazachenko, D. V. (Mater. 4th Nauchn. Konf. Aspirantov Rostov-na-Donu Univ., Rostov-on-Don, USSR, 1962, pp. 156/7; C.A. **60** [1964] 13124).
[10] Agrawal, Y. K.; Sharma, T. P. (J. Indian Chem. Soc. **54** [1977] 771/3).

[11] Tserkovnitskaya, I. A.; Charykov, A. K. (Izv. Vysshikh Uchebn. Zavedenii Khim. Khim. Tekhnol. **7** [1964] 544/50; C.A. **62** [1965] 3385).
[12] Wendlandt, W. W. (Anal. Chem. **29** [1957] 800/2).
[13] Moorthy, V. K.; Kulkarni, A. K. (Trans. Indian Ceram. Soc. **22** [1963] 116/29; C.A. **61** [1964] 10423).

15.6.4.3 Thorium Salicylates and a Salicylato Complex

Information on the known compounds is summarised in Table 55.

Table 55
Thorium Salicylates (=2-hydroxybenzoates) and Salicylato Complex.

$Th(2\text{-}OC_6H_4COO)(2\text{-}HOC_6H_4COO)_2$	[7]
$Th(OH)_2(2\text{-}HOC_6H_4COO)_2$ (?)	[8]
$ThO(2\text{-}HOC_6H_4COO)_2$	[1, 5]; solubility in saturated aqueous $2\text{-}HOC_6H_4COOH$ at pH ca. 2, 96°C, 3.7 mg Th/L [2]
$ThO(2\text{-}HOC_6H_4COO)_2\cdot 3H_2O$	differential thermogram [5] density, d_{obs} = 2.081 g/cm³ [5]
$ThO(2\text{-}HOC_6H_4COO)_2\cdot 4H_2O$ (?)	[4]
$ThCl(2\text{-}HOC_6H_4COO)_3$	[6]
$ThCl_2(2\text{-}HOC_6H_4COO)_2$	light brown, decomposes at 142 to 144°C [3]
$Na_2[ThO(2\text{-}OC_6H_4COO)_2]$	[1]

A salicylate(=2-hydroxybenzoate) of composition $Th(2\text{-}OC_6H_4COO)(2\text{-}HOC_6H_4COO)_2$ is reported to be precipitated when aqueous solutions of $Th(NO_3)_4$ and $2\text{-}HOC_6H_4COONa$ (molar ratio = 1:4 to 1:6) are mixed [7] (but see below). A product of composition $Th(2\text{-}OC_6H_4COO)_2$ (possibly $ThO(2\text{-}HOC_6H_4COO)_2$, see below) may be formed when the molar ratio Th:$2\text{-}HOC_6H_4COONa$ = 1:2 to 1:3 [7]. These products require further investigation. Another report suggests that $Th(OH)_2(2\text{-}HOC_6H_4COO)_2$ is precipitated which is potentially extractible into the higher alcohols [8].

$ThO(2\text{-}HOC_6H_4COO)_2$ is precipitated when $2\text{-}HOC_6H_4COOH$ is added to a weakly acidic solution of a Th^{IV} salt [1] and when aqueous 0.02 M solutions of $Th(NO_3)_4$ and $2\text{-}HOC_6H_4COONa$ are mixed in the molar ratio 1:4 at pH 2.40. The colloidal precipitate can be dried at 105°C [5]. Precipitation with the molar ratio = 1:3 and at pH 2.24 yields the trihydrate, $ThO(2\text{-}HOC_6H_4COO)_2\cdot 3H_2O$, in which the three H_2O molecules are evidently strongly held, for they are not removed at 120°C. The differential thermogram shows 3 endotherms (at 226, 265, and 273°C) which appear to be related to loss of H_2O and decomposition [5]. It has been suggested that anhydrous $ThO(2\text{-}HOC_6H_4COO)_2$ is precipitated initially in the above procedure, and that this is gradually converted to the trihydrate when left in contact with the supernatant [5]. A product which may be the tetrahydrate, $ThO(2\text{-}HOC_6H_4COO)_2\cdot 4H_2O$, also appears to be formed in the $Th(NO_3)_4$–$2\text{-}HOC_6H_4COONa$–H_2O system [4].

$ThO(2\text{-}HOC_6H_4COO)_2$ is partially hydrolysed on the addition of aqueous alkali; the anhydrous compound is reported to be readily soluble in acetone, ethanol, and ether [1]. Its solubility in H_2O and in aqueous $2\text{-}HOC_6H_4COONH_4$ has been studied at varying pH and temperature using tracer techniques [2]. The dipole moment of $ThO(2\text{-}HOC_6H_4COO)_2$ in dioxane solution is 3.43 Debye [5].

The chloride salicylate, $ThCl(2\text{-}HOC_6H_4COO)_3$, is formed in the reaction of $ThCl_4$ with $2\text{-}HOC_6H_4COOH$ by the procedure used to prepare the succinate, $Th(OOCCH_2CH_2COO)_2$ (p. 104) [6], whereas the product obtained by heating $ThCl_4$ with $2\text{-}HOC_6H_4COOH$ at 45 to 50°C has the composition $ThCl_2(2\text{-}HOC_6H_4COO)_2$. This is soluble in hot H_2O and in benzene or ether [3]. The solid is described as light brown in colour, presumably because of some decomposition.

The salicylato complex salt, $Na_2[ThO(2\text{-}OC_6H_4COO)_2]$, is obtained when $ThO(2\text{-}HOC_6H_4COO)_2$ (p. 130) is treated with aqueous $2\text{-}HOC_6H_4COONa$. It is not very stable and hydrolyses even when a large excess of Na salicylate is present [1].

There is solvent extraction evidence for the formation of a complex salicylate of composition $Th(2\text{-}HOC_6H_4COO)_4 \cdot 2\text{-}HOC_6H_4COOH$ [9] but this product was not isolated.

References for 15.6.4.3:

[1] Zvyagintsev, O. E.; Sudarikov, B. N. (Zh. Neorgan. Khim. **1** No. 1 [1956] 69/75; Russ. J. Inorg. Chem. **1** No. 1 [1956] 75/82).
[2] Zvyagintsev, O. E.; Sudarikov, B. N. (Zh. Neorgan. Khim. **2** No. 1 [1957] 128/37; Russ. J. Inorg. Chem. **2** No. 1 [1957] 196/213).
[3] Jaura, K. L.; Bajwa, P. S. (J. Sci. Ind. Res. [India] B **20** [1961] 391/4).
[4] Kazachenko, D. V. (Mater. 4th Nauchn. Konf. Aspirantov Rostov. Gos. Univ., Rostov, USSR, 1962, pp. 156/7; C.A. **60** [1964] 13124).
[5] Kovalenko, K. N.; Kazachenko, D. V.; Ivanova, E. M. (Zh. Neorgan. Khim. **7** [1962] 2340/4; Russ. J. Inorg. Chem. **7** [1962] 1213/6).
[6] Prasad, S.; Kumar, S. (J. Indian Chem. Soc. **39** [1962] 444/6).
[7] Agarwal, R. P.; Mehrotra, R. C. (Indian J. Chem. **2** [1964] 142/5).
[8] Tserkovnitskaya, I. A.; Charykov, A. K. (Izv. Vysshikh Uchebn. Zavedenii Khim. Khim. Tekhnol. **7** [1964] 544/50; C.A. **62** [1965] 3385).
[9] Hök-Bernström, B. (Acta Chem. Scand. **10** [1956] 174/85).

15.6.4.4 Thorium Salts of Other Hydroxybenzoic Acids

Information on the known compounds is summarised in Table 56.

Table 56
Thorium Salts of Other Hydroxybenzoic Acids.

3-hydroxybenzoates	
$Th(3\text{-}HOC_6H_4COO)_4$ (?)	[2]
$Th(OH)_2(3\text{-}HOC_6H_4COO)_2$	decomposes to ThO_2 at 550°C [8]; thermolysis curve illustrated in [8]
$Th(OH)_2(3\text{-}HOC_6H_4COO)_2 \cdot H_2O$	possibly $Th_2O(OH)_2(3\text{-}HOC_6H_4COO)_4(H_2O)_2 \cdot 2H_2O$; loses 1 H_2O/Th atom at 110°C, decomposes at 365, 408°C [3]; thermogram illustrated in [3] solubility in C_2H_5OH at 22°C: 5.27 g/100 g solvent [3]
4-hydroxybenzoates	
$Th(4\text{-}HOC_6H_4COO)_4$ (?)	[2]
$Th(OH)_2(4\text{-}HOC_6H_4COO)_2 \cdot 3H_2O$	possibly $ThO(4\text{-}HOC_6H_4COO)_2(H_2O) \cdot 3H_2O$ or $Th_2O(OH)_2(4\text{-}HOC_6H_4COO)_4(H_2O)_2 \cdot 4H_2O$; loses 2 H_2O/Th atom at 110°C, decomposes at 370, 405°C [3]; thermogram illustrated in [3] solubility in C_2H_5OH at 22°C: 10.12 g/100 g solvent [3]

References for 15.6.4.4 on p. 133

Table 56 (continued)

2,4-dihydroxybenzoate (β-resorcylate)		
$Th(OH)_2(2,4\text{-}(HO)_2C_6H_3COO)_2$	[1]	
digallate (tannate)		
(digallic acid = gallic acid 3-monogallate = $3,4,5\text{-}(HO)_3C_6H_2C(O)\text{-}3\text{-}OC_6H_2\text{-}4,5\text{-}(OH)_2COOH$ = $C_{14}H_{10}O_9$)		
$Th(OH)(C_{14}H_9O_9)_3$		decomposes >50°C, rapidly at >215°C, to ThO_2 at 565°C [6]; thermal decomposition curve illustrated in [6]

The precipitation of a 3-hydroxybenzoate, possibly $Th(3\text{-}HOC_6H_4COO)_4$, is mentioned in [2], but the preparative details are not clear. The basic 3-hydroxybenzoate, $Th(OH)_2(3\text{-}HOC_6H_4COO)_2 \cdot H_2O$, is obtained as a gelatinous precipitate, which becomes crystalline when finely ground, on mixing aqueous solutions of $Th(NO_3)_4$ and the acid (molar ratio = 1:4). An alternative formulation, $Th_2O(OH)_2(3\text{-}HOC_6H_4COO)_4(H_2O)_2 \cdot 2H_2O$ is also given and the compound may be a polymer [3]. The optimum pH for the precipitation of Th by 3-hydroxybenzoic acid is reported to be 3.5 to 6.0 [5], but the composition of the precipitate was not given.

The analogous 4-hydroxybenzoate, $Th(OH)_2(4\text{-}HOC_6H_4COO)_2 \cdot 3H_2O$, is obtained as a crystalline precipitate on mixing aqueous solutions of $Th(NO_3)_4$ and the acid (molar ratio = 1:4). Alternative formulations, $ThO(4\text{-}HOC_6H_4COO)_2(H_2O) \cdot 3H_2O$ and $Th_2O(OH)_2(4\text{-}HOC_6H_4COO)_4(H_2O)_2 \cdot 4H_2O$, are also given, and the latter could be a hydroxo-bridged dimer [3]. The 3- and 4-hydroxybenzoates of composition $Th(OH)_2(HOC_6H_4COO)_2 \cdot xH_2O$ are very sparingly soluble in H_2O and are sparingly soluble in acetone, dioxane, and ether [3].

The 2,4-dihydroxybenzoate (= β-resorcylate), $Th(OH)_2(2,4\text{-}(HO)_2C_6H_3COO)_2$, is precipitated when a hot 2% aqueous solution of the acid is added to a hot (90°C) aqueous solution of a Th salt (neutral to Congo Red). The composition of the precipitate is apparently insufficiently definite for the method to be used for the gravimetric determination of Th by direct weighing of the precipitate [1].

3,4,5-trihydroxybenzoic acid (gallic acid = $C_7H_6O_5$) in slight excess quantitatively precipitates Th^{IV} from a solution at pH ~3, thus providing a method for separating Th and U [10].

A Th digallate (= tannate; tannic acid = gallic acid 3-monogallate = $3,4,5\text{-}(HO)_3C_6H_2C(O)\text{-}3\text{-}OC_6H_2\text{-}4,5\text{-}(OH)_2COOH = C_{14}H_{10}O_9$) is precipitated when a hot 5% aqueous solution of the acid is added to a boiling aqueous solution of a Th^{IV} salt to which solid NH_4Cl, followed by CH_3COOH to neutrality, has been added. The precipitate was dissolved in hot dilute HCl, then reprecipitated [4], but the composition of the precipitate was not reported. However, it appears to be $Th(OH)(C_{14}H_9O_9)_3$ [6]. The precipitation is complete in the pH range 3.9 to 9.1 in a NH_3–NH_4Cl buffer, or at pH 8.2 to 9.4 in the presence of fluoride [7].

The thermal decomposition at 600, 800, and 1000°C of an unspecified Th tannate (precipitated from aqueous $Th(NO_3)_4$) has been reported in an investigation of the powder characteristics of ThO_2 produced by a variety of methods [9].

References for 15.6.4.4:

[1] Datta, S. K. (J. Indian Chem. Soc. **32** [1955] 687/93).
[2] Tserkovnitskaya, I. A.; Charykov, A. K. (Izv. Vysshikh Uchebn. Zavedenii Khim. Khim. Tekhnol. **7** [1964] 544/50).
[3] Kovalenko, K. N.; Kazachenko, D. V. (Zh. Neorgan. Khim. **10** [1965] 927/33; Russ. J. Inorg. Chem. **10** [1965] 501/5).
[4] Venkataramaniah, M.; Satyanarayanamurthy, T. K.; Raghava Rao, B. S. V. (J. Indian Chem. Soc. **27** [1950] 81/6).
[5] Deshmukh, G. S.; Xavier, J. (J. Indian Chem. Soc. **29** [1952] 911/4).
[6] Wendlandt, W. W. (Anal. Chim. Acta **18** [1958] 316/20).
[7] Yoshimura, C.; Kiboku, M. (Bunseki Kagaku **7** [1958] 445/9; C.A. **1960** 7428).
[8] Wendlandt, W. W. (Anal. Chem. **29** [1957] 800/2).
[9] Moorthy, V. K.; Kulkarni, A. K. (Trans. Indian Ceram. Soc. **22** [1963] 116/29).
[10] Tong, Chu; Wong, Yinwei; Go, Fuhsian (Huaxue Xueboa **1981** 114/9; C.A. **98** [1983] No. 172144).

15.6.4.5 Thorium Compounds with Substituted Hydroxybenzoic Acids

Information on the known compounds is summarised in Table 57.

Table 57
Thorium Compounds with Substituted Hydroxybenzoic Acids.

2-hydroxy-5-aminobenzoate (= 5-aminosalicylate; acid = 2-HO-5-$H_2NC_6H_3COOH$ = $C_7H_7NO_3$)	
$Th(OH)_3(C_7H_6NO_3)\cdot 3H_2O$	[5, 12]; very pale yellow [6]; IR spectrum illustrated in [14, 15]
5-bromosalicylates (acid = $C_7H_5O_3Br$)	
$Th(C_7H_4O_3Br)_4$	decomposes >200°C, to ThO_2 at 635°C thermal decomposition curve illustrated in [9]
$Th(OH)_2(C_7H_4O_3Br)_2$(?)	[17]
$Th(OH)_2(C_7H_4O_3Br)_2\cdot 3H_2O$	[5]
3,5-diiodosalicylate (acid = $C_7H_4O_3I_2$)	
$Th(OH)(C_7H_3O_3I_2)_3$	very pale yellow [3]
3-nitrosalicylates (acid = $C_7H_5NO_5$)	
$Th(OH)(C_7H_4NO_5)_3$ (?)	[17]
$Th(OH)_2(C_7H_4NO_5)_2$ (?)	[17]
5-nitrosalicylates	
$Th(OH)(C_7H_4NO_5)_3$ (?)	[17]
$Th(OH)_2(C_7H_4NO_5)_2$ (?)	[17]
$Th(OH)_3(C_7H_4NO_5)$	[5]
$Th(OH)_2(C_7H_3NO_5)$	IR spectrum: ν(OH), 3650 to 3100 cm^{-1}, ν(CO), 1615 cm^{-1} [16]

References for 15.6.4.5 on p. 136

Table 57 (continued)

3,5-dinitrosalicylate (acid = $C_7H_4N_2O_7$)		
$Th(OH)_2(C_7H_3N_2O_7)_2$ (?)	[17]	
***o*-cresotate** (acid = 2-HO-3-$CH_3C_6H_3COOH$ = $C_8H_8O_3$)		
$Th(OH)_3(C_8H_7O_3)\cdot 2H_2O$	[10]	
vanillates (acid = 4-hydroxy-3-methoxybenzoic acid = 4-OH-3-$CH_3OC_6H_3COOH$ = $C_8H_8O_4$)		
$Th(C_8H_7O_4)_4$	decomposes >310°C, to ThO_2 at 385°C thermal decomposition curve illustrated in [8]	
$Th(OH)(C_8H_7O_4)_3\cdot 2H_2O$	[2]	
5-bromoresorcilate (acid = $C_7H_5O_4Br$ = 5-bromo-2,4-dihydroxy benzoic acid)		
$Th(OH)_2(C_7H_4O_4Br)_2$	[5]	
acetylsalicylate (acid = 2-$CH_3COOC_6H_4COOH$ = $C_9H_8O_4$)		
$Th(OH)_2(C_9H_7O_4)_2$	decomposes rapidly above 140°C, to ThO_2 at 665°C thermal decomposition curve illustrated in [9]	
3-oximinomethylsalicylate (acid = 2-HO-3-(CH:NOH)C_6H_3COOH = $C_8H_7NO_4$)		
$Th(C_8H_5NO_4)_2\cdot 4H_2O$	[13]	
2-thioxothiazolidinone-N-salicylate (= rhodanine-N-salicylate; acid = $C_{10}H_7NO_4S_2$)		
$Th(OH)_3(C_{10}H_6NO_4S_2)\cdot 2H_2O$	[7]	

The basic 5-aminosalicylate, $Th(OH)_3(C_7H_6NO_3)\cdot 3H_2O$ (acid = 2-HO-5-$H_2NC_6H_3COOH$ = $C_7H_7NO_3$; the definition of the acid does not seem to be clear in the original papers) is precipitated when a hot 2% aqueous solution of the acid is added to a hot (90°C) solution of a Th^{IV} salt (neutral to Congo Red) containing added solid CH_3COONH_4 [5, 6]. The optimum pH for precipitation is reported to be 5.0 to 5.4 [15]. The precipitate can be dried to constant weight at 105 to 110°C [5, 6]. The IR spectrum of the compound indicates that the Th atom is bonded to the amino N and carboxylate O atoms [12]. The composition, particularly the H_2O content, of this product is not sufficiently definite for the determination of Th by direct weighing [5, 6], but the precipitation can be used for the indirect determination of Th by bromination or iodination of the acid anion [6]. Precipitation with this acid can also be used to separate Th from the cerite lanthanides [4].

The analogous 4-aminosalicylic acid has been used for the estimation of Th by precipitation from a hot (80°C) solution of a Th^{IV} salt (optimum pH = 1.3 to 3.5) on addition of a 1% aqueous solution of the acid, followed by ignition to ThO_2 [11].

The basic 5-bromosalicylate, $Th(OH)_2(C_7H_4O_3Br)_2\cdot 3H_2O$ (acid = $C_7H_5O_3Br$), is precipitated under the same conditions as the 5-aminosalicylate [5] but the procedure was later reported to yield $Th(C_7H_4O_3Br)_4$ [9]. The addition of aqueous $K(C_7H_4O_3Br)$ to an aqueous solution of a Th^{IV} salt (molar ratio = 2:1 or 4:1) yields a precipitate in which the ratio Th:$(C_7H_4O_3Br)$ = 1:2 [17] and this is presumably $Th(OH)_2(C_7H_4O_3Br)_2$.

A basic 3,5-diiodosalicylate, $Th(OH)(C_7H_3O_3I_2)_3$ (acid = $C_7H_4O_3I_2$), is precipitated when a hot 1% aqueous solution of $Na(C_7H_3O_3I_2)$ is added to aqueous $Th(NO_3)_4$. After digestion for 5 min the precipitate was washed with hot water, then hot acetone and dried to constant weight at 105 to 110°C. The precipitation can be used to separate Th from monazite sand [3]. The initial precipitate could be $Th(C_7H_3O_3I_2)_4$ which is partially hydrolysed by the washing procedure.

The addition of aqueous $K(C_7H_4NO_5)$ (= potassium 3-nitrosalicylate) to an aqueous solution of a Th^{IV} salt precipitates products in which the ratio $Th:C_7H_4NO_5 = 1:2$ when the molar ratio $Th:K(C_7H_4NO_5) = 1:2$, and $Th:C_7H_4NO_5 = 1:3$ when the molar ratio $Th:K(C_7H_4NO_5) = 1:5$ [17]. These are presumably $Th(OH)_2(C_7H_4NO_5)_2$ and $Th(OH)(C_7H_4NO_5)_3$, respectively. The same procedure with potassium 5-nitrosalicylate at the corresponding molar ratios yields products of the same composition as the 3-nitrosalicylates [17]. However, the addition of a hot 2% aqueous solution of 5-nitrosalicylic acid to an aqueous solution of a Th^{IV} salt under the conditions used for the preparation of the 5-bromosalicylate (see p. 134) yields a product of composition $Th(OH)_3(C_7H_4NO_5)$ which can be dried to constant weight at 105 to 110°C. The composition of this product is not sufficiently definite for the estimation of Th by direct weighing of the precipitate [5]. The thermal decomposition curve of the precipitate obtained by this route (composition not stated) shows that the first weight loss occurs at 40°C and that decomposition is explosive above 300°C [9]. The IR spectrum of a product described as $Th(OH)_2(C_7H_3NO_5)$ has also been reported. It was obtained in the course of a conductometric/potentiometric study of the reaction of Th^{IV} with 5-nitrosalicylic acid in aqueous solution [16].

A 3,5-dinitrosalicylate, which is presumably $Th(OH)_2(C_7H_3N_2O_7)_2$, is precipitated by the procedure used for the preparation of the 3- and 5-nitrosalicylates (see above) using the molar ratios $Th:K(C_7H_3N_2O_7) = 1:2$ and 1:4 [17].

The 2-hydroxy-3-methylbenzoate (= *o*-cresotate; acid = $2\text{-}HO\text{-}3\text{-}CH_3C_6H_3COOH = C_8H_8O_3$), $Th(OH)_3(C_8H_7O_3)\cdot 2H_2O$, is precipitated when an aqueous solution of $Th(NO_3)_4$, neutral to Congo Red and buffered with CH_3COONa/CH_3COOH, is treated with aqueous $NaC_8H_7O_3$; a small quantity of CH_3COONH_4 was added to improve filterability. The optimum pH range for precipitation is 3.72 to 5.2 and the precipitate can be weighed directly for the determination of Th [10].

Vanillic acid (= $4\text{-}HO\text{-}3\text{-}CH_3OC_6H_3COOH = C_8H_8O_4$) precipitates Th quantitatively at ca. pH 3.6 (solution neutral to Congo Red and containing added NH_4Cl) when added to a boiling aqueous solution of $Th(NO_3)_4$. The product was originally reported to be $Th(OH)(C_8H_7O_4)_3 \cdot 2H_2O$ [2] but it is apparently $Th(C_8H_7O_4)_4$ [8].

The basic 5-bromoresorcylate (acid = $C_7H_5O_4Br$), $Th(OH)_2(C_7H_4O_4Br)_2$, is precipitated by a hot 2% aqueous solution of $C_7H_5O_4Br$ under the same conditions as the 5-aminosalicylate (p. 134); the composition of this product is insufficiently definite for the determination of Th by direct weighing of the precipitate [5].

Th is precipitated from a boiling aqueous solution of a Th^{IV} salt (neutral to Congo Red and containing some added CH_3COONH_4) on addition of a boiling 1% aqueous solution of acetylsalicyclic acid (= $2\text{-}CH_3COOC_6H_4COOH$). A final pH of 4.0 to 4.4 is recommended for the separation of Th from the cerite lanthanides by this method [1]. The composition of the precipitate was not given in [1], but in later work it was reported to be $Th(OH)_2(CH_3COOC_6H_4COO)_2$ [9].

Aqueous 2% solutions of the Na salts of *o*-acetylcresotic (= $2\text{-}CH_3COO\text{-}3\text{-}CH_3C_6H_3COOH$) and *o*-benzoyl-cresotic (= $2\text{-}C_6H_5COO\text{-}3\text{-}CH_3C_6H_3COOH$) acids precipitate Th completely from aqueous $Th(NO_3)_4$ (neutralised with dilute aqueous NH_3 and buffered to pH 4 to 5 with CH_3COONa/CH_3COOH). The precipitation has been used for the gravimetric determination of Th as ThO_2 following ignition of the precipitate [18], but the compositions of the two precipitates were not recorded.

References for 15.6.4.5 on p. 136

The 3-oximinomethylsalicylate (acid = 2-HO-3-(CH:NOH)C_6H_3COOH = $C_8H_7NO_4$), $Th(C_8H_5NO_4)_2 \cdot 4H_2O$, is precipitated when an aqueous solution of $Th(NO_3)_4$ (neutral to Congo Red and containing added CH_3COONH_4) is treated with an excess of the acid. The mixture was heated to boiling and kept on a water bath for 5 min. The optimum pH for precipitation is 4 to 5. The composition of the precipitate may vary and ignition to ThO_2 is recommended for the gravimetric estimation of Th [13].

The basic Th^{IV} salt of rhodanine-N-salicylic acid (= 2-thioxothiazolidinone-N-salicylic acid = $C_{10}H_7NO_4S_2$), $Th(OH)_3(C_{10}H_6NO_4S_2) \cdot 2H_2O$, is precipitated when a 1.7- to 10-fold excess of the acid, as a 1% solution in ethanol, is added to an aqueous solution of $Th(NO_3)_4$ (neutral to Congo Red). The mixture was heated on a water bath for 15 min, then left to stand overnight. The precipitate can be dried at 110 to 115°C [7].

References for 15.6.4.5:

[1] Lakshmana Rao, B. R.; Raghava Rao, B. S. V. (J. Indian Chem. Soc. **27** [1950] 569/72).
[2] Krishnamurty, K. V. S.; Purushottam, A. (Recl. Trav. Chim. **71** [1952] 671/5).
[3] Datta, S. K.; Banerjee, G. (J. Indian Chem. Soc. **32** [1955] 167/72).
[4] Datta, S. K.; Banerjee, G. (J. Indian Chem. Soc. **32** [1955] 231/3).
[5] Datta, S. K. (J. Indian Chem. Soc. **32** [1955] 687/93).
[6] Datta, S. K.; Banerjee, G. (Anal. Chim. Acta **13** [1955] 23/7).
[7] Rout, M. K. (J. Indian Chem. Soc. **33** [1956] 683/4).
[8] Wendlandt, W. W. (Anal. Chim. Acta **17** [1957] 295/9).
[9] Wendlandt, W. W. (Anal. Chim. Acta **18** [1958] 316/20).
[10] Srivastava, T. N.; Agrawal, S. P.; Aggarwal, R. C. (Z. Anal. Chem. **169** [1959] 254/7).

[11] Subbanna, V. V.; Bhattacharya, A. K. (J. Sci. Ind. Res. [India] B **19** [1960] 409).
[12] Geleceanu, I.; Lapitskii, A. V. (Dokl. Akad. Nauk SSSR **144** [1962] 573/5; Dokl. Chem. Proc. Acad. Sci. USSR **142/147** [1962] 460/2).
[13] Ray, A. K. (Anal. Chim. Acta **28** [1963] 580/3).
[14] Galateanu, I. (Acad. Rep. Populare Romine Studii Cercetari Chim. **11** [1963] 239/46).
[15] Galateanu, I. (Oesterr. Chemiker-Ztg. **66** [1965] 275/85).
[16] Khadikar, P. V.; Sarwate, A. G. (Sci. Cult. [Calcutta] **37** [1971] 256).
[17] Pande, C. S.; Misra, G. N. (Indian J. Chem. **11** [1973] 292/3).
[18] Kudesia, V. P.; Srivastava, S. K.; Sharma, L. M. (Acta Cienc. Indica **3** [1977] 128/30).

15.6.4.6 Thorium Compounds with Aminobenzoic Acids

Information on the known compounds is summarised in Table 58.

Table 58
Thorium Compounds with Aminobenzoic Acids.

2-aminobenzoates (anthranilates; acid = 2-$H_2NC_6H_4COOH$ = $C_7H_7NO_2$)	
$Th(C_7H_6NO_2)_4$	pink [8] IR spectrum; ν(NH), 3325 cm^{-1}, Δν(NH), 125 cm^{-1} [6], $\nu_{as}(NH_2)$, 3400 cm^{-1}, $\nu_s(NH_2)$, 3300 cm^{-1} [8], $\nu_{as}(CO)$, 1670 cm^{-1} [8], 1510 cm^{-1} [6], $\nu_s(CO)$, 1395 cm^{-1} [8]
$Th(OH)(C_7H_6NO_2)_3$	IR spectrum; $\nu_{as}(CO)$, 1430 cm^{-1} (?) [7]

Table 58 (continued)

$Th(OH)_2(C_7H_6NO_2)_2$	[1, 2]; IR spectrum; $\nu_{as}(CO)$, 1425 cm^{-1} (?) [7]
$Th(OH)_3(C_7H_6NO_2)$	decomposes to ThO_2 at 450°C [11] thermolysis curve illustrated in [11]
$ThCl_2(C_7H_6NO_2)_2$	IR spectrum; $\nu(NH)$, 3170, 3100 cm^{-1}, $\nu(CO)$, 1615 cm^{-1}, $\nu(ThO)$, 388 cm^{-1}, $\nu(ThN)$, 335 cm^{-1}, $\nu(ThCl)$, 250 cm^{-1} [10] molar conductivity in $(CH_3)_2SO = 48.0\ \Omega^{-1}\cdot cm^2\cdot mol^{-1}$ [10]
$Th(OO)(C_7H_6NO_2)_2$	IR spectrum; $\nu(NH)$, 3125, 3090 cm^{-1}, $\nu(CO)$, 1590 cm^{-1}, $\nu_1(OO)$ (Raman) = 820 cm^{-1} [10] molar conductivity in $(CH_3)_2SO = 4.40\ \Omega^{-1}\cdot cm^2\cdot mol^{-1}$ [10]
2-amino-5-iodobenzoate (= 5-iodoanthranilate; acid = $C_7H_6NO_2I$)	
$Th(OH)(C_7H_5NO_2I)_3$	yellowish white [2]
N-phenyl-2-aminobenzoate (= N-phenylanthranilate; acid = $C_{13}H_{11}NO_2$)	
$Th(C_{13}H_{10}NO_2)_4$	light yellow [4], yellow [8] IR spectrum; $\nu_{as}(CO)$, 1585 cm^{-1}, $\nu_s(CO)$, 1400 cm^{-1} [8]
3-aminobenzoates	
$Th(OH)_3(C_7H_6NO_2)$	[5]
$Th(CH_3COO)_2(C_7H_6NO_2)_2$	decomposes to ThO_2 at ca. 850°C [9] IR spectrum; $\nu_{as}(CO)$, 1545 cm^{-1}, $\nu_s(CO)$, 1370 cm^{-1} [9]
4-aminobenzoates	
$ThO(OH)(C_7H_6NO_2)\cdot 4H_2O$	[1]; probably $Th(OH)_3(C_7H_6NO_2)\cdot 3H_2O$
$Th(OH)_3(C_7H_6NO_2)$	[5]; decomposes >75°C, to ThO_2 at 465°C thermal decomposition curve illustrated in [3]
$Th(CH_3COO)_2(C_7H_6NO_2)_2$	decomposes to ThO_2 at ca. 850°C [9] IR spectrum; $\nu_{as}(CO)$, 1600 cm^{-1}, $\nu_s(CO)$, 1380 cm^{-1} [9]

The 2-aminobenzoate, $Th(C_7H_6NO_2)_4$ (acid = anthranilic acid = 2-$H_2NC_6H_4COOH$ = $C_7H_7NO_2$) has been obtained by adding the calculated amount of $C_7H_7NO_2$ in ether to a solution of hydrated $Th(NO_3)_4$ in ethanol, followed by adjusting the pH to 4.0 with ethanolic ammonia. After refluxing the mixture on a steam bath for 1 h the solution was concentrated until the product separated. It is a non-electrolyte in methanol. The shift in $\nu(NH)$ in the IR spectrum ($\Delta\nu(NH)$, 125 cm^{-1}) suggests that the Th atom is bonded to the amine N atom [6]. However, the IR spectrum of the product of the same composition obtained by heating $Th(CH_3COO)_4$ (p. 47) with the stoichiometric quantity of $C_7H_7NO_2$ in ethanol under reflux, followed by addition of petroleum ether (40 to 60°C) to precipitate $Th(C_7H_6NO_2)_4$, does not show any shift in $\nu(NH)$, so that the Th atom is probably not bonded to the amino N atom; the frequencies of $\nu_{as}(CO)$ and $\nu_s(CO)$ suggest that only one carboxylate O atom of each $C_7H_6NO_2$ group is coordinated to Th [8], which seems rather unusual. The pink colour of this preparation may indicate that some decomposition has occurred.

References for 15.6.4.6 on p. 139

The basic anthranilates, $Th(OH)(C_7H_6NO_2)_3$ and $Th(OH)_2(C_7H_6NO_2)_2$, are reported to be formed on addition of the stoichiometric quantities of aqueous $K(C_7H_6NO_2)$ to an aqueous solution of a Th^{IV} salt [7]. The latter is precipitated when a boiling 1% aqueous solution of $C_7H_7NO_2$ is added to a boiling aqueous solution of a Th^{IV} salt which is just acid to Congo Red and to which some aqueous CH_3COONH_4 has been added [1, 2]. The precipitation can be used for the separation of Th from the cerite lanthanides, but the composition of the precipitate is not sufficiently definite for the gravimetric determination of Th, for which ignition to ThO_2 is recommended [1].

$ThCl_2(C_7H_6NO_2)_2$ has been prepared by adding the stoichiometric quantity of $C_7H_7NO_2$ in CH_3CN to a suspension of $ThCl_4$ in the same solvent. It can be recrystallised from hot acetone by partial evaporation and addition of ether while the solution cools [10]. The relatively high molar conductivity of $ThCl_2(C_7H_6NO_2)_2$ in $(CH_3)_2SO$ probably results from partial displacement of Cl^- ion (e.g. to $[Th(C_7H_6NO_2)_2Cl\{(CH_3)_2SO\}]^+$); the IR spectrum indicates that the amino N and carboxylate O atoms are coordinated to the Th atom [10]. The analogous peroxo compound, $Th(OO)(C_7H_6NO_2)_2$, is obtained by dissolving $ThCl_2(C_7H_6NO_2)_2$ in a 1:1 mixture of acetone and methanol containing a small quantity of pyridine, then adding H_2O_2 [10].

Halogeno-2-amino benzoic acids are reported to form basic compounds analogous to the basic 2-aminobenzoates (see above) [1] but no details of their preparation or composition are given in [1].

The 2-amino-5-iodobenzoate (acid = $C_7H_6NO_2I$), $Th(OH)(C_7H_5NO_2I)_3$, is precipitated when a 1% aqueous solution of Na (or NH_4)$C_7H_5NO_2I$ is added to a boiling aqueous solution of a Th^{IV} salt (neutral to Congo Red). The optimum pH for precipitation is $\geqq 3.5$(NH_4 salt) or 4(Na salt). The precipitate can be dried to constant weight at 105 to 110°C, but the composition is not sufficiently definite for the determination of Th by direct weighing and it is best to ignite the precipitate to ThO_2 for analytical purposes [2].

N-Phenyl-2-aminobenzoic acid (= $C_{13}H_{11}NO_2$) precipitates $Th(C_{13}H_{10}NO_2)_4$ quantitatively when a 1% ethanol solution of the acid is added to aqueous $Th(NO_3)_4$ at pH 1.5 to 3.4. The precipitate can be dried in air at 110°C. The precipitation can be used for the estimation of Th by direct weighing of the precipitate and, at pH ca. 2.0 to 2.5, for the separation of Th from U^{VI}, Ce^{III}, Al, and Fe [4]. The compound has also been prepared by heating $Th(CH_3COO)_4$ (p. 47) with the stoichiometric quantity of $C_{13}H_{11}NO_2$ in ethanol under reflux; the product is precipitated from the resulting solution on the addition of petroleum ether (40 to 60°C) [8].

The basic 3-aminobenzoate, $Th(OH)_3(C_7H_6NO_2)$, is precipitated from aqueous solutions of Th^{IV} salts at pH 3.3 to 6.0 [5]. The mixed acetate/3- and 4-aminobenzoates, $Th(CH_3COO)_2(C_7H_6NO_2)_2$, are precipitated when $Th(CH_3COO)_4$ (p. 47) is heated with the appropriate acid (molar ratio = 1:4) in ethanol under reflux [9].

A basic 4-aminobenzoate, formulated as $ThO(OH)(C_7H_6NO_2)\cdot 4H_2O$, is precipitated when a boiling 1% aqueous solution of $C_7H_7NO_2$ is added to a boiling solution of a Th^{IV} salt which is just acid to Congo Red and to which a little aqueous CH_3COONH_4 has been added. The precipitation can be used for the gravimetric determination of Th following ignition to ThO_2 and for the separation of Th from the cerite lanthanides [1]. The precipitate is probably $Th(OH)_3(C_7H_6NO_2)\cdot 3H_2O$, as indicated in [3]. $Th(OH)_3(C_7H_6NO_2)$ is reported to be precipitated from aqueous solutions of Th^{IV} salts at pH > 5 [5].

A 1:1 complex with 2-[(2-carboxyphenyl)azo]-4,5-diphenyl-imidazole is reported to be formed in aqueous solution [12], but the compound was not isolated.

References for 15.6.4.6:

[1] Murthy, D. S. N.; Raghava Rao, B. S. V. (J. Indian Chem. Soc. **27** [1950] 459/61).
[2] Datta, S. K.; Banerjee, G. (J. Indian Chem. Soc. **31** [1954] 779/83).
[3] Wendlandt, W. W. (Anal. Chim. Acta **17** [1957] 295/9).
[4] Banerjea, D.; Suryanarayana, S. V. (Z. Anal. Chem. **202** [1964] 161/4).
[5] Tserkovnitskaya, I. A.; Charykov, A. K. (Vysshikh Uchebn. Zavedenii Khim. Khim. Tekhnol. **7** [1964] 544/50).
[6] Deshpande, Y. H.; Rao, V. R. (J. Indian Chem. Soc. **48** [1971] 247/8).
[7] Pande, C. S.; Misra, G. N. (J. Indian Chem. Soc. **51** [1974] 835/6).
[8] Singh, M.; Singh, R. (J. Indian Chem. Soc. **56** [1979] 136/7).
[9] Singh, M.; Singh, R. (J. Indian Chem. Soc. **56** [1979] 1249/51).
[10] Westland, A. D.; Tarafder, M. T. H. (Inorg. Chem. **21** [1982] 3228/32).

[11] Wendlandt, W. W. (Anal. Chem. **29** [1957] 800/2).
[12] Hammam, A. M. (J. Electrochem. Soc. India **28** [1979] 217/20).

15.6.4.7 Thorium Compounds with Halogenobenzoic Acids

Information on the few known compounds is collected in Table 59.

Table 59
Thorium Compounds with Halogenobenzoic Acids.

2-chlorobenzoates (acid = $C_7H_5O_2Cl$)	
$Th(OH)(C_7H_4O_2Cl)_3 \cdot 4H_2O$ (?) [1, 2]	
$Th(OH)_2(C_7H_4O_2Cl)_2$	decomposes >100°C, to ThO_2 at 505°C thermal decomposition curve illustrated in [3]
2-iodobenzoate (acid = $C_7H_5O_2I$)	
$Th(OH)(C_7H_4O_2I)_3$	thermal decomposition curve given in [5]
3-iodobenzoates	
$Th(C_7H_4O_2I)_4$	decomposes >300°C, to ThO_2 at 540°C thermal decomposition curve illustrated in [3]
$Th(OH)_2(C_7H_4O_2I)_2$ (?)	pale pink [6]
2,3,5-triiodobenzoate (acid = $C_7H_3O_2I_3$)	
$Th(OH)_2(C_7H_2O_2I_3)_2$	red [6]

The addition of a boiling aqueous solution of 2-chlorobenzoic acid (= $C_7H_5O_2Cl$) to a boiling neutral aqueous solution of a Th^{IV} salt is reported to yield a basic product, $Th(OH)(C_7H_4O_2Cl)_3 \cdot 4H_2O$ [1]. However, a later paper [3] reports the product to be $Th(OH)_2(C_7H_4O_2Cl)_2$. The precipitation with this acid can be used for the separation of Th from the cerite lanthanides and for the gravimetric determination of Th following ignition of the precipitate to ThO_2 [1]. Th can also be separated from U by a double precipitation with $C_7H_5O_2Cl$ at pH 2.6 to 2.8 [2].

References for 15.6.4.7 on p. 140

2-Iodobenzoic acid ($=C_7H_5O_2I$) precipitates Th from aqueous solution at pH>2.0 [4]; the optimum pH for the precipitation is 2.2 to 3.5, the product being the basic compound $Th(OH)(C_7H_4O_2I)_3$ [5]. $C_7H_5O_2I$ has been proposed as a reagent for the gravimetric determination of Th as ThO_2 after ignition of the precipitate [5].

The addition of a hot 1% aqueous solution of sodium 3-iodobenzoate to aqueous $Th(NO_3)_4$ (neutral to Congo Red) is reported to precipitate the basic compound, $Th(OH)_2(C_7H_4O_2I)_2$ [6]; the precipitate was washed with hot H_2O and then hot ethanol. This procedure was later reported to yield $Th(C_7H_4O_2I)_4$ [3], a difference which may be the result of partial hydrolysis in the washing of the precipitate.

The basic 2,3,5-triiodobenzoate (acid $=C_7H_3O_2I_3$), $Th(OH)_2(C_7H_2O_2I_3)_2$, is precipitated and washed in the same way as the 3-iodobenzoate [6]. Precipitation with sodium 3-iodo- or 2,3,5-triiodobenzoate can be used for the separation of Th from monazite sand [6].

References for 15.6.4.7:

[1] Lakshmana Rao, B. R.; Raghava Rao, B. S. V. (J. Indian Chem. Soc. **27** [1950] 457/8).
[2] Murthy, T. K. S.; Lakshmana Rao, B. R.; Raghava Rao, B. S. V. (J. Indian. Chem. Soc. **27** [1950] 610/2).
[3] Wendlandt, W. W. (Anal. Chim. Acta **17** [1957] 295/9).
[4] Kuang Chang; Lian-Yao Su (K'o Hsueh T'ung Pao [Kexue Tongbao] **1963** No. 7, p. 55; C.A. **60** [1964] 6212).
[5] Chang, J.; Shu, L.-Y. (Acta Sci. Nat. Scholar. Super. Sinens. Chim. Chim. Tech. **1** [1966] 11/6; N.S.A. **23** [1969] No. 19623).
[6] Datta, S. K.; Banerjee, G. (J. Indian Chem. Soc. **32** [1955] 167/72).

15.6.4.8 Thorium Compound with 2-Mercaptobenzoic Acid

The only recorded compound with 2-mercaptobenzoic acid (= thiosalicylic acid = HSC_6H_4-COOH) is the basic salt, $Th_2(OH)_2(SC_6H_4COO)_3 \cdot 3H_2O$, which is precipitated when a 0.5M solution of HSC_6H_4COOH in 0.1M NaOH (pH = 4 to 4.5) is added to an aqueous solution of $Th(NO_3)_4$ (0.02M, pH = 1.6). The IR spectrum of the compound is illustrated in [2]. See also [1, 3]. The Th atom is probably bonded to the SH and COO groups [1].

References for 15.6.4.8:

[1] Geleceanu, I.; Lapitskii, A. V. (Dokl. Akad. Nauk SSSR **144** [1962] 573/5; Dokl. Chem. Proc. Acad. Sci. USSR **142/147** [1962] 460/2).
[2] Galateanu, I. (Acad. Rep. Populare Romine Studii Cercetari Chim. **11** [1963] 239/46).
[3] Galateanu, I. (Oesterr. Chemiker-Ztg. **66** [1965] 275/85).

15.6.4.9 Thorium Compounds with Other Substituted Benzoic Acids

Information on the known compounds is summarised in Table 60.

A 2-nitrobenzoate (acid $=C_7H_5NO_4$) of composition $Th(OH)_{0.5}(C_7H_4NO_4)_{3.5}$ is precipitated when aqueous $Th(NO_3)_4 \cdot 5H_2O$ is mixed with aqueous $Na(C_7H_4NO_4)$ (molar ratios = 1:2 and 1:8). The product may be $Th(OH)_2(C_7H_4NO_4)_2 \cdot 3Th(C_7H_4NO_4)_4$ [10].

The 3-nitrobenzoate, probably $Th(C_7H_4NO_4)_4 \cdot 4H_2O$, is precipitated when the aqueous acid is added to aqueous $Th(NO_3)_4$ and the mixture is heated to 80°C [1]. Anhydrous $Th(C_7H_4NO_4)_4$ is obtained by drying the precipitate at 70 to 153°C [2, 3]. Precipitation with 3-nitrobenzoic acid can be used for the gravimetric determination of Th by weighing either as $Th(C_7H_4NO_4)_4$ [2, 7] or as ThO_2 following ignition [2]. Th can be separated from U by a double precipitation with 3-nitrobenzoic acid at pH 2.6 to 2.8 [5]. See also [18] for earlier work.

Table 60
Thorium Compounds with Other Substituted Benzoic Acids.

2-nitrobenzoate (acid = $C_7H_5NO_4$)	
$Th(OH)_{0.5}(C_7H_4NO_4)_{3.5}$	decomposes >200°C in air, oxidises at 415°C [10]
3-nitrobenzoates	
$Th(C_7H_4NO_4)_4$	[2, 7]; decomposes >153°C, explosively at 300 to 305°C thermal decomposition curve illustrated in [3]
$Th(C_7H_4NO_4)_4 \cdot 4H_2O$ (?)	[1]
4-nitrobenzoate	
$Th(OH)_2(C_7H_4NO_4)_2 \cdot H_2O$	decomposes >140°C in air, oxidises at 420°C [10] density, $d_{obs} = 2.70$ g/cm³ [10] solubility product, $S = [Th^{4+}][OH^-]^2[C_7H_4NO_4^-]^2 = 8 \times 10^{-34}$ [10]
2-methylbenzoates (*o*-toluates; acid = $2\text{-}CH_3C_6H_4COOH = C_8H_8O_2$)	
$Th(C_8H_7O_2)_4$	decomposes >185°C, to ThO_2 at 440°C thermal decomposition curve illustrated in [8]
$ThO(C_8H_7O_2)_2$	IR spectrum; $\nu_{as}(CO)$, 1600 cm⁻¹, $\nu_s(CO)$, 1510 cm⁻¹ [11]
2-methoxybenzoate (*o*-anisate; acid = $2\text{-}CH_3OC_6H_4COOH = C_8H_8O_3$)	
$ThO(C_8H_7O_3)_2$	IR spectrum; $\nu_{as}(CO)$, 1540, 1510 cm⁻¹, $\nu_s(CO)$, 1400 cm⁻¹ [11]
4-methoxybenzoates (*p*-anisates)	
$Th(C_8H_7O_3)_4$	[6]; decomposes >255°C, to ThO_2 at 475°C thermal decomposition curve illustrated in [8]
$Th(CH_3COO)_2(C_8H_7O_3)_2$	IR spectrum; $\nu_{as}(CO)$, 1600, 1545 cm⁻¹, $\nu_s(CO)$, 1470, 1425 cm⁻¹ [11]
fluorescein complex (fluorescein = 3′,6′-dihydroxyspiro[isobenzofuran-1(3H)-9′-[9H]-xanthen]-3-one = $C_{20}H_{12}O_5$)	
$Na_2Th_2(C_{20}H_{10}O_5)_5 \cdot 11H_2O$ (?)	orange; loses weight at 60°C, decomposes at 360°C [11] IR spectrum; $\nu(OH)$ of H_2O, 3420 cm⁻¹, $\delta(H_2O)$/ring, 1630 cm⁻¹, $\nu_{as}(CO)$, 1570 cm⁻¹, $\nu_s(CO)$, 1385 cm⁻¹, $\nu(CO)$, 1320 cm⁻¹, $\nu(COC)$, 1295 cm⁻¹ [13]
eosin complex (eosin = tetrabromofluorescein, disodium salt = $Na_2C_{20}H_6O_5Br_4$)	
$H_{14}Th_2(C_{20}H_6O_5Br_4)_{11} \cdot 37H_2O$	red; IR spectrum, $\nu_{as}(CO)$, 1580 cm⁻¹, $\nu(CO) + \nu(C_6H_5O^-)$, 1455 cm⁻¹, $\nu_s(CO)$, 1345 cm⁻¹, $\nu(COC)$, 1250 cm⁻¹ [12]

References for 15.6.4.9 on p. 143

Table 60 (continued)

anilates, $Th(OH)_2\{2\text{-}(CONHR)C_6H_4COO\}_2 \cdot 2H_2O$	
$R = C_6H_5$	solubility, 2.1 g/L in H_2O at 40°C [16]
$R = 2\text{-}CH_3C_6H_4$	solubility, 2.5 g/L in H_2O at 40°C [16]
$R = 4\text{-}CH_3C_6H_4$	solubility, 2.2 g/L in H_2O at 40°C [16]
$R = 2\text{-}CH_3OC_6H_4$	solubility, 1.0 g/L in H_2O at 40°C [16]
$R = 1\text{-}C_{10}H_6$ (α-naphthyl)	solubility, 0.5 g/L in H_2O at 40°C [16]
anilates, $ThO\{2\text{-}CONHR)C_6H_4COO\}_2 \cdot 2H_2O$	
$R = 3\text{-}C_6H_4NO_2$	yellow, solubility, 1.5 g/L in H_2O at 40°C [16]
$R = 4\text{-}C_6H_4NO_2$	yellowish white, solubility, 1.3 g/L in H_2O at 40°C [16]

A basic 4-nitrobenzoate, $Th(OH)_2(C_7H_4NO_4)_2 \cdot H_2O$, is precipitated when aqueous $Th(NO_3)_4$ is mixed with aqueous $Na(C_7H_4NO_4)$. It is sparingly soluble in acetone, but is insoluble in benzene, chloroform, dioxane, ether, or methanol [10].

Th is precipitated quantitatively from hot (90°C) aqueous solution on addition of a saturated aqueous solution of 3,5-dinitrobenzoic acid at pH 2.5 to 3. The precipitate was ignited to ThO_2 for the gravimetric estimation of Th, and the composition of the precipitate was not reported [9].

The addition of a boiling 0.5% aqueous solution of *o*-toluic acid ($= 2\text{-}CH_3C_6H_4COOH = C_8H_8O_2$) to a boiling aqueous solution of a Th^{IV} salt (neutral to Congo Red) precipitates Th quantitatively. The precipitation can be used to separate Th from the cerite lanthanides at pH 3 to 4 [14]. The composition of the precipitate is not given in [14], but it has been shown to be $Th(C_8H_7O_2)_4$ [8]. The basic compound, $ThO(C_8H_7O_2)_2$, has been prepared by heating $Th(CH_3COO)_4$ (p. 47) with $C_8H_8O_2$ (molar ratio = 1:4) in ethanol under reflux. The IR spectrum of the compound indicates that the carboxylate groups are bidentate [11].

The basic *o*-anisate (acid $= 2\text{-}CH_3OC_6H_4COOH = C_8H_8O_3$), $ThO(C_8H_7O_3)_2$, has been prepared by heating $Th(CH_3COO)_4$ (p. 47) with $C_8H_8O_3$ (molar ratio = 1:4) in ethanol under reflux. The same procedure yields $Th(CH_3COO)_2(C_8H_7O_3)_2$ in the case of 4-methoxybenzoic acid [11]. The IR spectra of $ThO(C_8H_7O_3)_2$ and $Th(CH_3COO)_2(C_8H_7O_3)_2$ indicate that the carboxylate groups of the *o*- and *p*-anisate anions appear to be acting as bidentate ligands [11]. A study of the solvent extraction of Th by *o*-anisic acid has been reported [17], but the extracted species was not identified.

The addition of dilute aqueous anisic acid (the *p*-isomer according to the m.p. of the pure acid used) to a boiling aqueous solution of a Th salt (just acid to Congo Red) precipitates $Th(C_8H_7O_3)_4$ (after drying to constant weight at 105°C). The precipitation can be used for the separation of Th from the cerite lanthanides in monazite and for the gravimetric determination of Th as ThO_2 following ignition [6]. Th can be separated from U by a precipitation procedure in which a 0.75% boiling aqueous solution of *p*-anisic acid is added to a boiling aqueous solution containing the two elements which has been made neutral or nearly neutral to Congo Red and to which NH_4Cl has been added [15].

A boiling aqueous solution of veratric acid ($= 3,4\text{-}(CH_3O)_2C_6H_3COOH$) precipitates Th from aqueous solutions of its salts, and trimethylgallic acid ($= 3,4,5\text{-}(CH_3O)_3C_6H_2COOH$) also precipitates Th from a boiling aqueous solution which is neutral to Congo Red [4]. The compositions of these two precipitates are not stated in [4], but the precipitations can be used for the separation of Th from the cerite lanthanides and for the estimation of Th [4], presumably after ignition to ThO_2.

The addition of a 1% solution of the anilic acids 2-(CONHR)C_6H_4COOH in ethanol to a hot aqueous solution of $Th(NO_3)_4$ precipitates the basic compounds $Th(OH)_2${2-(CONHR)C_6H_4COO}$_2$ ·$2H_2O$ (R = C_6H_5, 2- or 4-$CH_3C_6H_4$, 2-$CH_3OC_6H_4$ and 1-$C_{10}H_6$ (α-naphthyl)) or ThO{2-(CONHR)-C_6H_4COO}$_2$·$2H_2O$ (R = 3- or 4-$C_6H_4NO_2$) [16].

The addition of 0.02M eosin (see Table 60, p. 141) to an aqueous 0.02M solution of a Th^{IV} salt (pH = 3.0 to 4.0) yields a precipitate of composition $H_{14}Th_2(C_{20}H_6O_5Br_4)_{11}\cdot 37H_2O$ which is soluble in dimethylformamide and in dimethyl sulfoxide, but is insoluble in other common organic solvents [12]. A similar procedure with fluorescein (see Table 60, p. 141) yields a precipitate of composition $Na_2Th_2(C_{20}H_{10}O_5)_5\cdot 11H_2O$ which is sparingly soluble in dimethylformamide and in dimethyl sulfoxide, but is insoluble in other organic solvents. The relatively high decomposition temperature (360°C) of this product may indicate that the compound is polymeric [13].

References for 15.6.4.9:

[1] Osborn, G. H. (Analyst **73** [1948] 381/4).
[2] Dupuis, T.; Duval, C. (Compt. Rend. **228** [1949] 401/2).
[3] Dupuis, T.; Duval, C. (Anal. Chim. Acta **3** [1949] 589/98).
[4] Venkataramaniah, M.; Satyanarayanamurthy, T. K.; Raghava Rao, B. S. V. (J. Indian Chem. Soc. **27** [1950] 81/6).
[5] Murthy, T. K. S.; Lakshmana Rao, B. R.; Raghava Rao, B. S. V. (J. Indian Chem. Soc. **27** [1950] 610/2).
[6] Krishnamurty, K. V. S.; Raghava Rao, B. S. V. (J. Indian Chem. Soc. **28** [1951] 261/4).
[7] Dutt, N. K.; Chowdhury, A. K. (Anal. Chim. Acta **12** [1955] 515/8).
[8] Wendlandt, W. W. (Anal. Chim. Acta **17** [1957] 295/9).
[9] Lavanchy, A. M. (Anales Fac. Quim. Farm. Univ. Chile **11** [1959] 54/8; C.A. **1960** 22168).
[10] Kovalenko, K. N.; Lutsenko, L. V. (Zh. Neorgan. Khim. **12** [1967] 2989/93; Russ. J. Inorg. Chem. **12** [1967] 1582/4).

[11] Singh, M.; Singh, A. (J. Indian. Chem. Soc. **56** [1979] 26/7).
[12] Agarwal, B.; Agarwala, B. V.; Dey, A. K. (Rev. Chim. Minerale **16** [1979] 535/42).
[13] Agarwal, B.; Agarwala, B. V.; Dey, A. K. (J. Indian Chem. Soc. **57** [1980] 130/3).
[14] Lakshmana Rao, B. R.; Raghava Rao, B. S. V. (J. Indian Chem. Soc. **27** [1950] 569/72).
[15] Krishnamurty, K. V. S.; Raghava Rao, B. S. V. (Recl. Trav. Chim. **70** [1951] 421/4).
[16] Datta, S. K. (Z. Anal. Chem. **148** [1955] 267/73).
[17] Hök-Bernström, B. (Acta Chem. Scand. **10** [1956] 174/85).
[18] Neish, A. C. (J. Am. Chem. Soc. **26** [1904] 780/93).

15.6.4.10 Thorium Benzene-(di and poly)carboxylates

Information on the known compounds is summarised in Table 61, p. 144.

The anhydrous phthalate, $Th(C_8H_4O_4)_2$, has been prepared by heating the dihydrate (see below) at 131°C [10]; see also [5]. It is also obtained by heating $ThCl_4$ in ether with phthalic acid in the same way as the succinate, $Th(OOCCH_2CH_2COO)_2$ (p. 104) [6] and by heating $Th(CH_3COO)_4$ (p. 47) with the stoichiometric quantity of $C_8H_6O_4$ in ethanol under reflux [12]. The IR spectrum of the compound indicates that the COO groups are bidentate [12].

References for 15.6.4.10 on pp. 145/6

Table 61
Thorium Benzene-(di and poly)carboxylates.

phthalates (= benzene-1,2-dicarboxylates)	
$Th(C_8H_4O_4)_2$	[2, 5, 6, 10]; IR spectrum: $\nu_{as}(CO)$, 1570 cm^{-1}, $\nu(CO)$, 1410 cm^{-1} [12]
$Th(C_8H_4O_4)_2 \cdot 2H_2O$	loses $2H_2O$ at 131°C, thermogram illustrated in [10] IR spectrum in the range 4000 to 400 cm^{-1} TG, DTG, and DTA curves illustrated in [13] activation energy of dehydration, E = 35.7 kJ/mol [13] density, d_{obs} = 2.286 g/cm^3 [10]
isophthalate (= benzene-1,3-dicarboxylate)	
$Th(C_8H_4O_4)_2 \cdot 2H_2O$	IR spectrum in the range 4000 to 400 cm^{-1} mentioned, TG, DTG, and DTA curves illustrated in [13] activation energy of dehydration, E = 44.2 kJ/mol [13]
terephthalates (= benzene-1,4-dicarboxylates)	
$Th_3(OH)_5(C_8H_4O_4)_2(C_8H_5O_4)_3 \cdot 5H_2O$	IR spectrum in the range 4000 to 400 cm^{-1} mentioned, TG, DTG, and DTA curves illustrated in [13] activation energy of dehydration, E = 24.4 kJ/mol [13]
$Th_5(OH)_{14}(C_8H_4O_4)_3 \cdot 7H_2O$	IR spectrum in the range 4000 to 400 cm^{-1} mentioned, TG, DTG, and DTA curves illustrated in [13] activation energy of dehydration, E = 35.3 kJ/mol [13]
tetrachlorophthalate	
$Th(C_8Cl_4O_4)_2$	solubility in H_2O at 20 ± 1°C = 2.27×10^{-7} g-atoms Th/L of solution at pH 6.20; 1.59×10^{-7} g-atoms Th/L of solution at pH 6.17; solubility product at pH 6.5, $S = [Th^{4+}][C_8Cl_4O_4^{2-}]^2 = 2.88 \times 10^{-20}$ [4]
benzene-1,2,4,5-tetracarboxylate (= pyromellitate; acid = 1,2,4,5-$C_6H_2(COOH)_4$ = $C_{10}H_6O_8$)	
$[Th(C_{10}H_2O_8)]_n$	stable to 405°C [8, 9]; thermogram illustrated in [9]
aurin tricarboxylates (acid = $O:C_6H_3(COOH):C\{C_6H_3(OH)(COOH)\}_2$ = $C_{22}H_{14}O_9$ = 4,4″-dihydroxy fuchsone tricarboxylic acid-(3,3′,3″))	
$(ThO)_2(OH)(C_{22}H_{11}O_9) \cdot L \cdot xH_2O$	L = C_5H_5N, x = 18; L = 2- or 3-$CH_3C_5H_4N$, x = 14; L = 4-$CH_3C_5H_4N$, x = 11; L = 2,4-$(CH_3)_2C_5H_3N$, x = 17; see Th E, pp. 12, 14
$(ThO)_3(OH)(C_{22}H_9O_9) \cdot 2L \cdot 15H_2O$	L = $C_5H_{11}N$ (piperidine); see Th E, pp. 15/6

The hydrated phthalate, $Th(C_8H_4O_4)_2 \cdot 2H_2O$, is precipitated quantitatively at pH 1 to 2 from boiling aqueous $Th(NO_3)_4$ on addition of a hot aqueous solution of the acid [5]. It has also been prepared by adding solid $Th(NO_3)_4 \cdot 5H_2O$ to hot aqueous $C_8H_6O_4$ (molar ratio = 1:4), then boiling the mixture for 5 min [10]. The dihydrate has also been prepared by adding an equivalent amount of 0.2M aqueous NH_4 benzene dicarboxylate (pH = 4.5) to hot aqueous 0.1M $Th(NO_3)_4$ solution. The precipitate formed was heated in the mother liquor at 353 to 363 K for 15 min and then washed with hot water and dried at 303 K [3].

The compound is almost insoluble in acetone, benzene, dioxane, ethanol, ether, methanol, or water [10]. It is sparingly soluble in water [13]. Other reports indicate that although aqueous phthalic and 3-nitrophthalic acids precipitate Th from aqueous solution, precipitation is incomplete [1]. Evidence for an insoluble phthalate, $Th(C_8H_4O_4)_2$ (probably hydrated), which dissolves in aqueous $K_2(C_8H_4O_4)$, was obtained in a solution chemistry study [2] and the precipitation curve for the titration of aqueous $Th(NO_3)_4$ against aqueous $K_2(C_8H_4O_4)$ has been reported [3]. Solubility studies of the system $Th(NO_3)_4$–$K_2(C_8H_4O_4)$–H_2O have provided evidence for the precipitation of $Th(C_8H_4O_4)_2$ (probably hydrated) and of $Th(OH)_x(C_8H_4O_4)_{(4-x)/2}$ (x unspecified) [11].

The dihydrate of the isophthalate, $Th(C_8H_4O_4)_2 \cdot 2H_2O$, was prepared in the same way as the phthalate. It is sparingly soluble in water [13].

Phthalate and isophthalate dihydrates are dehydrated between 343 and 453 K. The anhydrous compounds are stable up to 643 to 723 K. On further heating in air they decompose to a mixture of $ThOCO_3$ and carbon and finally to ThO_2 [13].

The terephthalate $Th_3(OH)_5(C_8H_4O_4)_2(C_8H_5O_4)_3 \cdot 5H_2O$ is prepared in the same way as the phthalate or isophthalate (see above). The basic compound $Th_5(OH)_{14}(C_8H_4O_4)_3 \cdot 7H_2O$ is obtained by adding NH_4 terephthalate to a solution of $Th(NO_3)_4$ at 291 K; this terephthalate is amorphous. Both compounds are sparingly soluble in water. On heating, the pentahydrate is firstly dehydrated, then decarboxylated to $Th_3(OH)_5(C_8H_4O_4)_2(C_7H_5O_2)_3$ and finally decomposes to ThO_2 with the intermediate formation of $ThOCO_3$. The heptahydrate is dehydrated on heating at 333 to 543 K. The anhydrous basic compound then decomposes forming ThO_2 [13].

Tetrachlorophthalic acid (= $C_8H_2Cl_4O_4$) precipitates Th quantitatively from aqueous solution (pH 1.0 to 1.1) at 70 to 80°C [1]. The solubility of the product obtained by mixing aqueous $Th(NO_3)_4 \cdot 4H_2O$ with the calculated quantity of $C_8H_2Cl_4O_4$ at pH 6.50 has been reported. The precipitate appears to be $Th(C_8Cl_4O_4)_2$ [4], although no analytical data are given. Precipitation of Th with this acid at pH 1.0 to 1.1 can be used to separate Th from Sc, which precipitates at pH 2.4 to 4.4 [7].

The polymeric pyromellitate, $[Th(C_{10}H_2O_8)]_n$, is precipitated when a dilute aqueous solution of $C_{10}H_6O_8$ is added dropwise to a boiling aqueous solution of $ThCl_4$ [9].

Complexes of Th aurintricarboxylates (aurintricarboxylic acid = O : C_6H_3(COOH) : C{C_6H_3-(OH)(COOH)}$_2$ = $C_{22}H_{14}O_9$) with neutral donor ligands are described in "Thorium" Suppl. Vol. E. They are included in Table 61 for completeness.

References for 15.6.4.10:

[1] Gordon, L.; Vanselow, C. H.; Willard, H. H. (Anal. Chem. **21** [1949] 1323/5).
[2] Bobtelsky, M.; Bar-Gadda, I. (Bull. Soc. Chim. France **1953** 382/6).
[3] Težak, B. (Proc. 1st Intern. Conf. Peaceful Uses At. Energy, Geneva 1955, Vol. 7, pp. 401/6).
[4] Wenger, P. E.; Kapetanidis, I. (Recl. Trav. Chim. **79** [1960] 567/73).
[5] Bogdan, E.; Ungureanu-Vicol, O. (Anal. Stiinte Univ. Al. I. Cuza Sect. I Iasi **6** [1960] 967/73).
[6] Prasad, S.; Kumar, S. (J. Indian Chem. Soc. **39** [1962] 444/6).
[7] Shu-Chuan Liang, Shui-Chieh Hung (Hua Hsueh Hsueh Pao [Huaxue Xuebao] **28** [1962] 139/47; C.A. **58** [1963] 9625).

[8] Tomic, E. A. (Am. Chem. Soc. Div. Polym. Chem. Preprints **4** [1963] 237/42; C.A. **62** [1965] 659).
[9] Tomic, E. A. (J. Applied Polym. Sci. **9** [1965] 3745/52).
[10] Kovalenko, K. N.; Kazachenko, D. V.; Vishnevetskaya, A. N. (Zh. Neorgan. Khim. **11** [1966] 1626/30; Russ. J. Inorg. Chem. **11** [1966] 869/71).
[11] Bilinski, H. (Croat. Chem. Acta **38** [1966] 71/81).
[12] Singh, M.; Singh, A. (J. Indian Chem. Soc. **56** [1979] 1249/51).
[13] Brzyska, W.; Karasiński, S. (J. Therm. Anal. **32** [1987] 55/9).

15.6.4.11 Thorium Salts of Other Aromatic Dicarboxylic Acids

Information on the known compounds and their properties is summarised in Table 62.

Table 62
Thorium Salts of Other Aromatic Dicarboxylic Acids.

2-carboxyphenoxyacetate (acid = 2-(HOOC)$C_6H_4OCH_2COOH$ = $C_9H_8O_5$)

$Th(C_9H_6O_5)_2 \cdot H_2O$	[10]

N-carboxymethylanthranilates (= phenylglycine-*o*-carboxylates)
(acid = 2-($HOOCCH_2NH$)C_6H_4COOH = $C_9H_9NO_4$ = N-(2-carboxyphenyl)glycine)

$Th(C_9H_7NO_4)_2 \cdot 2H_2O$	[1]
$Th(OH)_3(C_9H_8NO_4)$	decomposes > 240°C, to ThO_2 at 530°C thermal decomposition curve illustrated in [8]

phenylglycine-*p*-carboxylate (acid = N-(4-carboxyphenyl)glycine)

$Th(OH)(C_9H_8NO_4)_3 \cdot 2H_2O$	[6]

6,6′-dihydroxy-3,3′-thiodibenzoate (= 5,5′-thiodisalicylate)
(acid = $S(5\text{-}C_6H_3\text{-}2\text{-}(OH)COOH)_2$ = $C_{14}H_{10}O_6S$)

$Th(C_{14}H_8O_6S)_2 \cdot 4H_2O$	yellow, m.p. with decomposition 280°C [13] IR spectrum; ν_{as}(CO), 1500 cm^{-1}, ν_s(CO), 1610 cm^{-1}, δ(OCO) + ν(CO), 1150 cm^{-1}, ν(ThO), 435 cm^{-1} [13]; see also [14] diamagnetic susceptibility, $\chi_{mol} = 286.7 \times 10^{-6}$ [13]

diphenates (= 2,2′-biphenyldicarboxylates)
(acid = $HOOCC_6H_4C_6H_4COOH$ = $C_{14}H_{10}O_4$)

$Th(C_{14}H_8O_4)_2$	[2, 5, 7]; solubility in H_2O, ca. 0.1 g/L at 60°C [3]
$Th(C_{14}H_8O_4)_2 \cdot 2H_2O$	IR spectrum; ν(CO), 1580 cm^{-1}, ν(ThO), 380 cm^{-1} [15]
$Th(OH)_3(C_{14}H_9O_4)$	decomposes > 375°C, to ThO_2 at 575°C thermal decomposition curve illustrated in [8]

The 2-carboxyphenoxyacetate, $Th(C_9H_6O_5)_2 \cdot H_2O$ (acid = 2-(HOOC)$C_6H_4OCH_2COOH$ = $C_9H_8O_5$), is precipitated quantitatively from hot aqueous solutions of Th^{IV} salts (pH 4.2) on the addition of a hot 2% aqueous solution of the acid [9, 10]. The precipitation has been tested for the possible separation of Th from the lanthanides and some solubility data are given in [10].

A phenylglycine-*o*-carboxylate, $Th(C_9H_7NO_4)_2 \cdot 2H_2O$ (acid = 2-($HOOCCH_2NH$)C_6H_4COOH = $C_9H_9NO_4$ = N-(2-carboxyphenyl)glycine), is precipitated when a slight excess of a boiling 1% aqueous solution of the acid is added to a boiling aqueous solution of a Th^{IV} salt which is just neutral to Congo Red [1]. The optimum pH range is given as 4.4 to 5.2 in [1], but pH 3.5 is also apparently satisfactory [6]. The precipitate can be air-dried at 105 to 110°C, but the composition is not definite enough for the determination of Th by direct weighing except after ignition to ThO_2 [1]. The precipitation can be used to separate Th from the lanthanides [1]. In a later paper [8], the precipitate obtained by the method described in [1] is reported to be $Th(OH)_3(C_9H_8NO_4)$.

Phenylglycine-*p*-carboxylic acid (= N-(4-carboxyphenyl)glycine) gives a precipitate of composition $Th(OH)(C_9H_8NO_4)_3 \cdot 2H_2O$ under the conditions used to prepare the *o*-carboxylate [6].

5,5′-Thiodisalicylic acid (= 6,6′-dihydroxy-3,3′-thiodibenzoic acid = S(5-C_6H_3-2(OH)COOH$)_2$ = $C_{14}H_{10}O_6S$) precipitates Th quantitatively from aqueous solutions of $ThCl_4$ at pH 5.0 to 7.0 [11] and stability constants for complex formation with this acid are reported in [12]. The precipitate obtained when a 0.1M ethanol solution of the acid is added to an aqueous solution of a Th^{IV} salt, followed by heating under reflux for 1 h, has the composition $[Th(C_{14}H_8O_6S)_2(H_2O)_4]$ after drying at 80°C. The IR spectrum of this product has been interpreted on the basis of monodentate carboxylate groups [13], but the frequencies assigned to $\nu_{as}(CO)$ and $\nu_s(CO)$ (see Table 62) may be transposed. See also [14].

The anhydrous diphenate, $Th(C_{14}H_8O_4)_2$ (acid = $HOOCC_6H_4C_6H_4COOH$ = $C_{14}H_{10}O_4$), is reported to be precipitated from a hot aqueous solution of a Th^{IV} salt (neutral to Congo Red) on the addition of the acid. It is also precipitated in the cold in the presence of CH_3COONH_4 [2]. Th is not precipitated from acid media, so that the acid can be used to separate Th from Zr, which precipitates at pH < 2 [2, 4], and from the cerite lanthanides [2, 5], from U [2] and from Ti and Fe [4]. Precipitation of Th at pH 4.5 to 8.6 from boiling aqueous solution, followed by drying at 110 to 115°C for 1 to 2 h, can be used for the direct gravimetric determination of Th by weighing as $Th(C_{14}H_8O_4)_2$ [3, 5]. The precipitation of this compound at pH 4 to 5 has also been noted in a solution chemistry study in which the first stage of the reaction between aqueous $Th(NO_3)_4 \cdot 4H_2O$ and $C_{14}H_{10}O_4$ is apparently the formation of a soluble species of composition $[Th_2(C_{14}H_8O_4)]^{6+}$ [7]. The product of a precipitation carried out by the method given in [2, 3, 5] is also reported to be $Th(OH)_3(C_{14}H_9O_4)$ [8], which suggests that the primary product is rather easily hydrolysed. However, a hydrate of composition $Th(C_{14}H_8O_4)_2 \cdot 2H_2O$ is apparently precipitated when an aqueous solution of a Th^{IV} salt, mixed with a slight excess of $C_{14}H_{10}O_4$ in 40% aqueous dioxane, is adjusted to pH 8 to 10 and the mixture is heated under reflux for 30 min. This product behaves as a non-electrolyte in dimethylformamide [15].

References for 15.6.4.11:

[1] Datta, S. K.; Banerjee, G. (J. Indian Chem. Soc. **31** [1954] 149/52).
[2] Banerjee, G. (Naturwissenschaften **42** [1955] 417).
[3] Banerjee, G. (Z. Anal. Chem. **147** [1955] 404/9).
[4] Banerjee, G. (Z. Anal. Chem. **147** [1955] 409/15).
[5] Banerjee, G. (Z. Anal. Chem. **148** [1955] 105/10).
[6] Datta, S. K. (Z. Anal. Chem. **159** [1958] 241/9).
[7] Bobtelsky, M.; Ben-Bassat, A. H. I. (Bull. Soc. Chim. France **1958** 1138/44).

[8] Wendlandt, W. W. (Anal. Chim. Acta **18** [1958] 316/20).
[9] Datta, S. K. (Z. Anal. Chem. **174** [1960] 104/8).
[10] Datta, S. K. (Z. Anal. Chem. **174** [1960] 109/18).

[11] Good, M. L.; Srivastava, S. C. (Talanta **12** [1965] 181/3).
[12] Kumar, A. N.; Srivastava, P. C.; Nigam, H. L. (Indian J. Chem. **9** [1971] 488/9).
[13] Nigam, H. L.; Pandeya, K. B.; Srivastava, P. C. (Current Sci. [India] **40** [1971] 600/1).
[14] Srivastava, P. C.; Nigam, H. L.; Pandeya, K. B. (Fert. Technol. **13** [1976] 170/1; C.A. **86** [1977] No. 163049).
[15] Sharma, C. L.; Jain, P. K. (Indian J. Chem. A **15** [1977] 1110/2).

15.6.4.12 Thorium Naphthoates

The known compounds are listed in Table 63 together with the available physical data.

Table 63
Thorium Naphthoates.

1-hydroxy-2-naphthoates (acid = 1-$HOC_{10}H_6$-2-COOH = $C_{11}H_8O_3$)	
$Th(OH)_2(C_{11}H_7O_3)_2$	decomposes rapidly >310°C, to ThO_2 at 485°C thermal decomposition curve illustrated in [6]
$Th(OH)_3(C_{11}H_7O_3)\cdot 2H_2O$	[3]
$Th(CH_3COO)_2(C_{11}H_7O_3)_2$	light brown [9] IR spectrum: ν(CO), 1625 cm^{-1}, ν_{as}(CO), 1575, 1550, 1495 cm^{-1}, ν_s(CO) 1410 cm^{-1} [9]
1-hydroxy-4-bromo-2-naphthoate (acid = $C_{11}H_7O_3Br$)	
$Th(OH)_3(C_{11}H_6O_3Br)\cdot 2H_2O$	orange [3]
1-hydroxy-4-nitro-2-naphthoate (acid = $C_{11}H_7NO_5$)	
$Th(OH)_3(C_{11}H_6NO_5)\cdot 2H_2O$	yellow [3]
3-hydroxy-2-naphthoates	
$Th(C_{11}H_7O_3)_4$	yellowish white [7]
$Th(CH_3COO)_2(C_{11}H_7O_3)_2$	light yellow [9] IR spectrum: ν(CO), 1640, 1610 cm^{-1}, ν_{as}(CO), 1575, 1550, 1520 cm^{-1}, ν_s(CO), 1340 cm^{-1} [9]
$Th(C_{11}H_6O_3)_2$	greenish [1]
$Th(C_{11}H_6O_3)_2\cdot 2H_2O$	yellow-green [7]
$Th(C_{11}H_6O_3)_2\cdot 4H_2O$	whitish yellow [7]
$Th_3(C_{11}H_6O_3)_6\cdot C_5H_5N\cdot 5H_2O$	see Th E, pp. 12, 14

Table 63 (continued)

2-hydroxy-3-naphthoates	
$Th(C_{11}H_6O_3)_2 \cdot 4H_2O$	[8]
$ThO(C_{11}H_6O_3) \cdot 3H_2O$	[8]
3-amino-2-naphthoate (acid = $C_{11}H_9NO_2$)	
$Th(C_{11}H_8NO_2)_4$	IR spectrum: ν(NH), 3380 cm^{-1}, ν(CO), 1640 cm^{-1}, ν(C=C), 1610 cm^{-1}, ν_{as}(CO), 1540 cm^{-1}, ν_s(CO) = ca. 1390 cm^{-1}, ν(ThN), 535 cm^{-1} [10]
2,3,6,7-naphthalenetetracarboxylate (acid = 2,3,6,7-$C_{10}H_4(COOH)_4$ = $C_{14}H_8O_8$)	
$[Th(C_{14}H_4O_8)]_n$	decomposes at 360 [12] to 380°C [13] thermogram illustrated in [13]

The basic 1-hydroxy-2-naphthoate, $Th(OH)_3(C_{11}H_7O_3) \cdot 2H_2O$ (acid = $C_{11}H_8O_3$), is quantitatively precipitated when a hot 0.5% solution of the acid (or 1% solution of its Na salt) is added to a hot aqueous solution of a Th^{IV} salt which is neutral to Congo Red, followed by addition of 2% aqueous CH_3COONH_4 to pH 4.2 to 5.2. The 1-hydroxy-4-bromo- and 1-hydroxy-4-nitro-2-naphthoates, $Th(OH)_3L \cdot 2H_2O$, are precipitated in the same way at pH 3.6 to 5.2 and 3.9 to 5.2, respectively, using a hot 1% aqueous solution of the Na salts of the acids [3]. These products can be dried at ca. 100°C and the precipitation can be used to separate Th from the cerite lanthanides and from monazite sand, but the procedure is only suitable for the gravimetric determination of Th after igniting the precipitates to ThO_2 [3]; see also [2]. The 1-hydroxy-2-naphthoate obtained in this way was later reported [6] to have the composition $Th(OH)_2(C_{11}H_7O_3)_2$.

The mixed acetate/1-hydroxy-2-naphthoate, $Th(CH_3COO)_2(C_{11}H_7O_3)_2$, is precipitated when $Th(CH_3COO)_4$ (p. 47) is heated with the acid in ethanol under reflux for 4 h [9]. The product is described as light brown, which may indicate that some decomposition had occurred.

The anhydrous 3-hydroxy-2-naphthoate, $Th(C_{11}H_6O_3)_2$, is obtained when the hydrated pyridine complex, $Th_3(C_{11}H_6O_3)_6 \cdot C_5H_5N \cdot 5H_2O$, is heated above 150°C, and preferably at 170 to 180°C [1]. The dihydrate, $Th(C_{11}H_6O_3)_2 \cdot 2H_2O$, is obtained by heating the tetrahydrate. The latter is precipitated when aqueous $Th(SO_4)_2$ or $Th(NO_3)_4$ is added to the stoichiometric quantity of the acid in hot water, adjusted to pH 5 with aqueous NH_3. The dihydrate is soluble in acetone [7]. Precipitation from aqueous $Th(NO_3)_4$ at pH 6 to 7 (molar ratio, Th : $C_{11}H_8O_3$ = 1 : 4) gives $Th(C_{11}H_7O_3)_4$, which is soluble in acetone, but insoluble in other organic solvents [7].

The mixed acetate/3-hydroxy-2-naphthoate, $Th(CH_3COO)_2(C_{11}H_7O_3)_2$, is obtained in the same way as the corresponding 1-hydroxy-2-naphthoate (see above) [9].

Th is precipitated quantitatively by 2-hydroxy-3-naphthoic acid at pH 3.5 [4] or 4.3 to 5.3 [5], and by 4,6-dinitro- at pH 3 [4], or pH 4.3 to 5.3 [5], 1-bromo- [4], at pH 3.8 to 5.3 [5], 1,6-diiodo- [4], at pH 3.8 to 4.3 [5] and 1-nitroso-2-hydroxy-3-naphthoic acid at pH 2 to 5 [4], or pH 3.8 to 4.3 [5], and by 2-acetyl-3-naphthoic acid [4], at pH 4.3 to 5.3 [5], but the compositions of the precipitates were not reported. Precipitation with these acids can be used for Th analysis, following ignition to ThO_2, and for the separation of Th from Zr, the lanthanides, and U [5].

The 2-hydroxy-3-naphthoate, $Th(C_{11}H_6O_3)_2 \cdot 4H_2O$, is precipitated when the Na salt of the acid is added to aqueous $Th(NO_3)_4$. Addition of aqueous NH_3 to the reaction mixture yields a precipitate of composition $ThO(C_{11}H_6O_3) \cdot 3H_2O$ (or $Th(OH)_2(C_{11}H_6O_3) \cdot 2H_2O$) [8].

The 3-amino-2-naphthoate, $Th(C_{11}H_8NO_2)_4$ (acid $= C_{11}H_9NO_2$), has been prepared by treating a solution of a Th^{IV} salt (solvent unspecified) with an aqueous ethanolic solution of the Na salt of the acid. The compound has a limited solubility in dimethylformamide and dimethyl sulfoxide, and is insoluble in common organic solvents. It is a non-electrolyte in dimethylformamide. The large shift in ν(NH) in the IR spectrum (ca. 100 cm^{-1}) may indicate that the Th atom is bonded to the amino N atom [10].

The addition of hot aqueous 1,2- or 1,8-naphthalic acid to an aqueous solution of a Th^{IV} salt yields a gelatinous precipitate of unknown composition [11].

2,3,6,7-Naphthalene tetracarboxylic acid ($= C_{14}H_8O_8$) forms a polymeric compound of composition $[Th(C_{14}H_4O_8)(H_2O)_2]_n$ when a boiling dilute aqueous solution of the acid is added to hot aqueous $ThCl_4$; the final pH is 6. This dehydrates at 100 to 250°C [13]. See also [12].

References for 15.6.4.12:

[1] Poni, M.; Cernătescu, R. (Ann. Sci. Univ. Jassy I **30** [1944/47] 219/28; C.A. **1949** 2538).
[2] Datta, S. K. (J. Indian Chem. Soc. **33** [1956] 257/60).
[3] Datta, S. K. (J. Indian Chem. Soc. **33** [1956] 394/8).
[4] Datta, S. K. (J. Indian Chem. Soc. **34** [1957] 238/44).
[5] Datta, S. K. (J. Indian Chem. Soc. **34** [1957] 531/6).
[6] Wendlandt, W. W. (Anal. Chim. Acta **18** [1958] 316/20).
[7] Bogdan, E.; Bold, A. (Acad. Rep. Populare Romine Filiala Iasi Studii Cercetari Stiinte Chem. **13** [1962] 49/58).
[8] Agarwal, R. P.; Mehrotra, R. C. (Indian J. Chem. **2** [1964] 138/41).
[9] Singh, M.; Singh, R. (J. Indian Chem. Soc. **53** [1976] 437/8).
[10] Biradar, N. S.; Mallur, N. B. (J. Karnatak Univ. Sci. **21** [1976] 191/5; C.A. **88** [1978] No. 130098).

[11] Gordon, L.; Vanselow, C. H.; Willard, H. H. (Anal. Chem. **21** [1949] 1323/5).
[12] Tomic, E. A. (Am. Chem. Soc. Div. Polym. Chem. Preprints **4** [1963] 237/42; C.A. **62** [1965] 659).
[13] Tomic, E. A. (J. Appl. Polym. Sci. **9** [1965] 3745/52).

15.6.4.13 1-Anthraquinonyloxamates

1-Anthraquinonyloxamic acid ($=(C_{14}H_7O_2)NHC(O)COOH$) and 4-nitro-1-anthraquinonyloxamic acid ($C_{16}H_8N_2O_7$) yield brownish yellow Th compounds when a solution of a Th salt and the acid (molar ratio = 1 : ca. 2) in ethanol is heated under reflux for ca. 15 min. The products separate when the solution is cooled. The complex with 1-anthraquinonyloxamic acid absorbs at 360 nm, the complex with its 4-nitro-derivative at 350 and 450 nm (probably in ethanol). In the IR spectra both show a feature at ca. 1630 cm^{-1} (ν(CO)) [1]. No analytical data or suggested formulae for the products are given in [1].

Reference for 15.6.4.13:

[1] Idriss, K. A.; Seleim, M. M.; Abu Zubri, A. Z. (Indian J. Chem. A **17** [1979] 532/4).

15.6.5 Thorium Compounds with Heterocyclic Carboxylic Acids

15.6.5.1 Thorium O-Heterocyclic Carboxylate

A thorium compound with coumarin-3-carboxylic acid is reported to have a superior toxicity to rats and mice, but only a minor toxicity to domestic animals [1]. The composition of this compound is not recorded and no other compounds with O-heterocyclic carboxylic acids appear to have been reported.

Reference for 15.6.5.1:

[1] Kitigawa, H.; Iwaki, R.; Saito, H.; Takaeda, R.; Suzuki, T.; Turuya, M.; Inazuka, N. (Nippon Oyo Dobutsu Konchu Gakkaishi **7** [1963] 97/101, 157/8; C.A. **62** [1965] 3354).

15.6.5.2 Thorium Pyridinemonocarboxylates

Information on the known compounds is summarised in Table 64.

Table 64
Thorium Pyridinemonocarboxylates.

pyridine-2-carboxylates (= α-picolinates)	
$Th(NC_5H_4\text{-2-COO})_4$	decomposes above 500°C [5], to ThO_2 at ca. 850°C [1] IR spectrum: $\nu_{as}(CO)$, 1640 cm^{-1} [5], 1625 cm^{-1} [1], $\nu_s(CO)$ 1340 cm^{-1}, $\nu(ThN)$, 470 cm^{-1}, $\nu(ThO)$, 415 cm^{-1} [5] molar conductivity at 36°C in $CH_3OH = 7.9$, in $H_2O = 42\ \Omega^{-1}\cdot cm^2\cdot mol^{-1}$ [5]
$ThCl_2(NC_5H_4\text{-2-COO})_2$	IR spectrum: $\nu(CO)$, 1655 cm^{-1}, $\nu(ThO)$, 395 cm^{-1}, $\nu(ThN)$, 350 cm^{-1}, $\nu(ThCl)$, 225 cm^{-1} [4] molar conductivity in $(CH_3)_2SO = 43.05\ \Omega^{-1}\cdot cm^2\cdot mol^{-1}$ [4] ^{13}C NMR spectrum in [4]
$Th(OO)(NC_5H_4\text{-2-COO})_2$	IR spectrum: $\nu(CO)$, 1650 cm^{-1} Raman spectrum, $\nu_1(OO)$, 822 cm^{-1} [4] molar conductivity in $(CH_3)_2SO = 0.60\ \Omega^{-1}\cdot cm^2\cdot mol^{-1}$ [4] ^{13}C NMR spectrum in [4]
$Th(acac)(CH_3COO)_2(NC_5H_4\text{-2-COO})\cdot 2H_2O$	see Th E, pp. 134/5
$Th(acac)_2(CH_3COO)(NC_5H_4\text{-2-COO})\cdot 2H_2O$	see Th E, pp. 134/5
$Th(acac)_3(NC_5H_4\text{-2-COO})\cdot H_2O$	see Th E, pp. 134/5

References for 15.6.5.2 on p. 153

Table 64 (continued)

N-oxo-pyridine-2-carboxylate	
$Th(ONC_5H_4\text{-}2\text{-}COO)_4$	decomposes above 400°C [5] IR spectrum: $\nu_{as}(CO)$, 1635 cm^{-1}, $\nu_s(CO)$, 1335 cm^{-1}, $\nu(NO)$, 1210, ($\nu(NO)$, 1265 cm^{-1} in the free ligand), $\nu(ThO)$, 370 cm^{-1} [5] molar conductivity at 37°C in CH_3OH = 8.6, in H_2O = 76 $\Omega^{-1}\cdot cm^2\cdot mol^{-1}$ [5]
pyridine-3-carboxylate (= nicotinate)	
$Th(CH_3COO)_2(NC_5H_4\text{-}3\text{-}COO)_2$	decomposes to ThO_2 at ca. 850°C [1] IR spectrum: $\nu(CO)$, 1700, 1605, 1560 cm^{-1} [1]
pyridine-4-carboxylates (= isonicotinates)	
$Th(OH)_2(NC_5H_4\text{-}4\text{-}COO)_2\cdot 8H_2O$	[2]; IR spectrum: $\nu_{as}(CO)$, 1620 cm^{-1}, ν(ring, C:C, C:N), 1580 cm^{-1}, π(C:C, CH), 1040, 840, 750 cm^{-1} [3] binding energy, $Th5d_{5/2}$ = 86.8 eV, N1s = 399.5 eV [3]
$Th(NO_3)_4(NC_5H_4\text{-}4\text{-}COOH)\cdot 2H_2O$	[3]

The pyridine-2-carboxylate (= α-picolinate), $Th(NC_5H_4\text{-}2\text{-}COO)_4$, has been prepared by heating $Th(CH_3COO)_4$ (p. 47) with the stoichiometric quantity of α-picolinic acid in ethanol under reflux, then adding petroleum ether (40 to 60°C) to precipitate the product [1]. It is also obtained as a precipitate when a concentrated solution of the acetylacetonate, $Th(acac)_4$, in acetone is heated under reflux with the stoichiometric quantity of the acid in the same solvent [5]. It is reported to be insoluble in the common organic solvents [1] but is soluble in H_2O and moderately soluble in polar solvents such as ethanol, methanol, dimethylformamide and dimethyl sulfoxide. It can be recrystallised from methanol [5]. The IR spectrum of the compound indicates that the carboxylate group is bidentate [1] and that the pyridine N atom is bonded to the Th atom [5].

$ThCl_2(NC_5H_4\text{-}2\text{-}COO)_2$ is obtained by treating a suspension of $ThCl_4$ in CH_3CN with the stoichiometric quantity of the acid in the same solvent. It can be recrystallised from hot acetone by partial evaporation of the solution, followed by addition of ether while the solution cools. The relatively high conductivity of the compound in $(CH_3)_2SO$ is probably due to displacement of Cl^- by the solvent; the ^{13}C NMR spectrum of the compound in $(CH_3)_2SO$ indicates that the carboxylate group is not displaced by the solvent. The IR spectrum of the compound indicates that the pyridine N atom and the carboxylate O atoms are coordinated to the Th atom [4].

The peroxide compound, $Th(OO)(NC_5H_4\text{-}2\text{-}COO)_2$, is prepared by dissolving $ThCl_2(NC_5H_4\text{-}2\text{-}COO)_2$ in a 1:1 mixture of acetone and methanol containing a small amount of pyridine, then adding H_2O_2 [4].

The N-oxo-pyridine-2-carboxylate, $Th(ONC_5H_4\text{-}2\text{-}COO)_4$, is precipitated when a concentrated solution of $Th(acac)_4$ in acetone is heated under reflux with the stoichiometric quantity of the acid in the same solvent. It shows the same solubility behaviour as the α-picolinate (see above). The IR spectrum of the compound indicates that the N-oxo group is bonded to the Th atom [5].

The pyridine-3-carboxylate (= nicotinate), $Th(CH_3COO)_2(NC_5H_4\text{-}3\text{-}COO)_2$, has been prepared from $Th(CH_3COO)_4$ and the acid (molar ratio = 1:4) in the same way as the picolinate (see above). Its IR spectrum indicates that the carboxylate group is bidentate [1].

Pyridine-4-carboxylic acid (= isonicotinic acid) can act as a neutral N-donor ligand, forming the complex $Th(NO_3)_4 \cdot (NC_5H_4\text{-}4\text{-}COOH) \cdot 2H_2O$ when a mixture of aqueous $Th(NO_3)_4$ and a hot aqueous solution of the acid (molar ratio = 1:2 or 1:4) is evaporated to small volume and cooled [3]. The basic compound, $Th(OH)_2(NC_5H_4\text{-}4\text{-}COO)_2 \cdot 8H_2O$, is prepared by dissolving freshly precipitated Th hydroxide in a boiling aqueous solution of the acid (molar ratio = 1:4, final pH = ca. 5). The product separates when the solution is cooled. Its IR spectrum indicates coordination via the carboxylate group only [3].

References for 15.6.5.2:

[1] Singh, M.; Singh, R. (J. Indian Chem. Soc. **56** [1979] 136/7).
[2] Molodkin, A. K.; Ivanova, O. M.; Balakaeva, T. A.; Belyakova, Z. V. (Zh. Neorgan. Khim. **25** [1980] 158/62; Russ. J. Inorg. Chem. **25** [1980] 84/6).
[3] Molodkin, A. K.; Balakaeva, T. A.; Salyn, Ya. V.; Ivanova, O. M.; Belyakova, Z. V.; Kolesnikova, L. E. (Zh. Neorgan. Khim. **25** [1980] 1902/6; Russ. J. Inorg. Chem. **25** [1980] 1054/7).
[4] Westland, A. D.; Tarafder, M. T. H. (Inorg. Chem. **21** [1982] 3228/32).
[5] Ta, N. C.; Sarkar, A. R.; Mukherjee, T. K. (Transition Metal Chem. [Weinheim] **9** [1984] 29/31).

15.6.5.3 Thorium Pyridinedicarboxylates

Information on these compounds is summarised in Table 65.

Table 65
Thorium Pyridinedicarboxylates.

pyridine-2,6-dicarboxylates	
$[Th(NC_5H_3(COO)_2)_2]_n$	decomposes at 250 to 500°C [4] IR spectrum: $\nu_{as}(CO)$, 1670 cm^{-1} [1], 1660 cm^{-1} [3], 1580 cm^{-1} [1], $\nu_s(CO)$, 1390, 1340 cm^{-1} [1], $\nu(ThO)$, 400 cm^{-1}, $\nu(ThN)$, 325 cm^{-1} [3] molar conductivity in $(CH_3)_2SO$ = 7.8 $\Omega^{-1} \cdot cm^2 \cdot mol^{-1}$ [3] ^{13}C NMR spectrum reported in [3]
$Th(NC_5H_3(COO)_2)_2 \cdot 3H_2O$	loses $3H_2O$ at 150°C; thermogravimetric analysis reported in [4] IR spectrum: $\nu_{as}(CO)$, 1655 cm^{-1}, $\nu_s(CO)$, 1330 cm^{-1}, $\nu(ThN)$, 490 cm^{-1}, $\nu(ThO)$, 425 cm^{-1} [4]

References for 15.6.5.3 on p. 156

Table 65 (continued)

$[Th(NC_5H_3(COO)_2)_2(H_2O)_4]$	loses $4H_2O$ at 115°C [1] crystallographic data; monoclinic, space group C2/c-C_{2h}^6 (No. 15), lattice parameters, a=30.57(2), b=8.86(1), c=20.49(2) Å, β=92.51(3)°; density, d_{obs}=2.37 g/cm³ [1] IR spectrum: ν_{as}(CO), 1625 cm^{-1}, ν_s(CO), 1370 cm^{-1} [1]
$Th(NC_5H_3(COO)_2)_2 \cdot 4\,ur$	see Th E, pp. 39, 42
$Th(NC_5H_3(COO)_2)_2 \cdot (C_6H_5)_3AsO \cdot xH_2O$	x=0 or 3; see Th E, p. 99
$Th(OO)(NC_5H_3(COO)_2) \cdot H_2O$	IR spectrum: ν_{as}(CO), 1650 cm^{-1} Raman spectrum: ν_1(OO), 840 cm^{-1} [3] molar conductivity in $(CH_3)_2SO$ = 0.22 $\Omega^{-1} \cdot cm^2 \cdot mol^{-1}$ [3] ^{13}C NMR spectrum reported in [3]
$Th(CH_3COO)_2(NC_5H_3(COO)_2) \cdot 4H_2O$	IR spectrum: ν_{as}(CO), 1590 cm^{-1}, ν_s(CO), 1420 cm^{-1} (CH_3COO groups); ν_{as}(CO), 1657, 1635 cm^{-1} ($NC_5H_3(COO)_2$ group) [2]
$Th(CF_3COO)_2(NC_5H_3(COO)_2) \cdot 2H_2O$	IR spectrum; ν_{as}(CO), 1690 cm^{-1}, ν_s(CO), 1403 cm^{-1} (CF_3COO groups); ν_{as}(CO), 1658, 1630 cm^{-1} ($NC_5H_3(COO)_2$ group) [2]
$Th(CH_3COO)(CCl_3COO)(NC_5H_3(COO)_2) \cdot 2H_2O$	IR spectrum: ν_{as}(CO), 1585 cm^{-1}, ν_s(CO), 1420 cm^{-1} (CH_3COO group); ν_{as}(CO), 1640 cm^{-1}, ν_s(CO), 1398 cm^{-1} (CCl_3COO group); ν_{as}(CO), 1655, 1629 cm^{-1} ($NC_5H_3(COO)_2$ group) [2]
$Th(CH_3COO)(NC_5H_4\text{-}2\text{-}COO)(NC_5H_3(COO)_2) \cdot 4H_2O$	IR spectrum: ν_{as}(CO), 1570 cm^{-1}, ν_s(CO), 1415 cm^{-1} (CH_3COO group), ν_{as}(CO), 1643 cm^{-1} (NC_5H_4-2-COO group), ν_{as}(CO), 1660, 1633 cm^{-1} ($NC_5H_3(COO)_2$ group) [2]
$[As(C_6H_5)_4][Th(NC_5H_3(COO)_2)_3]$	IR spectrum: ν_{as}(CO), 1655 cm^{-1}, ν_s(CO), 1345 cm^{-1} [1]
$[As(C_6H_5)_4][Th(NC_5H_3(COO)_2)_3] \cdot 3H_2O$	loses $3H_2O$ at 120°C [1] IR spectrum: ν_{as}(CO), 1645 cm^{-1}, ν_s(CO), 1360 cm^{-1} [1]

The anhydrous pyridine-2,6-dicarboxylate, $Th(NC_5H_3(COO)_2)_2$, has been prepared by heating the trihydrate at 50 to 150°C [4] or the tetrahydrate at 115°C [1], and by adding the stoichiometric quantity of $NC_5H_3(COOH)_2$ in CH_3CN to a suspension of $ThCl_4$ in the same solvent [3]. It can be recrystallised from hot acetone by partial evaporation of the solution and addition of ether while the acetone solution cools [3]. The product obtained by heating the tetrahydrate is reported to be polymeric; it rehydrates on standing in air for a few hours. The

IR spectrum of the anhydrous compound suggests the presence of both unidentate and bridging bidentate carboxylate groups [1].

The trihydrate is prepared by heating $Th(acac)_4$ with the acid in a 1:1 mixture of acetone and methanol under reflux. It is soluble in water but only sparingly soluble in the common organic solvents. Its IR spectrum indicates that the Th atom is coordinated to the pyridine N atom [4].

The tetrahydrate is precipitated when the stoichiometric quantity of $Th(NO_3)_4 \cdot 5H_2O$ in methanol is added to a saturated solution of $NC_5H_3(COOH)_2$ in the same solvent. It can be recrystallised from hot water [1]. The 10-coordinate Th atom in the structure of $[Th(NC_5H_3(COO)_2)_2(H_2O)_4]$ is bonded to the O atoms of the four H_2O molecules and to both pyridine N atoms as well as to four O atoms of the monodentate carboxylate groups. The coordination geometry is a bicapped square antiprism in which the two N atoms are located outside the mid-points of the two opposite rectangular faces of the prism (**Fig. 11**) [1].

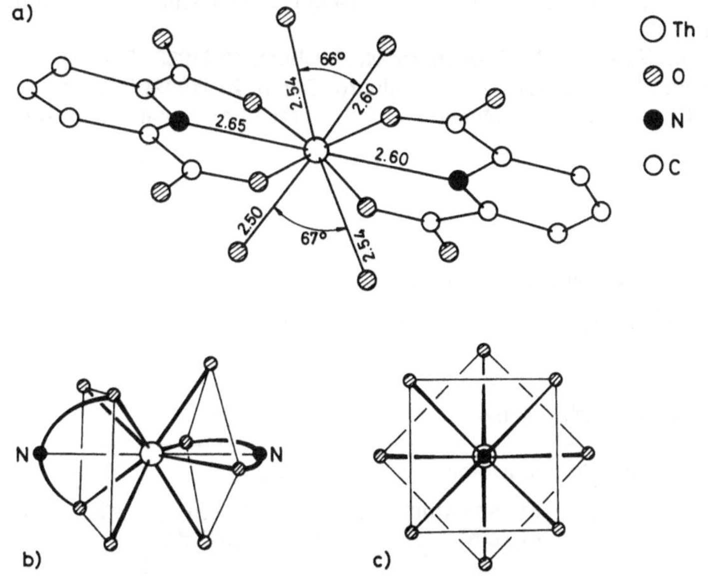

Fig. 11. The molecular structure of $[Th\{NC_5H_3(COO)_2\}_2(H_2O)_4]$ projected on the ac plane (a) and schematic drawings of the coordination polyhedron about the Th atom (b, c) [1]. Some distances and angles are shown in figure a).

The peroxide compound, $Th(OO)(NC_5H_3(COO)_2) \cdot H_2O$, has been prepared by heating a solution of $Th(NC_5H_3(COO)_2)_2$ in a 1:1 mixture of acetone and methanol containing a small amount of pyridine and added H_2O_2. After boiling for 5 min the solution was cooled, whereupon the product separated [3].

The mixed acetate compound, $Th(CH_3COO)_2(NC_5H_3(COO)_2) \cdot 4H_2O$, has been obtained by heating a 1:1 mixture of $Th(acac)_4$ and $NC_5H_3(COOH)_2$ in ethanol under reflux. The product separates when the solution is evaporated to ca. 20% of the original volume. The acetate groups result from decomposition of the coordinated acetylacetonate groups via an intermediate such as $Th(acac)_2(NC_5H_3(COO)_2)$. The IR spectrum of the compound suggests that the CH_3COO groups are bridging bidentate and that the N atom and two O atoms of the

References for 15.6.5.3 on p. 156

$NC_5H_3(COO)_2$ group are bonded to the Th atom [2]. When a solution of CF_3COOH, CCl_3COOH, or NC_5H_4-2-COOH in benzene is added dropwise to a suspension of $Th(CH_3COO)_2(NC_5H_3(COO)_2) \cdot 4H_2O$ in the same solvent (molar ratio = 1:1 in each case), the compound slowly dissolves. The mixed complexes $Th(CF_3COO)_2(NC_5H_3(COO)_2) \cdot 2H_2O$, $Th(CH_3COO)(CCl_3COO)(NC_5H_3(COO)_2) \cdot 2H_2O$ and $Th(CH_3COO)(NC_5H_4\text{-2-}COO)(NC_5H_3(COO)_2) \cdot 4H_2O$ precipitate when the resulting solutions are heated under reflux for 3 h [2].

The hydrated anionic complex salt, $[As(C_6H_5)_4]_2[Th(NC_5H_3(COO)_2)_3] \cdot 3H_2O$, is obtained when an aqueous solution of $Th(NO_3)_4$ or $ThCl_4$ is added to an excess of $NC_5H_3(COOH)_2$ and $[As(C_6H_5)_4]Cl$ in water. Crystals of the compound separate on standing overnight. It dehydrates on heating at 120°C [1].

References for 15.6.5.3:

[1] Degetto, S.; Baracco, L.; Graziani, R.; Celon, E. (Transition Metal Chem. [Weinheim] **3** [1978] 351/4).
[2] Aly, M. M.; El-Awad, A. M. (J. Inorg. Nucl. Chem. **42** [1980] 567/71).
[3] Westland, A. D.; Tarafder, M. T. H. (Inorg. Chem. **21** [1982] 3228/32).
[4] Ta, N. C.; Sarkar, A. R.; Mukherjee, T. K. (Transition Metal Chem. [Weinheim] **9** [1984] 29/31).

15.6.5.4 Thorium Quinolinecarboxylates

Information on the known compounds is summarised in Table 66.

Table 66
Thorium Quinolinecarboxylates.

quinoline-2-carboxylate (quinaldinic acid = $C_{10}H_7NO_2$)	
$Th(acac)_2(C_{10}H_6NO_2)_2$	IR spectrum: $\nu_{as}(CO)$, 1630 cm^{-1}, $\nu_s(CO)$, 1360 cm^{-1}, $\nu(ThN)$, 470 cm^{-1}, $\nu(ThO)$, 420, 390 cm^{-1} [3]
quinoline-2,3-dicarboxylates (acid = $C_{11}H_7NO_4$)	
$Th(C_{11}H_5NO_4)_2$	[3]
$Th(C_{11}H_5NO_4)_2 \cdot 2H_2O$	dehydrated at 50 to 160°C, decomposes > 290 to 500°C; thermogravimetric analysis reported in [3] IR spectrum: $\nu_{as}(CO)$, 1645, 1630 cm^{-1}, $\nu_s(CO)$, 1335 cm^{-1}, $\nu(ThN)$, 475 cm^{-1}, $\nu(ThO)$, 430 cm^{-1} [3]
5,6-benzoquinaldinate (acid = $C_{14}H_9NO_2$)	
$Th(C_{14}H_8NO_2)_4$	stable to 150°C [1]

The mixed acetylacetonate/quinoline-2-carboxylate (quinaldinate; acid = $C_{10}H_7NO_2$), $Th(acac)_2(C_{10}H_6NO_2)_2$, has been prepared by heating $Th(acac)_4$ with the stoichiometric quantity of $C_{10}H_7NO_2$ in acetone under reflux. It is insoluble in water and sparingly soluble in the common organic solvents. Its IR spectrum indicates that the quinoline N atom is bonded to the

Th atom [3]. Th is precipitated when a 2% aqueous solution of the acid is added to an aqueous solution of $Th(NO_3)_4$ and the mixture is neutralised to Congo Red. The precipitation can be used for the separation of Th from Zr [2], but the composition of the precipitate has not been reported.

The quinoline-2,3-dicarboxylate (acid = $C_{11}H_7NO_4$), $Th(C_{11}H_5NO_4)_2 \cdot 2H_2O$, is precipitated when $Th(acac)_4$ is heated with the acid in a 1:1 v/v mixture of acetone and methanol. It is insoluble in water and in the common organic solvents; it can be dehydrated at 50 to 160°C. The IR spectrum of the hydrate indicates that the quinoline N atom is bonded to the Th atom. The compound may be polymeric [3].

The 5,6-benzoquinaldinate (acid = $C_{14}H_9NO_2$), $Th(C_{14}H_8NO_2)_4$, is precipitated quantitatively when an ethanol solution of the acid is added dropwise to a boiling aqueous acid solution of a Th^{IV} salt (at pH 3). The precipitate can be dried at 110°C and the procedure can be used for the gravimetric determination of Th by direct weighing of the precipitate or of ThO_2 following ignition [1].

References for 15.6.5.4:

[1] Majumdar, A. K.; Banerjee, S. (Anal. Chim. Acta **14** [1956] 306/10).
[2] Majumdar, A. K.; Sen Gupta, J. G. (Z. Anal. Chem. **162** [1958] 262/5).
[3] Ta, N. C.; Sarkar, A. R.; Mukherjee, T. K. (Transition Metal Chem. [Weinheim] **9** [1984] 29/31).

15.6.5.5 Thorium Compounds with Benzylpenicillinic Acid (Penicillin G)

A compound of composition $ThO(NO_3)(C_{16}H_{17}N_2O_4S) \cdot 3H_2O$ (benzylpenicillinic acid = $C_{16}H_{18}N_2O_4S$) is precipitated when a methanol solution of $Th(NO_3)_4$ is added gradually to a solution of the potassium salt of the acid in the same solvent [1]. A mixed chlorotetracycline (aureomycin)-penicillin G complex, $Th(C_{22}H_{20}ClN_2O_8)(C_{16}H_{17}N_2O_4S) \cdot HCl \cdot 6H_2O$, is included in Th E, pp. 141/2.

Reference for 15.6.5.5:

[1] Taguchi, K. (Chem. Pharm. Bull. [Tokyo] **8** [1960] 205/11; C.A. **1961** 6454).

15.6.5.6 Thorium Thiophene-2-carboxylate

The only recorded thorium thiophene-2-carboxylate appears to be the mixed acetate compound, $Th(CH_3COO)(SC_4H_3COO)_3$. This is precipitated when a suspension of $Th(CH_3COO)_4$ (p. 47) in ethanol is mixed with a solution of the acid in the same solvent and the mixture is heated under reflux. It decomposes above 300°C. The IR spectrum of the compound exhibits features assigned to $\nu_{as}(CO)$ at 1520 and 1420 cm^{-1}, $\nu_s(CO)$ at 1390 cm^{-1} and $\nu(CS)$, which appears at 1105 cm^{-1} in the free acid, is shifted to 1125 cm^{-1} in the spectrum of the salt, which may indicate that the S atom is coordinated to the Th atom [1].

Reference for 15.6.5.6:

[1] Singh, M.; Singh, R.; Singh, R.; Sandhu, S. S. (J. Indian Chem. Soc. **53** [1976] 1226/7).

15.7 Thorium Thiolate and Dithiolate

15.7.1 Thorium Thiolate (Mercaptide)

Information on the only recorded compound, the mercaptobenzothiazolide, is summarised in Table 67.

Table 67
Thorium Thiolate.

thorium mercaptobenzothiazolide (mercaptobenzothiazole = $C_6H_4(N{:}C(SH)S) = C_7H_5NS_2$)	
$Th(C_7H_4NS_2)_4$	some decomposition (?) at >35°C [2], stable to 120°C [1]; decomposes to ThO_2 above 910°C; thermolysis curve illustrated in [2] IR spectrum: ν(NCS), 1502 cm^{-1}, spectrum illustrated in [4]

$Th(C_7H_4NS_2)_4$ (mercaptobenzothiazole = $C_6H_4(N{:}C(SH)S) = C_7H_5NS_2$) is precipitated when aqueous $Na(C_7H_4NS_2)$ is added to aqueous $Th(NO_3)_4$. The precipitate can be dried at 105 to 110°C and the precipitation can be used for the gravimetric determination of Th by direct weighing as $Th(C_7H_4NS_2)_4$ [1]. The beginning of decomposition above 35°C which was noted in a thermolysis study [2] may be due to the removal of water from the precipitate. The compound is insoluble in molten biphenyl at 250°C [3]. Mercaptobenzothiazole in the thioimine form does not appear to react with Th salts [4].

References for 15.7.1:

[1] Spacu, G.; Pirtea, T. I. (Acad. Rep. Populare Romine Bul. Stiinte Ser. Mat. Fiz. Chim. **2** [1950] 669/76).
[2] Wendlandt, W. W. (Anal. Chem. **29** [1957] 800/2).
[3] Baldwin, W. H. (ORNL-2864 [1960]; C.A. **1960** 11731).
[4] Galateanu, I. (Acad. Rep. Populare Romine Studii Cercetari Chim. **11** [1963] 239/46).

15.7.2 Thorium Dithiolate

The precipitate which is obtained when an aqueous solution of a Th salt is added to a solution of ethane-1,2-dithiol (= $HSCH_2CH_2SH$) in aqueous NaOH is probably the dithiolate $Th(SCH_2CH_2S)_2$. The product was identified as such by physicochemical methods [1], but it was not analysed.

Reference for 15.7.2:

[1] Saxena, R. S.; Bhatia, S. K. (J. Indian Chem. Soc. **51** [1974] 660/2).

15.8 Thorium Compound with a Thiocarboxylic Acid

Crystallographic data for the only recorded compound are summarised in Table 68.

Table 68
Thorium Compound with a Thiocarboxylic Acid.

dithiooxalate	
$Th(H_2O)_6(S_2C_2O_2)_4Ni_2 \cdot 6.5H_2O$	crystallographic data: monoclinic, space group P2/c-C^4_{2h} (No. 13), lattice parameters, a = 26.27(1), b = 10.770(1), c = 13.476(2) Å, β = 100.75(2)°; Z = 4 [1]

A dithiooxalate of composition $Th(H_2O)_6(S_2C_2O_2)_4Ni_2 \cdot 6.5H_2O$ is precipitated when a hot (50°C) dilute (ca. 10^{-2} M) aqueous solution of $Th(NO_3)_4$ is added to a 10^{-2} M aqueous solution of $K_2Ni(S_2C_2O_2)_2$, also at 50°C. The coordination polyhedron of the 10-coordinate Th atom in this compound can be represented as a tetradecahedron, with the Th atom coordinated to four O atoms from two dithiooxalate groups and six O atoms from six H_2O molecules (**Fig. 12**). Bond distances (in Å) for $Th(H_2O)_6(S_2C_2O_2)_4Ni_2 \cdot 6.5H_2O$ (Fig. 12) are shown in Table 69 [1].

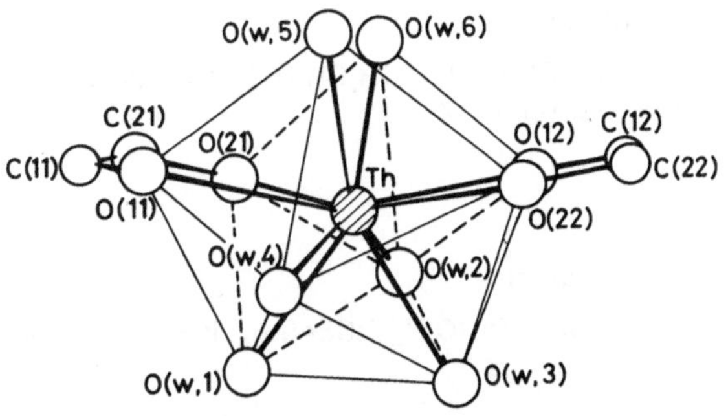

Fig. 12. The tetradecahedral coordination polyhedron of Th in the dithiooxalate, $Th(H_2O)_6(S_2C_2O_2)_4Ni_2 \cdot 6.5H_2O$ [1]; w denotes oxygen atoms from water.

Table 69
Bond distances (in Å) for $Th(H_2O)_6(S_2C_2O_2)_4Ni_2 \cdot 6.5H_2O$ [1].

Th–O(w,1): 2.49(2)	Th–O(w,6): 2.64(3)
Th–O(w,2): 2.51(3)	Th–O(11): 2.61(3)
Th–O(w,3): 2.49(3)	Th–O(21): 2.56(3)
Th–O(w,4): 2.56(3)	Th–O(12): 2.51(3)
Th–O(w,5): 2.48(3)	Th–O(22): 2.62(3)

There are two kinds of tunnel in the structure which are partially occupied by inserted H_2O molecules; these allow easy diffusion of the H_2O molecules and the instability of the compound can be related to this highly porous feature of the lattice [1].

Reference for 15.8:

[1] Trombe, J. C.; Gleizes, A.; Galy, J. (Nouv. J. Chim. **9** [1985] 55/63).

15.9 Thorium Carbamates, Oxothio- and Dithiocarbamates

These compounds are discussed in Th E, pp. 172/4. The known compounds are listed in Table 70.

Table 70
Thorium Carbamates, Oxothio- and Dithiocarbamates.

carbamates	
$Th(R_2NCOO)_4$	R = CH_3, C_2H_5, or *i*-C_4H_9; see Th E, pp. 172, 174
oxothiocarbamates	
$Th(R_2NCOS)_4$	R = CH_3 or C_2H_5; see Th E, pp. 172, 174
dithiocarbamates	
$Th(R_2NCSS)_4$	R = CH_3 or C_2H_5
$ThL\{(C_2H_5)_2NCSS\}_2$	H_2L = 2-$HOC_6H_4CH{:}NCH_2CH_2N{:}CHC_6H_4$-2-OH; see Th E, pp. 172/4

15.10 Thorium Diselenocarbamate

The only recorded compound, $Th\{(C_2H_5)_2NCSeSe\}_4$, is discussed in Th E, pp. 173/4.

Physical Constants and Conversion Factors

Avogadro constant N_A (or L) $= 6.02214 \times 10^{23}$ mol^{-1}
Faraday constant $F = 9.64853 \times 10^{4}$ C/mol
molar gas constant $R = 8.31451$ J·mol^{-1}·K^{-1}
molar volume (ideal gas) $V_m = 2.24141 \times 10^{1}$ L/mol
(273.15 K, 101325 Pa)

Planck constant $h = 6.62608 \times 10^{-34}$ J·s
elementary charge $e = 1.60218 \times 10^{-19}$ C
electron mass $m_e = 9.10939 \times 10^{-31}$ kg
proton mass $m_p = 1.67262 \times 10^{-27}$ kg

1 kg = 2.205 pounds
1 m = 3.937×10^{1} inches = 3.281 feet
1 m^3 = 2.642×10^{2} gallons (U.S.)
1 m^3 = 2.200×10^{2} gallons (Imperial)

Force	N	dyn	kp
1 N	1	10^{5}	1.019716×10^{-1}
1 dyn	10^{-5}	1	1.019716×10^{-6}
1 kp	9.80665	9.80665×10^{5}	1

Pressure	Pa	bar	kp/m^2	at	atm	Torr	lb/in^2
1 Pa = 1 N/m^2	1	10^{-5}	1.019716×10^{-1}	1.019716×10^{-5}	9.86923×10^{-6}	7.50062×10^{-3}	1.450378×10^{-4}
1 bar = 10^6 dyn/cm^2	10^{5}	1	1.019716×10^{4}	1.019716	9.86923×10^{-1}	7.50062×10^{2}	1.450378×10^{1}
1 kp/m^2 = 1 mm H_2O	9.80665	9.80665×10^{-5}	1	10^{-4}	9.67841×10^{-5}	7.35559×10^{-2}	1.422335×10^{-3}
1 at (technical)	9.80665×10^{4}	9.80665×10^{-1}	10^{4}	1	9.67841×10^{-1}	7.35559×10^{2}	1.422335×10^{1}
1 atm = 760 Torr	1.01325×10^{5}	1.01325	1.033227×10^{4}	1.033227	1	7.60×10^{2}	1.469595×10^{1}
1 Torr = 1 mmHg	1.333224×10^{2}	1.333224×10^{-3}	1.359510×10^{1}	1.359510×10^{-3}	1.315789×10^{-3}	1	1.933678×10^{-2}
1 lb/in^2 = 1 psi	6.89476×10^{3}	6.89476×10^{-2}	7.03069×10^{2}	7.03069×10^{-2}	6.80460×10^{-2}	5.17149×10^{1}	1

Work, Energy, Heat	J	kW·h	kcal	Btu	eV
1 J = 1 W·s = 1 N·m = 10^7 erg	1	2.778×10^{-7}	2.39006×10^{-4}	9.4781×10^{-4}	6.242×10^{18}
1 kW·h	3.6×10^{6}	1	8.604×10^{2}	3.41214×10^{3}	2.247×10^{25}
1 kcal	4.1840×10^{3}	1.1622×10^{-3}	1	3.96566	2.6117×10^{22}
1 Btu (British thermal unit)	1.05506×10^{3}	2.93071×10^{-4}	2.5164×10^{-1}	1	6.5858×10^{21}
1 eV	1.602×10^{-7}	4.450×10^{-14}	3.8289×10^{-11}	1.51840×10^{-10}	1

1 cm^{-1} = 1.239842×10^{-4} eV 1 Hz = 4.135669×10^{-15} eV
1 hartree = 27.2114 eV 1 eV ≙ 23.0578 kcal/mol

Power	kW	hp	kp·m·s^{-1}	kcal/s
1 kW = 10^3 J	1	1.35962	1.01972×10^{2}	2.39006×10^{-1}
1 hp (horsepower, metric)	7.3550×10^{-1}	1	7.5×10^{1}	1.7579×10^{-1}
1 kp·m·s^{-1}	9.80665×10^{-3}	1.333×10^{-2}	1	2.34384×10^{-3}
1 kcal/s	4.1840	5.6886	4.26650×10^{2}	1

References:

International Union of Pure and Applied Chemistry, Manual of Symbols and Terminology for Physicochemical Quantities and Units, Pergamon, London 1979; Pure Appl. Chem. **51** [1979] 1/41.
The International System of Units (SI), National Bureau of Standards Spec. Publ. 330 [1972].
Landolt-Börnstein, 6th Ed., Vol. II, Pt. 1, 1971, pp. 1/14.
ISO Standards Handbook 2, Units of Measurement, 2nd Ed., Geneva 1982.
Cohen, E. R., Taylor, B. N., Codata Bulletin No. 63, Pergamon, Oxford 1986.

Key to the Gmelin System of Elements and Compounds

System Number	Symbol	Element
1		Noble Gases
2	H	Hydrogen
3	O	Oxygen
4	N	Nitrogen
5	F	Fluorine
6	**Cl**	**Chlorine**
7	Br	Bromine
8	I	Iodine
8a	At	Astatine
9	S	Sulfur
10	Se	Selenium
11	Te	Tellurium
12	Po	Polonium
13	B	Boron
14	C	Carbon
15	Si	Silicon
16	P	Phosphorus
17	As	Arsenic
18	Sb	Antimony
19	Bi	Bismuth
20	Li	Lithium
21	Na	Sodium
22	K	Potassium
23	NH_4	Ammonium
24	Rb	Rubidium
25	Cs	Caesium
25a	Fr	Francium
26	Be	Beryllium
27	Mg	Magnesium
28	Ca	Calcium
29	Sr	Strontium
30	Ba	Barium
31	Ra	Radium
32	**Zn**	**Zinc**
33	Cd	Cadmium
34	Hg	Mercury
35	Al	Aluminium
36	Ga	Gallium

System Number	Symbol	Element
37	In	Indium
38	Tl	Thallium
39	Sc, Y La–Lu	Rare Earth Elements
40	Ac	Actinium
41	Ti	Titanium
42	Zr	Zirconium
43	Hf	Hafnium
44	Th	Thorium
45	Ge	Germanium
46	Sn	Tin
47	Pb	Lead
48	V	Vanadium
49	Nb	Niobium
50	Ta	Tantalum
51	Pa	Protactinium
52	**Cr**	**Chromium**
53	Mo	Molybdenum
54	W	Tungsten
55	U	Uranium
56	Mn	Manganese
57	Ni	Nickel
58	Co	Cobalt
59	Fe	Iron
60	Cu	Copper
61	Ag	Silver
62	Au	Gold
63	Ru	Ruthenium
64	Rh	Rhodium
65	Pd	Palladium
66	Os	Osmium
67	Ir	Iridium
68	Pt	Platinum
69	Tc	Technetium[1]
70	Re	Rhenium
71	Np, Pu . . .	Transuranium Elements

Examples shown with arrows beside the tables: HCl → 6 Cl; $ZnCl_2$ → 32 Zn; $CrCl_2$ → 52 Cr; $ZnCrO_4$ → 52 Cr.

Material presented under each Gmelin System Number includes all information concerning the element(s) listed for that number plus the compounds with elements of lower System Number.

For example, zinc (System Number 32) as well as all zinc compounds with elements numbered from 1 to 31 are classified under number 32.

[1] A Gmelin volume titled "Masurium" was published with this System Number in 1941.

A Periodic Table of the Elements with the Gmelin System Numbers is given on the Inside Front Cover